MULTIPURPOSE TREE SPECIES:

Research – Retrospect and Prospect

K. R. Solanki
A. K. Bisaria
A. K. Handa

AGROBIOS (INDIA)

Published by:
AGROBIOS (INDIA)
Agro House, Behind Nasrani Cinema
Chopasani Road, Jodhpur 342 002
E. mail: agrobiosindia@gmail.com

AGROBIOS (INDIA)

ISBN No. (10): 81-7754-045-9
ISBN No. (13): 9788177540451

Published by: *Dr. Updesh Purohit for Agrobios (India), Jodhpur*
Lasertypeset at: *Shriya Computers and Printers, Jodhpur*
Cover Design by: *Reena*
Printed at: *Agrobios Digitals, Jodhpur*

Foreword

Research on Multipurpose Tree Species is the basic input to the sustainable development of profitable agroforestry enterprise. Ever since the initiation of agroforestry research, many strides have been made to define, characterize, test and apply Multipurpose Tree and Shrub through diagnosis and design exercise and examining them for their suitability, growth, development and biomass production. Sufficient information is now available about MPTS for different agroecological zones of the country and for the different farming systems.

National Research Centre for Agroforestry, Jhansi, organized a National Symposium on Multipurpose Tree Species for Agroforestry Systems during July 11-13, 1998, in which more than 100 delegates from different states of Indian Union belonging to State Agricultural Universities, ICAR Institutes, State Forest Department and Non-Government Organizations participated to update the information on MPTS.

I am immensely pleased to know that the organizers of the National Symposium have compiled and edited the research and review papers presented in the symposium in the form of a book entitled "**Multipurpose Tree Species Research – Retrospect and Prospect**". I am confident that this publication will inspire research and development efforts on the subject, would make available knowledge for the benefit of the researchers, development agencies and policy makers and serve the users as a ready reference. The organizers of the symposium and the editors deserve due appreciation for bringing out this valuable publication.

(G. B. Singh)
Deputy Director General
Natural Resource Management
Indian Council of Agricultural Research
New Delhi – 110 001

Preface

Farmers have nurtured trees on their farms, pasture land and around homes for millennia to satisfy their livelihood needs. These trees were later on defined as multipurpose trees which provide fodder, firewood and also replenish nutrients in soil. Many strides have been made to define characterize, evaluate multipurpose tree species for various agroforestry system under different agro-ecological zones of the world. Since the initiation of agroforestry research in India efforts are going on for identifying promising multipurpose trees and shrubs through diagnosis and design exercise and evaluating them for their suitability pertaining to growth, development, biomass production and natural resource conservation. Now voluminous data is available about multipurpose tree species for all the twenty one agro-ecological zones of the Indian Union.

In view of continuous reduction in forest cover, demographic pressure for livelihood needs and importance of trees, National Research Centre for Agroforestry, Jhansi organized a National Symposium on "**Multipurpose Tree Species for Agroforestry System**" during 11-13 June, 1998 to collect information on MPTS in relation to agroforestry systems and to provide a platform to discuss various aspects of MPTS and exchange of scientific knowledge among scientific community, foresters, planner and NGOs.

This volume brings together a vast and useful collection of papers by experts in agroforestry Emphasis has been given to many problems and challenges facing their sustainable development such as evaluation, management and ecology. Overall the papers reflects both progress and degree of comprehension achieved in understanding, management, utilization and conservation of MPTS.

We owe a specific debt of gratitude of Dr. R. S. Paroda, Secretary, DARE, Govt. of India and Director General, ICAR, New Delhi for encouragement. We record our sincere gratitude to Dr. G. B. Singh, Deputy Director General (Natural Resource Management), ICAR, Krishi Bhavan, New Delhi his keen interest, valuable suggestions and writing forward. Thanks are also due to Dr. P. S. Pathak, ADG (AF.) ICAR, Krishi Bhavan, New Delhi for his peliotropic help. We with to record our sincere thanks to scientists of NRCAF, Jhansi especially Dr. P. Rai, Pr. Scientist (Agronomy), Dr. K. S. Dadhwal, Pr. Scientist (Soil) and Dr. V. K. Gupta, Pr. Scientist (Plant Breeding) for their suggestions and cooperation. Thanks are also due to Dr. Rajeev Tiwari, Technical Officer, Mrs. Uma, Sr. Technical Assistant and Sh. Hoob Lal, Jr. Steno for their tireless efforts in bring out this volume.

It is fervently hoped that this volume will facilitate research on multipurpose tree species more vigorously and will serve the researchers, development agencies and policy makers as a ready reference.

K. R. Solanki
A. K. Bisaria
A. K. Handa

Contents

Inaugural Address

G. B. SINGH
Deputy Director General (NRM)
Indian Council of Agricultural Research, Krishi Bhawan, New Delhi-110001

Jhansi the historical city famous for the first freedom struggle (1857) of the nation waged under the dynamic leadership of the great iron Lady, the queen of Jhansi popularly known as *Jhansi Ki Rani* is the most appropriate place to hold this ***National Symposium on Multipurpose Tree Species for Agroforestry Systems*** particularly on the occasion of this Golden Jubilee year (1997-98) of India's independence.

Tree research has been on the mandate of many research institutes and universities of our country. There has been a swift shift in the research objectives of these organizations from commercial timber yielding species to the multipurpose tree species which are often needed in the rural set up, a positive pointers leading to the importance of Agroforestry Research in the country.

Art of agroforestry practice is ancient but its science is relatively recent. Its global recognition is due to demographic reasons for meeting the demands of food, fodder, fuel wood, and timber and herbal medicines. The history of agroforestry is as old as the origin of agriculture. Keeping in view the paramount importance of agroforestry, Indian Council of Agricultural Research launched All Indian Coordinated Research Project on Agroforestry (AICRPAF) with 12 centres spread over in different agro-ecological zones of country. Now there are 35 centres under AICRPAF. Later on it was decided to establish a National Research Centre for further augmentation of research culminating into establishment of National Research Centre for Agroforestry on 8th May 1988. The prime objective of NRCAF is to undertake the basic and strategic research in agroforestry. There are two major components in agroforestry system viz. tree and crops.

I am happy to be with you on this occasion and share with you some thoughts on the role of multipurpose tree species in agroforestry system in the national perspective. Multipurpose tree species are woody perennials that provide more than one product of direct or indirect benefits to mankind. These benefits could be as fuel, timber, pulp, fodder, food, medicines, oil, gum, etc. In the light of the aforesaid definition it may strictly be difficult to identify a tree species, which is not multipurpose. Multipurpose tree species in agroforestry system have specific significance. We have to search for such multipurpose tree species for the direct and/or indirect benefits to mankind, which are compatible with agricultural crops and can be introduced under various agroforestry systems.

India is a vast country with a total geographical area of 329 million ha which is about 2.0 per cent of the world but she has to provide food and fodder to huge human and livestock population of 960 and 490 millions which are likely to exceed 1.00 and 0.60 billion, respectively by 2000 A.D. In order to meet the increasing demand of our growing population we would require to boost the production of food grain, fuel wood for human consumption and green and dry fodder for livestock to the tune of 250, 350 and 2085 million tonnes, respectively besides 75 million m^3 of timber. Under the aforesaid situation we have no alternative but to follow the integration of multipurpose tree species with agricultural crops. Some fast growing and short duration multipurpose tree species have been identified for different agro-ecological zones.

1. *Populus deltoides*
2. *Eucalyptus tereticornis*
3. *Leucaena leucocephala*
4. *Erythrina indica*
5. *Ceiba pentandra*
6. *Albizia procera*
7. *Acacia auriculiformis*
8. *Ailanthus excelsa*
9. *Melia azedarach*

The demand for fuel fodder and herbal medicines is increasing but the land holding has reduced to 0.6 ha/capita. The reduction in arable land is the major problem, which can be solved to a considerable extent by planting suitable multipurpose tree species to fulfill the daily need of the rural society.

The traditional medicinal plants such as *Terminalia arjuna, T. belerica, T. chebula, Tamarindus indica, Turmeric* etc can be grown under various agroforestry systems. In this way rural population will not only fulfill their needs of medicine but also fetch additional income.

There is about 15 million ha. land under pasture which can be exploited for planting nitrogen fixing multipurpose tree species namely *Leucaena leucocephala, Albizia procera, A. lebbeck, A. amara, H. binata etc.* to fetch additional income from fodder and fuel and to improve soil fertility status through nitrogen fixation and litter fall and improving highly polluted environment.

The current concept of agroforestry aims at sustainable land use systems involving intimate and interacting association of agricultural/ horticultural/pasture and woody components viz. trees, shrubs, palms, climbers, bamboos, etc. Such systems involves two major objectives:

1. Promotion in productivity/unit land/time.
2. **Sustainability:** Which implies conservation or improvement of natural resources soil water, biodiversity and environment. Some multipurpose tree species are also capable for nitrogen fixation. There are *Leucaena leucocephala*, *Acacia albida*, *Albizia procera*, etc. are well known nitrogen fixing multipurpose tree species (NFTs).

Agroforestry is new synthesis of various existing discipline of science. It is of paramount importance to select the right multipurpose tree species to suit the planting sites, creation of seed orchard, germplasm bank, selection and vegetative propagation, tree improvement standardization of nursery planting and management practices. Determination of Photosynthetically Active Radiation (PAR) requirement of various agricultural crops and manipulation of tree canopy for optimizing crop and tree productivity. I am to emphasis that diversity in agroforestry system is essential for the ecological sustainability. The MPTS which have adverse effect through chemical ecology / allelopathy should be avoided. Physiological and biochemical investigation should be initiated to control the allelopathic toxins/chemicals.

RESEARCH AND DEVELOPMENT PRIORITIES

- ☐ Genetic upgrading of suitable tree species for faster biomass production and suitability for different stress situations.

- ☐ Suitable nursery and planting techniques to ensure proper establishment and cost reduction.
- ☐ Production and management techniques (pruning, lopping, optimum rotation length, grazing etc.) for enhancing and sustaining production.
- ☐ Introduction of top feed species to ensure fodder availability during lean periods.
- ☐ Nutrient dynamics studies under different management systems.
- ☐ Management ☐ environment and genotypes ☐ environment interaction including allelopathic/negative effect under different land use systems.
- ☐ Soil and water conservation techniques including use of water absorbing polymers to ensure protective irrigation source near the planting side.
- ☐ Tissue culture technique to exploit mass multiplication of elite clones of tree/shrubs/grasses.
- ☐ Availability of technical information and advice for selection of species, establishment and maintenance of plantations to the growers.
- ☐ Timely availability of inputs and adequate funds to carry out the operations regularly.
- ☐ Arrangements for marketing the produce.

I am sure that the participating galaxy of delegates from different states of the country will provide guidelines for selection and optimum utilization of vast array of MPTS in Agroforestry systems. I am confident that deliberations will be helpful in shaping some useful recommendations for scientists, foresters, farmers, developmental agencies, planners, policymakers and all those who are interested in green India and making it a self sufficient with sustainability.

Important Agroforestry Systems in Arid and Semi-Arid Regions of India

R. P. SINGH AND N. K. SHARMA*
61, Sardar Club Scheme, Jodhpur (Rajasthan)
**Agricultural Research Station, Keshwana, Jalore (Rajasthan)*

Land can be used to produce agricultural crops- "*agriculture*", animals and their products- "*animal husbandry*", and trees and their products- "*forestry*" (Tejwani, 1994). When these products are obtained by growing or raising of agricultural crops, animals and trees in combinations then the system is known as agroforestry. Number of definitions of agroforestry has been given by various workers. Lundgren and Raintree (1992) suggested that agroforestry is a collective name for land use systems and technologies where woody perennials (trees, shrubs, palms, bamboos, etc) are deliberately used on the same land management units as agricultural crops and/or animals, in some forms of spatial arrangement or temporal sequence. In agroforestry systems there are both ecological and economical interactions between the different components. According to Nair (1991) agroforestry:

- Is a collective name for land-use systems involving trees combined with crops and/or animals on the same unit of land;
- Combines production of multiple outputs with protection of the resource base;
- places emphasis on the use of indigenous, multipurpose trees and shrubs,
- Is particularly suitable for low input conditions and fragile environments;

- ☐ Involves the interplay of socio-cultural values more than in most other land-use systems;
- ☐ Is structurally and functionally more complex than monoculture.

An ideal agroforestry system should possess following three attributes (Nair, 1993).

1. **Productivity:** Agroforestry system should maintain or increase production as well as productivity of the land. Agroforestry can improve productivity in many different ways. These include increased output of tree products, improved yields of associated crops, reduction of cropping system inputs, and increased labour efficiency.
2. **Sustainability:** By conserving the production potential of the resource base, mainly through the beneficial effects of woody perennials on soils, agroforestry can achieve and indefinitely maintain conservation and fertility goals.
3. **Adaptability:** The fact that agroforestry is a relatively new word for an old set of practices means that, in some cases, agroforestry has already been accepted by the farming community. However, the implication here is that improved or new agroforestry technologies that are introduced into new areas should also conform to local farming practices.

In fact, cultivating trees and agricultural crops in intimate combination with one another is an ancient practice that farmers have used throughout the world. Most of the farmers combine agricultural crops with trees and livestock in their farming practices. Shifting cultivation, growing of tea and coffee under the shade of trees, intercropping under coconut, growing of vegetables, spices, fruit plants in kitchen garden is an age old practice prevailing in India and many other countries. Research on growing of the tea bushes under the shade of trees having been started by the British in the 19th century in India. Research on silvipasture was started in fifties (Tejwani, 1994).

Objectives of agroforestry: Rajora (1998) listed following five objectives in watershed areas:

1. **Biomass production:** The maximum production of biomass per unit area in time is the primary objective of agroforestry systems in the watershed areas.

2. **Soil conservation:** Compared to the permissible limit of soil loss 4.5 t/ha/year, the average soil displaced is around 16 t/ha/year. Watershed, especially sloping terrain, should be treated through agroforestry in watershed programme.
3. **Soil improvement:** Agroforestry practices enrich soil by nitrogen fixation and addition of organic matter. Agroforestry helps in meeting nutrient requirement of plants growing in association with trees and at the same time, the soil structure and infiltration rates are also improved.
4. **Agro based village industries:** Agroforestry in watershed areas help in promotion of agro based industries like herbal drugs, poultry, piggery, dairy, bee keeping, sericulture, mushroom cultivation etc.
5. **Moderation of microclimate:** The micro environs in the neighborhood of trees in moderated by adopting an agroforestry programme. The field crops in vicinity of trees receive multidirectional effects and benefits. The impact is more prominent in arid and semi-arid zones.

AGROFORESTRY SYSTEMS AND THEIR SUSTAINABILITY

An agroforestry system is a specific local example of a practice, characterized by environment, plant species and their arrangement, management, and socioeconomic functioning. Agroforestry systems can be categorized according to structural basis, functional basis, socioeconomic basis and ecological basis. Within an agroforestry system, there can be more than one agroforestry practices. The global inventory of tropical and subtropical agroforestry systems undertaken during 1982-1987 revealed that agroforestry systems existing in different areas are so complex and diverse. The ecological potential of an area has been found to be the prime factor that determines the adoption and spread of specific agroforestry systems. Socio-economic factors determine the complexity of the system and the degree of intensity of its management. Some approaches are common to many ecological regions but the nature of components that constitute the systems in specific region will vary depending upon site-specific factors. In different areas with identical agroclimatic conditions, factors such as human population pressure availability of labour and other production resources, proximity and accessibility to market sources, etc, are the main determinants of the types and forms of agroforestry systems. Irrespective of the socio-cultural differences in different geographical regions, the major types of agroforestry systems are structurally similar in areas with identical ecological characteristics.

Some systems and approaches are particularly appropriate for specific ecological conditions. For example, silvipastoral systems, windbreaks and shelterbelts, and multipurpose trees on farmlands, are the major specific approaches for arid and semi-arid lands, whereas, soil conservation hedges, silvipastoral combination, and plantation crop systems are common in the tropical highlands. Improved tree fallows in shifting cultivation areas, modified taungya, use of under exploited woody perennials, multilayer tree gardens, tree fodders and other aspects of silvipastoral management are some agroforestry approaches to multiple use management of tropical forests are the main example of agroforestry systems of tree + cattle/grass combinations, and intercropping systems involving hardwood and nut trees (Nair 1991).

There are some functional aspects of specific agroforestry practices that are necessary in specific agroecological situations. For example, the functional emphasis of agroforestry systems in sloppy lands is erosion control and soil conservation; in wind prone areas, woody species are successfully used in agroforestry practices of shelterbelts and windbreaks, the emphasis is on fuelwood production in areas with fuelwood shortage, there are also specific agroforestry approaches for the reclamation of degraded lands and other wastelands (such as highly eroded lands, area with high level of salinity/alkalinity, etc).

The success of any agroforestry system depends on the nature of components and their arrangement. In agroforestry systems, there are three basic sets of elements or components that are managed by the man, viz. woody perennials usually referred to as "*trees*", herbaceous plants (agricultural crops including pasture species) and animals. On the basis of these components,

Agroforestry systems are of 3 basic types:

1. Agrisilviculture (crop and trees)
2. Silvipastoral (pasture/animal + trees)
3. Agrisilvipastoral (crops + pasture/animals + trees)

If woody component is a horticultural/fruit plant then systems will be known as:

1. Agrihorticulture (crops and fruit plants)
2. Hortipastoral (Pasture/animals + fruit plants)
3. Agrihortipastoral (crops + pasture/animals + fruit plants)

When trees, fruit plants and crops are planted simultaneously on the same land then the system will be known as Horti-silvi pasture. In this system, tree species should be planted on the periphery of the orchard which would act as shelterbelt and windbreak. While planting trees for shelterbelt, selection of species should be on the basis of height. The tall species should be planted on the boundary, the middle row should consist of shrub, having less height and again the third row should be of tall species. These should be staggered rather than in a straight row (Vashishtha and Prasad, 1995). The wind velocity will be minimized and would minimize the damage caused by hot and cold winds prevalent in the arid region.

The most common agroforestry practices that constitute the diverse agroforestry systems throughout the tropics and their main characteristics are discussed below. These practices need to be improved scientifically and developed into modern agroforestry technology (Nair, 1991).

A brief description of some important agroforestry systems is as follows:

SHIFTING CULTIVATION

Shifting cultivation is an age-old practice, which is essentially prevalent in North-Eastern States and to some extent, in hills of Andhra Pradesh, Bihar, Madhya Pradesh, Orissa and Karnataka (Tejwani, 1994). Shifting cultivation consists of cultivation of crops on steep slopes of hills and hillocks, etc., by clearing certain trees, shrubs, bushes, etc, upto the stump level during winter. Useful trees and shrubs are left standing, and are some times highly pruned. The cut and fell materials are left for drying and these are burnt to make the land ready for sowing different crops before the rains. After 2 - 3 years, this area is abandoned and a new site is selected. The number of years before a community returns to the same patch of land will depend on the land available, the community and its population. The length of fallow phase is considered critical to the success and sustainability of the system. During this period the soil, having been depleted of its fertility during the cropping period, regain its fertility through the regenerative action of the woody vegetation (Nair, 1993). The yield of arable crops declines with reduction in cycle time.

This system of cultivation is mostly applied in tropical and subtropical climatic areas with high rainfall and high humidity. Forest prevalent in such areas is tropical evergreen, temperate evergreen, conifer and pine forests (Rajora, 1998). This system is very labour intensive and requires small quantities of inputs and capital. In shifting cultivation, there is huge loss of soil and run off. So, shifting cultivation should be controlled. Rajora (1998) suggested that the indiscriminate cleaning of slopes should be prohibited and no land with more than 20 per cent slope should be cleared for the cultivation.

TAUNGYA

Basically the word taungya was originated from Myanmar (Burma). It comprises of two words- "*taung*" means hill and "*ya*" means cultivation (Nair, 1993). Taungya is a system of establishing commercial forest plants in which agricultural crops are grown on temporary basis (during early establishment period) in between rows of the tree species. Wood production is the ultimate objective of the taungya system, the immediate motivation for practicing it is food production. In India, the system is prevailing in Kerala, Tamil Nadu, Uttar Pradesh, Orissa, Andhra Pradesh, Karnataka and N-E regions (Rajora, 1998). In south India, the system is known as "*Kumari*" cultivation.

The land is prepared and planting is done. In certain areas, crops are grown one year before of plantation. The land usually belongs to the forest department and arable farming is done by selected farmers. Department establishes the tree plantation and the cultivators, besides using intervening area for cultivation, provide the labour to take care of the plantation. All the inputs for crop production are provided by the farmers. This agreement would last for two or three years, during which time the forestry species would grow and expend its canopy. This type of approach has been recommended for community wasteland development in the semi-arid Aravalli hills in Rajasthan (Rajora, 1998).

The taungya system permits sustainable use of forestland for food production by landless people who would otherwise be engaged in forest encroachment. Taungya system is often cited as a popular and mostly successful agroforestry approach to establishing forest plantation (Nair, 1993).

The modified taungya system has effectively been used by the progressive farmers in irrigated areas of Uttar Pradesh, Gujarat, Haryana and Punjab for establishing poplar plantation. A private firm WIMCO has been working very extensively for popular plantation on the farmers fields.

ALLEY CROPPING

Alley cropping is planting perennial trees/shrubs yielding forage in rows at some distances in association with agricultural crops. It is an agroforestry practice, which emerged as an alternative to shifting cultivation in humid tropics. It is a system in which food crops are grown in alleys formed by hedge rows of trees and shrubs. The primary purpose of alley cropping is to maintain or increase crop yields by improvement of soil fertility (through legume tree component and mulching) and microclimate (Singh, 1995). Alley cropping with tree rows aligned along contours will prove to be an effective means of erosion control and also impart stability to production. The hedge rows are cut back at planting of arable crops and kept pruned during the cropping system so as to prevent shading and reduce competition with food crops. When there are no crops, the hedge rows are allowed to grow freely or cut to meet the fodder needs.

Multipurpose tree species (MPTS) like *Leucaena leucocephala, Calliandra calothyrsus, Sesbania aegyptica* are generally preferred because they impart flexibility to the system (Singh and Singh, 1988). *Colospermum mopane, Dichrostachys nutan, Zizyphus nummulria, Mauriatina, Morus alba, Capparis decidua* are the important shrubs suitable for the silvipastoral systems in arid and semi-arid regions of India. Fruit plants like *Prunus persica* (peach), *Pyrus communis* (pear), *Psidium guajava* (guava) are the important constituents of agroforestry systems in semi-arid regions. Ber and pomegranate are the common constituents of agroforestry in farmer's field of Rajasthan.

Hedge rows should be planted in the direction of East and West to minimize the effect of shade. Hedge rows of MPTS are periodically pruned to prevent shading and utilized for mulching/green manuring both for high as well as low rainfall zones in sloppy and plain areas (Chinnamani, 1989).

The perennial trees are planted thick enough to act as barrier for run off of excess water and soil erosion. The continuous barriers across the slope provide series of vegetative checks for control of soil erosion and run off water. In course of time, a soil bund is automatically formed which may take the first line to defence against erosion.

Alley cropping is being practiced in several watershed development programme (Itnal, 1988). However, wide spread of adoption of this practice will depend on the ability of alley cropping to stabilize production, in the final analysis, to pay economically. One of the strong arguments for alley cropping is its ability to produce useful material even in years of severe drought when agricultural crops fail to grow.

Four aspects of hedgerow management of *Leucaena leucocephala*. viz., choice of crop, alley width, cutting height, and frequency/interval of cutting have been researched by CRIDA, Hyderabad in greater details (Singh, 1995). The experience with regard to each of these are presented below:

1. Based on 3 years experimental results (1984 to 1986), it was seen that yield reduction was higher (61 to 62 per cent) in pigeonpea and castor than sorghum (48 per cent) and pearl millet (30 per cent).
2. Alley width between 3 to 4 m are severely competitive with crops, more so in a drought year. Irrespective of the crop, the decline in yield was more in case of narrow (3.6 m) row width than in wider (7.8 m) row width, the per cent reduction being more in case of castor and pigeonpea.
3. Hedge row height of 100 and 150 cm are severely competitive with crops, in regard to light and moisture. Shading at anthesis/grain formation affects the productivity of crops in alley adversely. Hedge row cutting height of 15 cm was found advantageous, resulting in yield advantage of 14 per cent in case of pearl millet.
4. Intervals of 40 to 50 days and cutting heights of 75 and 150 cm yielded as much as 50 per cent more foliage than 15 cm cutting height.

Management of crop grown in the alleys has not received due attention and so is the case with microclimate and tree interaction studies. Some of the important studies with *Leucaena leucocephala* are as follows:

Leucaena leucocephala (Subabul) is a tree legume which can be maintained as tree for fuelwood or pulpwood and poles, as fodder bush or live hedge to harvest fodder 5-8 times a year for over 20 years. Forage with 25-30 per cent protein and 65 to 70 per cent digestibility makes an excellent feed to increase the profitability of dry farming. When established as a fodder crop, the annual fodder yield ranges between 20 to 80 t green fodder or 6 to 18 t dry matter/ha. This wide range of variation in forage yield of *Leucaena* is due to variation in the agroclimatic conditions and crop management practices (Hedge and Brewbaker, 1988).

Several recommendations have been made regarding the height of cutting, in the range between 50 to 150 cm. In India, 75 to 100 cm height is found to be ideal for higher yields as well as for minimizing the labour cost of weeding and manual harvesting (Hegde, 1983).

Tejwani (1994) reported that fuelwood and fodder production of *Leucaena* is influenced by rainfall. Under rainfed conditions, it seems to yield more fodder and fuelwood in the 1000 to 1500 mm annual rainfall range than in high (1500 to 2000 mm) or low rainfall (500 to 750 mm).

Leucaena yields more fodder when it is closely spaced (75 □ 25 cm) than when widely spaced (375 □ 25 cm). When it is intercropped, its fodder and fuelwood yields generally tend to go down, but there is an indication that fertilizers applied to crops contribute to its growth. The intercrops yield less than pure crops. The yield and type of crop grown is determined by amount of rainfall (Khybri *et. al.*, 1988, Grewal *et. al.*, 1992; Prasad and Verma, 1987).

Khybri *et. al.*, (1992) reported that pure cropping of paddy and wheat was found much less economical than intercropping with *Leucaena* at Dehradun. Rana *et. al.*, (1988) reported the results of an experiment conducted with subabul at HAU, Hissar. Experiment was planted under three row spacing (15, 30 and 45 cm). Anjan grass was transplanted between the rows of subabul. Narrow row spacing of 90 cm and intra row spacing of 15 cm produced significantly higher green and dry fodder yield of subabul. However, higher green and dry forage yield of anjan grass was recorded in wider inter and intra row spacing. But total fodder yield of subabu l+ anjan grass was recorded highest in narrow spacing (90 □ 15 cm). However, the quality of anjan grass was better under wider inter and intra row spacing.

In another experiment, subabul was planted at five different spacing (1.0 □ 0.5, 1.0 □ 1.0, 1.5 □ 1.0, 2.0 □ 1.0, and 3.0 □ 1.0 m) under sole and alley cropping system in the sub-humid subtropical Gangetic alluvial soil of West Bengal during 1981 to 1984. The highest herbage productivity (dry matter) of subabul (17.8 t/ha/year in 5 cuts) was obtained from sole crop planted at 1.0 □ 0.5 m spacing and that of anjan was 5.0 t/ha/year obtained from sole as well as in alleys of subabul planted at 1.5, 2.0 and 3.0 m^2/plant spacing. When anjan grass was grown in between subabul rows established at 1.0 □ 1.0 m spacing, the herbage yield of anjan and subabul were 4.4 t and 13.5 t/ha, respectively, the corresponding land equivalent ration (LER) value and monetary advantage were 1.69 and Rs. 1537 per ha (Chatterjee *et. al.*, 1988).

Intercropping under Scattered Planted Trees

Reports are available on extensive intercropping systems in which a variety of locally adaptive multipurpose trees are widely scattered over farmlands. The species diversity in this system is very much related to ecological conditions. Intercropping under ecologically adapted, naturally grown, scattered trees is the simplest and most popular form of agroforestry.

Range areas where ecologically adapted woody perennials are scattered on the grassland is the best example of such type of agroforestry system. *Capparis decidua* (kair), *Zizyphus numularia* (bordi), *Prosopis cineraria* (khejri), *Salvadora* (jal), *Acacia nilotica* (deshi babul), *Prosopis cineraria* (vilayati babul) are very common tree species of Indian desert abundantly seen on the rangelands.

Prosopis cineraria (khejri) and *Zizyphus nummularia* (bordi) are common woody components for such type of agroforestry systems in Indian arid tract.

KHEJRI BASED AGROFORESTRY

Khejri *(Prosopis cineraria)* is a multipurpose tree species, known as *kalp vraksha* due to its variety of economic uses. The species is rarely destroyed by the farmers. Jindal (1995) reported that on deep alluvial soils, the number of khejri plants per hectare varies from 20 to 40 giving a 4 to 10 per cent cover. On very deep soils of Nagaur and Ganganagar, 40 to 120 trees/ha are kept with a rainfall range from

150 to 350 mm and even upto 150 trees/ha in south of Sikar where annual rainfall is around 400 mm. In Western part of Rajasthan where rainfall ranges from 80 to 180 mm (around Jaisalmer) only 10 trees/ha are found in the natural habitat. Shah (1957) recommended tree lands of khejri with the density of 50 trees/ha to be the best landuse for Rajasthan desert.

Khejri is considered to be the most companion woody component for agroforestry in arid and semi-arid regions of India. Khejri shows best compatibility with the crops growing in its vicinity. It enhances the under story photomass production by fixing atmospheric nitrogen through rhizobial activities and by adding organic matter through its falling leaves. Moreover, it also reduces moisture losses by checking evapotranspiration through shading (Paroda and Jatasra, 1985).

Singh and Lal (1969) and Shankar and Saxena (1976) reported higher yield of crops growing in the vicinity of khejri tree. Shankar and Saxena (1976) reported that khejri tree canopy had a boosting effect on the herbage yield, growth and botanical composition of natural pasture. Singh and Lal (1969) pointed out that small leaves of khejri crown produce no real shadowing effect. They further reported that the soil under khejri had higher content of organic matter, total N, available P and soluble calcium as compared to open field conditions. pH values of soil under khejri was near neutral (7.3). This is an important feature, as the role of soil reaction in controlling the availability of nutrients is well known. High organic matter content in the soil under khejri may be attributed to leaf fall and its faster mineralization perhaps due to favourable moisture condition that exist under khejri canopy (Gupta, 1975).

In well-planned agroforestry practices of agricultural crops and khejri, the trees are lopped in a systematic manner from mid November to mid January every year (Mann and Saxena, 1980, Saxena, 1980). In case, if farmer takes *rabi* crops in khejri field the lopping of khejri should be completed by the end of October or first week of November. Khejri tree remains dormant in the winter months. By mid February the trees start sprouting. The tender wings are cut during the rainy season, from July to September. Khejri tree attain full foliage by the end of September, when the kharif crops also start maturing. The best examples of this practice are found in the districts of Pali, Nagpur, Sikar, Jaipur and Jhunjhunu (Tejwani, 1994). The lopping of the tree is based on traditional knowledge acquired by farmers by trial and error through generations of experience.

In early stage of growth, khejri is a slow growing tree, generally it takes about 10-15 years to become a tree in 200-350 mm rainfall zone. The variations in growth arise due to the variations in soil depth and total annual rainfall. Tree growth as well as tree density are better in high rainfall areas (350-450 mm) than in low rainfall areas (100-200 mm) shallow soils.

In general, khejri tree provide leaf fodder from the 10th year onwards and may go on producing upto 200 years. A moderately sized tree yields nearly 25-30 kg of dry leaves per annum (Tejwani, 1994). Khejri is a multipurpose tree. It provides forage, fuelwood, small timber (but rarely harvested for timber), fruit, tannin, gum and medicine. Branches, twinges and other parts after lopping are used as fuelwood. Its thorny twinges and branches are used for fencing the fields. The green and dry pods are used as vegetable, and ripe pods are considered the dry fruit of desert.

BORDI BASED AGROFORESTRY

Bordi *(Zizyphus nummularia)* plays a vital role in pastroralsilviculture in arid zone of India. Village common grazing lands with shallow soils and kankar pan at 30-45 cm soil depth are generally dominated bordi shrubs. There are some big *Zizyphus* orans dedicated to deities in Bikaner, Jaisalmer and Jodhpur districts of Rajasthan. These organs have 70-75 per cent dominance of *Zizyphus* plants. The rest 10-25 per cent is generally shared by *P. cineraria, Balanites aegyptiaca* and *Capparis decidua*. Here the shrub density with clear bole and crown ranges from 190 to 120 plants per hectare (Saxena, 1981). Shrubs of bordi also have remarkable power of regeneration through root suckers. They start sprouting goats. Bordi starts bearing fruits after the second year of establishment. Yearly cut shrubs on cultivated fields generally do not bear fruits. A five year olds bush in a grassland may produce 3.5 to 4.0 kg of fruits, old well developed shrubs in grassland produce 3-5 kg of fruits in the 150-250 mm rainfall zone. Harvesting of *Zizyphus* leaves starts with the season of leaf fall in winter. It has been confirmed by the field studies that winter harvest gives maximum above ground biomass leaf and shoot production, productivity per day and other growth parameters of the shrubs (Saxena and Sharma, 1981). The 14 per cent density of bordi in an organ has been found ideal for obtaining the maximum yield of leaf fodder and grass as compared to 11 or 18 per cent shrub density.

Bordi grows abundantly and wild in arid and semi-arid zones. The harvesting of bordi shrubs coincides with the season of its leaf fall and after the kharif crops are harvested. In a good or normal rainfall year, the farmers get a double harvest, i.e., the crop as well as bordi leaves for fodder while in drought years, when the crop fails, the cultivator gets only leaf fodder.

The plants are cut to the ground level and heaped together in one bundle. For one week all the bundles are put at one place for drying. Once dried, the leaves easily separated from the twinges by beating with a stick. These leaves (locally known as pala) are collected and stored for feeding to cattle during lean periods (Saxena, 1981). The yearly cut shrubs in agricultural fields do not produce any fruit. The bordi shrubs has a remarkable power of regeneration through its root suckers. The cut stumps start coppicing by February or March. The sprouting continues through April to June, indicating that the plant growth is independent of surface soil moisture. The spring sprouts are grazed during the summer months (Tejwani, 1994).

Bordi is one of the most common shrubs which occupies all the habitats. Flat alluvial plains with clay loam to clay soils and a kankar pan at 45-60 cm show high shrubs density (4 per cent cover) in semi-arid regions. Shallow soils with kanka pan at 30 to 45 cm are generally dominated by bordi shrubs. The gravelly plains where sheet wash has exposed the kankar pan, are dominated by a community of *Z. nummularia* and *Capparis decidua*. There are about 120-150 bordi plants per hectare which contribute 8 per cent crown cover (Saxena, 1981). In piedomont plains and eroded rocky surface, bordi and kair are equally dominant in the community of bordi and kair. These lands mostly serve as common grazing grounds. Sandy, undulating, hummocky plains and interdunal plains have a few scattered shrubs of bordi. In protected forests on low hills and piedomont zone, bordi is an important associated of *Acacia senegal*. The shrub density varies from 80 to 350 plants/ha but with small crown cover (5-7 per cent. In unprotected areas, with a community of thor (*Euphorbia caducifolia*), bordi has negligible presence (10-15 shrub/ha) due to its severe exploitation (Saxena, 1981).

Inter Cropping with Regularly Planted Trees

This type of approach is most commonly observed in the cultivated areas. The planting consists of both annual crops and perennial trees. The perennial tree species are systematically planted in rows at a particular distance. The planting distance varies with the species, and with the purpose of planting. The inter-space is utilized for growing of annual/seasonal crops.

Certain tree species after the possibility of providing atleast a portion of optimum crop nutrients by natural leaf drop or by loppings for purposes of green leaf manuring. This includes both fixed nitrogen as well as other nutrients recycled from the deeper soil depth. This is essentially true with *Acacia albida* which when matures (after 3 years) is said to be deciduous in the kharif season. As such it offers less competition for light and moisture at the time when crops need them most. *A. albida* is a native of Africa and annual rainfall in its native zone ranges from 300 mm to 1800 mm. It can tolerate temperature exceeding 44 °C and mild frost too. The species it strongly resistant to termite attack (Hocking, 1993). In Africa, *A. albida* considered to be one of the most important trees for agroforestry. The leaves are produced at the end of rains and are retained throughout the hot dry season, providing an invaluable source of animal fodder when no other supply is available. The remaining nitrogen rich leaves are then shed at the start of the rains, enriching the soil with plant nutrients. Leaflessness during rains has the further advantages of minimizing competition for light and moisture with under storey grass and legumes. A more common management strategy is not to lop for leaf fodder, making use of only the pods for fodder and leaving the leaves for soil enrichment. Plant generally produces pods after 5 years.

Das (1995) reported that it is the fastest growing legume found in semi-arid lands, reaches 6.5 m height after 4 years and 10.5 m after 7.5 years. He further stated that under favourable conditions it can reach the height of 30 m. *A. albida* planted at CAZRI, Jodhpur attained a height of about 4.5 m after 7 years of establishment (Singh, 1995 b).

Now a days intercropping with poplar has been commercialized by the prosperous farmers of Uttar Pradesh, Gujarat, Haryana, and Punjab. Poplar are planted on irrigated fields at 4 □ 5 m or 5 □ 5 m distance. After tree plantation, the crops can be taken up to five years. Popular trees are ready to harvest between the tenth and twelfth years. At this time, the wood is considered suitable for plywood, matchwood and packing cases. The plantation yields about 100 cu m of wood or 20 cu m of plywood per hectare after 12 years of rotation (Tejwani, 1994). The WIMCO, a private firm is doing lot of plantation of popular on the farmers field.

Intercropping of arable crops under trees involves boundary planting or systematic line planting of trees on crop fields is a form of agroforestry, adopted by the farmers. Tree species like *Acacia* species, *Azadirachta indica, Dalbergia sissoo, Eucalyptus species, Albizia lebbeck, Leucaena leucocephala, Prosopis species* etc. are commonly planted on the farm boundary, bunds/strips. In arid and semi-arid conditions, *Hardwickia binata* and *Colospermum mopane* have been found the potential top feed tree species for agroforestry systems.

AGRISILVIHORTICULTURE

Agrihorticulture system is commonly observed in western Uttar Pradesh and many other parts of the country. Cultivation of arable crops in the orchard of *Mangifera indica (mango), Psidium guava* (guava), *Emblica officinalis* (aonla), *Prunus persica* (peach), *Prunus persica* (pear) etc. on the farmer's fields in general and during early establishment in particular is a common practices. Jamun plantation is very common on the field boundaries and in strips in the field. Farmers planted trees on the farm boundaries, which act as wind break and minimized the damage by reducing the wind velocity.

Hortipasture, another system of landuse is very much suitable for arid and semi-arid areas. Vashishtha and Prasad (1995) reported the results of an experiment on Hortipastoral system conducted at CAZRI, Jodhpur. They introduced *Cenchrus ciliaris* in an established orchard of jujube in between rows of 6 m distance. During third year 32.9 q/ha grass was harvested with 25 kg/ha grass seed in addition to 29 kg ber fruits per tree. They found that there was no adverse effect of grass on the yield of ber.

In another experiment conducted at Hyderabad, guava and custard apple used as fruit plant, *Stylosanthes hamata* as forage legume and *Cenchrus ciliaris* as forage grass. Plant height of guava was found more with *C. ciliaris* whereas in custard apple it was more in control. The yield of *C. ciliaris* in first year was 5.0 to 9.6 t/ha.

Hortipastoral systems produce more biomass per unit area. This system also meets the basic requirement of fruit, fuel and fodder for rural masses of arid zone. Vashishtha and Prasad (1995) emphasized that there is need to conduct intensive research on multilocation to generate information for this composite system. This system will not only provide the judicious use of available land resources but will also generate additional employment opportunities for rural people. Papaya is another fruit plant, which have been used for agroforestry under irrigated areas.

REFERENCES

1. Chatterjee, B. N.; Maiti, S.; Majhi, S. K.; and Pal, S. 1988. In: Proc. Natl. Rangeland Symposium, IGFRI, Jhansi, Nov. 9-12, 1987 (Eds.: Punjab Singh and P. S. Pathak), 473.
2. Chinnamani, S. 1989. In: National Symposia on Agroforestry Systems in India, CRIDA, Hyderabad, Jan. 11-13.
3. Das, R. B. 1995. In: Regional training cum workshop on Silvipastoral systems in arid and semi-arid ecosystems. CAZRI, Jodhpur, Nov. 15-Dec. 5.
4. Grewal, S. S.; Mittal, S. P.; Dayal, S. and Agnihotri, Y. 1992. Agroforestry systems 17: 183-191.
5. Gupta, J. P. 1975. Annual Report. CAZRI, Jodhpur, 127.
6. Hedge, N. G. and Brewbaker, J. L. 1988. In: Proc. National Rangeland Symposium, IGFRI, Jhansi, Nov. 9-12, 1987 (Eds.: Punjab Singh and P. S. Pathak), 473.
7. Hegde, N. G. 1993. In: *Leucaena* research in the Asia-pacific region, Proc. of the workshop held in Singapore, 23-26 November 1982. National Development Centre, Canada. 73-78.
8. Hocking, D. 1993. Trees for drylands Oxford. IBH Publishing Co. Pvt. Ltd., New Delhi
9. Itnal, C. J. 1988. In: Proc. of National Symposia on Recent Advances in Dryland Agriculture, held at CRIDA, Hyderabad from Sept. 13-16.561-570.
10. Jindal, S. K. 1995. In: Regional training cum workshop on Silvipastoral systems in arid and semi-arid ecosystems. CAZRI, Jodhpur, Nov. 15-Dec. 5.
11. Khybri, M. L.; Sewa Ram and Bhardwaj, 1988. In: Regional training cum workshop on Silvipastoral systems in arid and semi-arid ecosystems. CAZRI, Jodhpur, Nov. 15-Dec. 5.
12. Lundgren, M. L. and Raintrree, J. B. 1982. In: Agricultural Research for Development: Potentials and Challenges in Asia (ed. Nestel, B.) ISNAR, The Hague, The Netherlands. 37-49.

13. Mann, H. S. and Saxena, S. K. 1980. Khejri in the Indian desert. Monograph No. 11, CAZRI, Jodhpur, 78.
14. Nair, P. K. R. 1983. An introduction to agroforestry. Kluwer Academic Publishers, London, 499.
15. Paroda, R. S. and Jatasra, D. S. 1985. In: Dryland Resources and Technology (Eds.: Alam Singh and G. R. Chowdhary). Scientific publishers and Geo-environ Academia, Jodhpur, 256.
16. Prasad, A. and Verma, B. 1987. Annual Report, CSWCRTI, Dehradun.141-142.
17. Rajora, R. 1998. Integrated Watershed Management. Rawat Publications, Jaipur New Delhi, 616
18. Rana, D. S.; Lodhi, G. P.; Taneja, K. D.; and Jatasra, D. S. 1988. In: Proc. Natl. Rangeland Symposium, IGFRI, Jhansi, Nov. 9-12, 1987 (Eds.: Panjab Singh and PS Pathak). 473 p.
19. Saxena, S. K. 1989. In: Khejri (*Prosopis cineraria*) in Indian desert, (Eds.: H. S. Mann and S. K. Saxena) CAZRI, Jodhpur. 4-10.
20. Saxena, S. K. 1981. In: Bordi, Monograph No. 3, (Eds.: H. S. Mann and S. K. Saxena) CAZRI, Jodhpur.3-9.
21. Saxena, S. K. and Sharma, S. K. 1981. In: Bordi, Monograph No. 3, (Eds.: H. S. Mann and S. K. Saxena) CAZRI, Jodhpur. 25-30.
22. Shah, S. A. 1957. Indian Forester 83: 489-490.
23. Singh, R. P. and Singh, S. P. 1988. In: Dryland Resources and Technology (Eds.: Alam Singh, G. R. Chowdhary and D. Kumar), Geo Environ Academia, Jodhpur.
24. Singh, R. P. 1995. In: Regional training cum workshop on Silvipastoral systems in arid and semi-arid ecosystems. CAZRI, Jodhpur, Nov. 15 Dec. 5.
25. Singh, K. S. and Lal, P. 1969. Ann. Arid zone 8: 33-36.
26. Shankar, V.; Dadhich, N. K. and Saxena, S. K. 1976. Forage Research 2: 91-96.
27. Shankar, V. and Saxena, S. K. 1976. Indian Farming 26: 22.
28. Tejwani, K. G. 1994. Agroforestry in India. Oxford & IBH publishing Co. Pvt. Ltd., New Delhi, 233 p.
29. Vashishtha, B. B. and Prasad, R. N.1995. In: Regional training cum workshop on silvipastoral systems in arid and semi-arid ecosystems. CAZRI, Jodhpur, Nov. 15-Dec.

Multiple-use Trees : Some Socio-Economic Implications for Agroforestry Programme

JAGDISH C. KALLA
Consultant, Aravalli Project, Gurgaon (Haryana)

Trees have always been an important constituent of Indian farming. Integration of tree component in conjunction with corps have been preferred for this use potential viz. wood for implements and fruitful fruits and seeds and leaves for food, top feed for animals bark and roots for medicinal purposes, and gums and rains for industrial uses pruned some of the prominent mix of uses of plants.

Post independence period in India is marked specifically for efforts in enhancement of productivity per unit of land. Consequently monoculture with its higher intensity of use of land and other resources relegated trees to relative redundancy. Other factors like higher population growth lack of alternative employment and resultant shrinkage and fragmentation of holdings hastened the process of allocating available resources for intensive monoculture use. In long run, strides in productivity of monoculture product mix may have provided resilience to food productivity and thus security. This however has a no-so apparent costs pertaining to continued acceleration of degradation of natural resource base, jeopardizing the very sustainability of farming. The technology of farming has thus came one full circle to realise the importance of reintroduction of plant components in the farm product mix. The continued experimentation pertaining to spatial geometry of tree plantation have concomitantly necessitated continued search for suitable plant material. The suitability of plant material has to be conditioned by use of maximizing potentials of plants under constraining resource base.

The principle determinants of allocating lands to trees in conjunction with regular crops call for.

1. **Choice of Plants:** The choice has to be determined principally by the farmers. Almost all Agroforestry programmes in the country presuppose that what is provided by scientists must be acceptable to farmers. This presupposition at times, almost forces the scientist's or organization's choice on farm families without even taking into account their perceptions of the multiple utility of plants. Further, wherever this hurdle is crossed, the programme tends to ignore the women folks who may not have decision-making power in rural households but surely are the major executors and users. Unless the women folks on farm front are not empowered to express this choice matrix, the programme may lack the sustenance and continuity.

2. **Geometrical considerations:** Considerations of plant densities which preclude the interaction of natural resources like sunshine, wind and hygroscopy, would be self-defeating. The selection of trees, which under full-grown conditions do not compete with regular crops for sunshine, is a prime consideration, which merits attention of technologists and protagonists of agroforestry systems.

3. **Symbiotic Externalities:** Tree growth in and around farm fields entail some crowns which harbour birds. Experience and studies have shown that almost 13 per cent and sometimes more of the crop products gets lost due to birds. This externality could be contained if in selection crown light tree species are preferred over crown heavy species. Another externality incidental to the existence of trees is occurrence of pathogens and parasites, which are commonly injurious to both tree and regular crops. A judicious combination of treating trees to selection and pruning would go a long way in providing resource-consuming resilience to the economy of farmers.

4. **Socio economic considerations:** Farm size under Indian conditions continues to decline over time. Consequently income-generating and manpower absorption capacity is also on decline. The technological options thus have to be tailored by incorporating relevant tree components. The relevance of tree component would be ensured if the price and market link ups are explored before the product mix are realised.

5. **Conclusion and policy implications:** Agroforestry programmes have acquired renewed meaning under existing conditions. For sustainable development of farm sector, it is, therefore, important that incorporating tree components rationally makes the present allocation of agricultural lands. The uncertainties associated with monoculture production have resulted in to unbearable and despotic financial stresses. This has gone for enough to cause taking extreme steps of resorting to suicides. The monetary and fiscal incentive for promotion of diversification of land use by incorporation of trees and guarantees of price and market support are therefore necessary preconditions for the success of Agroforestry programmes.

Strategy for Promotion of Agroforestry in Dry Regions

N. G. HEGDE
BAIF, Warje, Pune (Maharashtra)

ABSTRACT

Maharashtra is one of the water deficit states in India and also has a sparse forest cover. As a result, farmers in many of the districts are faced with acute shortage of fodder, fuel, timber and other tree products. Agricultural productivity being low in dry regions, tree-based farming systems has good potential to provide a stable income to farmers, particularly during the dry period.

The major problems in promoting agroforestry are lack of awareness about the benefits, improper selection of tree species, non-availability of good quality planting material, and inadequate returns due to unfavourable conditions, poor marketing facilities and hindrances posed by Tree Acts. In the absence of extension and financial support, farmers lack initiative to grow trees under agroforestry systems. These problems can be resolved by selecting agroforestry systems which can yield higher returns, promote farmers' groups to develop necessary infrastructure for input supply, processing and marketing of produce and providing techno-management support through government and non-government agencies.

INTRODUCTION

Agroforestry is a good system to improve the land productivity and microclimate although the scope for improving the income is restricted to some extent in dry regions. Nevertheless, the trees can ensure an alternate source of income when arable crops fail during lean years. In Maharashtra state where large areas are dry, agroforestry can help to improve the environment while meeting the basic needs of fodder, fuel, timber and other products. This state situated in the West-Central Sector of the Indian sub-continent has total land areas of 30.758 million hectares, representing about 10% of the geographical area of the country. The state covers 7.0% forests, 11% sown area, 4.0% net irrigated and 6.0% cultivable irrigated land in the country (GOM, 1986).

The forest cover in the state includes 4.232 million ha reserved forests, 1.5564 million ha., Protected forests and 0.6034 million ha. Unclassified forests. About 10% (3.070 million ha) of the land areas is barren. However, the total wastelands including other fallows have been estimated at 6 million ha. The major forests in the state are the Sahyadri (Western Ghats) in the west, the Satpuda ranges in the north-west, the Melghat ranges in the North and the Ajanta ranges in the north-east. The projected annual demand of fuelwood in Maharashtra has been estimated at 17 million m^3m resulting in a wide gap between the demand and supply.

The livestock population of Maharashtra in 1992 was 36.391 million, representing 8.0 per cent of the total livestock in India. On an average, there were five units of livestock per ha of cropped area, while the carrying capacity was only three or six units of livestock under rainfed and irrigated conditions respectively. The cropping intensity being low, crop residues alone are not adequate to maintain the livestock and will therefore have to be supplemented by other sources (GOM, 1986).

In the absence of adequate grazing lands, it was uneconomical to cultivate fodder crops on private lands. It was observed that about 98% of the farmers did not make any attempt to produce fodder, was directly linked to the income generated in the form of sale of milk or ready market for forage. Fodder production was economical, only if the livestock maintained by farmers were efficient in production (Hegde, 1991a). In the absence of a well-planned livestock management policy, scarcity of fodder was anticipated to continue in Maharashtra State as well as in the country.

SOCIAL FORESTRY IN MAHARASHTRA

Social forestry was introduced during the second five year plan through a scheme of assistance to village *Panchayats* for afforestation and development of pastures. A forestry extension scheme was introduced during the fourth plan to take up tree planting on roadsides and canal banks. So far this scheme has covered about 1500 km long strips in the State. After the establishment of the social forestry department in 1982, various on-going tree-planting programmes were transferred to this department for implementation. Under the USAID aided social forestry project, the Government of Maharashtra set the objectives of increased supply of firewood, fodder, fruit, timber and other minor forest products in rural areas, while reducing the rate of deforestation.

The success of the social forestry department based on the rate of survival of the saplings was reported to be about 70 per cent. Without the involvement of the community members, these plantation required continuous supervision. Looking to the poor participation of the local people, the following recommendations were made for successful implementation of the social forestry programme (IRDE, 1990):

1. Emphasis should be given on motivation and extension activities.
2. Cultivation of fodder, bamboo and non-wood forest produce species to generate short term income before the trees matures.
3. Establishment of cooperatives to raise *kisan* nurseries and handle dairy and forestry produce.
4. Setting up of training and demonstration facilities for farmers and extension workers.

SCOPE FOR AGROFORESTRY

Considering the rainfall pattern, forest cover and fuelwood supply, three based farming systems can play a significant role in improving the farm productivity in the state. Trees established on agricultural fields should improve crop production and total return per unit area, while improving the ecosystem and microclimate.

Among many agroforestry systems, the scope in dry regions is limited to only a few models, which can generate substantial income. The most popular models presently practiced in certain dry regions of

Maharashtra are block plantation of *Eucalyptus, Jatropha curcus* fruit orchards and windbreaks. The first three models are intended to replace the traditional arable crops for enhancing income and the windbreak is expected to improve the agricultural production and the micro-climate.

POPULAR MPTS IN DRY REGION

Some of the important species presently planted under various agroforestry systems in dry regions of Maharahstra are listed below (Hegde, 1991b):

- **MPTS for Farm Forestry:** *Eucalyptus* spp., *Jatropha curcus, Dendrocalamus strictus, Albizia lebbeck, Leucaena leucocephala, Azadirachta indica.*
- **MPTS for Orchards:** *Mangifera indica, Annona squamosa, Zizyphus mauritiana, Emblica officinalis* and *Tamarindus indica.*
- **MPTS for Windbreaks:** *Acacia auriculiformis, A. nilotica* var. *cupressiformis, A. nilotica* var. *telia, Agave sisalana, Ailanthus excelsa, A. indica, Cassia siamea, Dalbergia sissoo, Eucalyptus* spp., *Derris indica, Gliricidia sepium, L. leucocephala* and *Parkinsonia aculeata.*

PROBLEMS OF PROMOTING AGROFORESTRY IN DRY REGIONS

Promotion of agroforestry particularly as windbreak and alley cropping in dry regions is difficult compared to farm forestry and establishment of fruit species. In case of farm forestry and orchards, farmers are prepared to invest a large amount hoping to generate high income. Farmers prefer those species, having better demand and ready outlet for the produce. As the block plantations have large tree population, farmers take more interest to manage such plantations unlike windbreaks and alley cropping. In spite of assured benefits from trees, agroforestry is not expanding fast, due to several problems. The major problems in dry region are:

- Lack of awareness among farmers.
- Choice of desired species.
- Non-availability of planting material,

- ☐ Inadequate returns.
- ☐ Unfavourable conditions and
- ☐ Local uses and market outlets.
- ☐ Hindrance of the tree acts.

STRATEGY FOR PROMOTION OF AGROFORESTRY

The problems affecting the promotion of agroforestry need to be solved through research back up, extension and policy matters. Various support required to promote agroforestry in dry regions of Maharashtra have been clubbed under the following sections.

Selection of Suitable Systems and Species

In dry areas, where tree growth is slow and biotic pressure in high, it is better to establish plants in small blocks with closer spacing and provide protection through the establishment of livehedges and cattle-proof trenches, at a low cost.

Selection of species should be based on their adaptability and utility. Preference may be given to non-browsing by nitrogen fixing tree species. The choice of nitrogen fixing trees fro dry regions is limited. Hence, search for additional tree species should be continued through on-farm trials. In areas where tree planting is not viable, hardier species like agave (*Agave sisalana*) can be establishment on contour bunds. This can help to conserve soil moisture and regenerate native grasses. This system would contribute to the income through grass and fibre from the second or third year. Subsequently, these species can be established in the plots. Short duration species such as pigeon pea (*Cajanus cajan*) and castor (*Racinus cumuinis*) can also be established on field bunds, while sowing arable crops at the onset of monsoon. These plants grow fast and produce a small quantity of biomass at the end of the rainy season.

Our present knowledge about the tree-crop interaction is very limited and adequate information is not available e to quantify the advantages and disadvantages of trees in cropping areas. It is also necessary to quantify the direct and indirect benefits of trees on soil and moisture conservation, soil productivity and the environment to motivate the farmers to adopt Agroforestry. Hence, on-farm research should be initiated to upgrade the technologies.

Various systems developed in consultation with the local farmers should be established for demonstration. The responsibility of maintaining such demonstrations should be entrusted to the land owners, by providing suitable training and incentives.

PROMOTION OF AGROFORESTRY GROUPS

It is necessary to organize groups of farmers who are interested in establishing agroforestry, preferably located in one block. Such a united effort will help in easy procurement of inputs, processing and marketing of the produce and collective protection of plantations, both from human and livestock interference. Selected members of this group can serve as trainers and motivators to promote agroforestry in the neighbouring villages. They can also represent their problems to the government to modify the Tree Acts, assure minimum support price for the produce and provide necessary technical and financial support.

TECHNO-MANAGEMENT SUPPORT

To encourage farmers to grow trees, government and non-government organisations should strengthen their field network for technical and extension services. While promoting tree planting, other related aspects of conservation and production should also be included in the programme. These include soil and water conservation, adoption of proper tillage operations and low external inputs such as biofertilizer, green manure, vermi-compositing and mulching. Such systems can be more attractive. Regular training based on local needs should be organised. This should be followed by awareness and extension activities to highlight the benefits of various trees. The extension wing can provide information regularly on new technologies, modification in the tree Acts and market demand supply situation. Such support can improve the profitability of agroforestry systems even in dry regions.

REFERENCES

1. GOM, 1986. Seventh. Five Year Plan, 1985-90. Planting Department, Government of Maharashtra, Bombay.
2. Hegde, N. G. 1991a. Yojana 35 (16): 27-29.

3. Hegde, N. G. 1991b. Impact of afforestation programme on Socio-economic transformation of the rural poor. Ph.D. Thesis submitted to Pune University.

4. IRDE, 1990. Recommendation of the social forestry seminar (USAID Research Project) held in Pune on July 10, Institute of Rural Development and Education, Pune: 6.

Grow Neem under Agroforestry Systems to Bring Prosperity to Farmers

R. DEB ROY
Ex. Director, NRCAF, Jhansi (UP), B-387 Sarita Vihar, New Delhi

INTRODUCTION

With fast increasing human and livestock population in our country which is likely to exceed 1020 and 600 m, respectively by 2000 A.D. the gap between the demand and the supply of fuelwood, fodder and timber besides, food is also widening. Despite the massive afforestation and tree planting programmes the forest cover or our country is still much below the national target of 33 per cent of the total geographical area. In order to meet the daily demand of our people in respect of fuelwood and fodder besides food and also to achieve the national target of $1/3^{rd}$ area under tree and grass cover we have no alternative but to involve our people including our farmers to grow more MPTS including neem tree under forestry and agroforestry systems.

Neem (*Azadirachta indica*) the *Kalp Vriksh* is held in great esteem in India from time immemorial and is variously called as "*Nature bitter flow*", the "*wonder tree*", "*Nature's gift to mankind*", "*The tree that purify*", "*The tree of the twentieth century*" and the "*tree for solving global problems*" (Saxena, 1993). At present we have about 19 m neem tree in the country, 50% of which is confined to U.P. (Deb Roy, 1997). Looking to the importance of this multipurpose tree both nationally and internationally we should aim at and plan to grow about 1000 m neem trees by the next 10-12 years to bring prosperity to our farmers and the country as a whole thereby also meeting the national target of $1/3^{rd}$ area under tree and grass cover.

USES OF NEEM

The commercial valued neem has been known to our country since the *vedic* times (about 1500 B.C.) Every part of neem tree, leaf, flower, fruit, seed, kernel, bark, wood, root etc. has been is use especially in "*Ayurvedic and or Unani*" systems of medicine and has been traded for various purposes. Today neem is being used in a large number of products such as pesticides and allied chemicals, medicines, plant nutrients, animal feeds, fuel, raw materials for industries etc. (Parmer and Ketkar, 1993).

Neem is a renewable source of various useful biofriendly products. A fully grown neem tree yields about 40-50 kg of fruits and 300-350 kg of leaves annually. Assuming that there are about 20 m neem tree and about 15 m fully grown tree we can expect annually about 6.0 m tonnes of leaves and about 0.87 m tonnes of fruits. The fruits contain 24% skin pulp, 47% pulp, 19% shell and 10% kernel and the kernel contain 30-45% oil. The annual income just from the sale of fruits and leaves would be about Rs. 5300 million and would generate huge rural man power for collection of fruits and leaves. Neem oil could be exported as it has got a large demand for eco-friendly pesticides, insecticides, neem coated fertilizer, medicine etc. and can bring foreign exchange.

Neem in Agroforestry

Neem can be introduced alongwith other MPTS in any of the different agroforestry systems The purpose of introducing neem in various agroforestry systems is to increase the population of neem to meet the target of 1000 m plants by next 10-12 years that India can produce enough raw material for neem based products.

Agrisilvicultural Systems

The purpose of introducing neem in this system is not only to increase biomass production but also to improve polluted environment and increase agrisilvicultural production by *in situ* uses of neem leaves and fruits by reducing demands of insecticides etc.

Agrisilvicultural studies carried out at NRCAF, Jhansi and elsewhere revealed that there is not much reduction in crop production even with 800 MPTS /ha in the initial years provided these MPTS are managed properly. In order to reduce light competition the canopies are manipulated annually by pruning and removing 40% plants thinning in the fourth year thereby producing extra leaf fodder for livestock and fuelwood. In case of neem the twigs could be utilized for brushing and cleaning teeth. The population of neem could be 25% of the MPTS to start with other MPTS such as *H. binata*, *A. cupressiformis, L. leucocephala, D. sissoo, Casuarina equisetifolia* or even *E. tereticornis* for quick return which gave maximum biomass of 147.5 t/ha through thinning followed by 23.2 and 20.7 t/ha respectively in *Leucaena* and *A. cupressiformis* (Deb Roy, 1994; 1995).

On 9th year with about 315 plant/ha the biomass production from pruning was as high as 10.6 kg/tree in *A. lebbeck* with 7.4 kg of fuelwood and 6.9 kg in *D. sissoo* and 2.2 kg in *M. azedarch* plus fruits, soybean production was highest (7.82 q/ha) in association with *A. nilotica* and is at par of control (without crops). Wheat production however was highest 17.0 q/ha in associaltion with *M. azedarach* compared to control (21.52 q/ha) (NRCAF, 1996-97). Poplar based agrisilvicultural studies in J & K., H.P., Punjab and Tarai areas of U.P. revealed that at 6-9 yrs the crops productivity was 75.8% compared to control (Dhanda and Verma, 1998). Bisaria *et. al.* (1996-97) recommend (400-800 tree/ha) of *H. binata* for optimum tree growth and crop production.

At NRCAF, Jhansi crops like urd/moong are grown under rainfed conditions with about 400 neem trees with pruning them regularly to minimise shade effect. Pruning not only reduces shade and provide green leaf fodder and fuelwood but also improve the quality of wood having longer clean bole. At the age of 10 years, about 25% plants should be removed retaining better ones and at 15 yrs only 150 plants/ha should be kept to the final rotation of 20-25 years. Vivekanandan (1998) reported from a drought prone village as Tamil Nadu where farmers grow, green gram, millet, sorghum, cotton with neem trees starting with more trees and gradually reduced to about 36 trees/ha. By careful pruning farmers keep the trees free of lateral branches upto a height of 3.5 m. Farmers protect the young trees with thorny branches of *P. juliflora* and with paddy straw and sprinkled with cowdung slurry to repel animals. They collect 50-60 kg/plant of seed (fruits) every year from 10th year and harvest them at 15-20 years As about 189 farmers growing neem in about 300 ha. They can afford to purchase small tractor which could plough field in between trees even upto 800-1000 trees/ha. They can easily start with 400 trees/ha pruning/lopping them regularly at alternate year to get regular flow of fruits which bring them good income. More emphasis needs to be given for further research to include neem alongwith other MPTS in agrisilvicultural systems.

Agrihorti-silvicultural System

Deb Roy (1994; 1995) and others (Tejwani, 1994; Grewal *et. al.*, 1992; Solanki, 1997) discussed in detail the potential of agrihortisilvicultural system. Recently Deb Roy (1997); Pathak (1997) and Vashistha (1997) also emphasized the role of neem trees in agrihortisilvicultural system for improving the economy of the farmers. Earlier Deb roy reported potential of introducing fast growing MPTS like Leucaena (200 plants/ha) in Agrihortisilvicultural system which gave annually 2.5-3.0 t/ha of nutritious green fodder besides 1.8-2.5 t/ha of dry fuelwood with 200-400 fruit trees/ha. Number of MPTS will depend on fruit trees and their standard spacing. In this system 25-30% ot the MPTS may be neem tree. So out of 200 MPTS 50-75 neem trees and 125 to 150 Leucaena or other fast growing NFTS. Here neem would provide valuable neem leaves for fodder and also for *in situ* mulching around fruit trees in the initial years and fruits in the later years. Leachate from leaves and fruit falling on the ground will help to protect the crops from nematodes, pests and disease. Collected fruits also give some additional income to the farmers annually. Neem trees will also act as wind breaks during hot summer and cold winter protecting the fruits trees besides the falling fruits and leaf biomass would improve the nutrient status of the soil. These aspects need further research.

For further results MPTS especially neem should be prunned annually from 2nd/3rd year so that it grows fast having lean bole so as to give less shade and better wood value. Neem started fruiting at the age of about 6 years after which trees may be lopped one third every year to get fruits as well as leaves besides fruit wood and twigs.

Commercial Crops

In commercial crops like Tea, Coffee, and Cocoa introduction of neem plants alongwith other MPTS may be tested as a shade tree besides amelioration of soil and protecting the crops from pests and disease.

Plantation Crops

Most important plantation crops in India are coconut and areca nut which are also grown with various crops like tapioca, yam, sweet potato, ginger, turmeric, pine apple, green gram, black gram, paddy guinea grass, cowpea etc. (Rethinum, 1993, Tejwani, 1994). Research to introduce neem in coconut and areca nut based agroforestry system needs to be taken up for improving the production in this system.

Hortisilvipastoral System

Under semi arid and arid rangelands there are large number of naturally growing *Zizyphus nummularia* (Borch) plants available which could be successfully budded with improved varieties of ber viz. sev, gola, banarasi karaka, umran etc. Fuel cum fodder trees like anjan *(H binata)* khejri (*P. cineraria),* subabul (*L. leucocephala*) and neem could be introduced if not already available alongwith perennial grass like *Cenchrus, Chyrysopogon, Lasiurus, Stylosanthes*. Fruiting of ber could start from the second year with 4 kg fruits/tree reaching 10 kg in the 4th year. It gives an income of Rs. 1600 from second year to Rs. 9880 per ha in the 4th year (Rai *et. al.*, 1993, Deb Roy, 1994). Mulching ber plants with neem leaves can improve growth and production; pruning/lopping of MPTS including neem provides green leaf fodder besides fuelwood. Neem fruits can also give additional income. Effect of *in situ* mulching of ber plants with neem leaves and the effect of neem fruits on the production of ber fruits needs to be tested experimentally.

Silvipastoral System

In marginal, submarginal and other degraded lands silvipasture has been found to be the most economic agroforestry system under arid and semi-arid conditions (Deb Roy and Pathak, 1983; Deb Roy, 1994; 1995; Pathak and Khan, 1990; Singh, 1990). Deb Roy *et. al.*, (1980; 1982) discussed in detail the forage production of 3.5 t/ha/annum from *A. tortilis and L. leucocephala* based silvipastoral system with *C. cililaris, C. setigerus and Stylosanthes* and fuelwood production of 7-8 t/ha/annum on a 13 year rotation. Rai *et. al.* (1991-92) reported forage production of 6.8 t/ha/annum with *A. amara, D. cinerea and L. leucocephala* 800 pl/ha with *Chrysopogon and Stylosanthes* species. Introducing young goat and sheep for grazing from second year revealed that the former gain body weight @ 41 g/h/day and the later 58 g/h/d without any concentrate.

Neem being a good fodder for goat, its introduction in the silvipastoral system would further increase livestock production besides providing fuelwood from annual lopping and thinning and fruit production from the $5^{th}/6^{th}$ year. Neem may constitute 25-30% of the total MPTS in silvipastoral system.

Shelterbelt and Windbreak

In arid and semi arid areas shelterbelts and wind break around the crop fields are planted to protect the crops from desiccating hot winds, three row of shelterbelt is quite affective in reducing the wind speed and wind erosion during hot summer and increasing crop production in arid areas (Gupta and Venkateswarlu, 1993). *A. tortilis, A. indica, E. camaldulensis, D. cinerea, A. amara* are used commonly in arid areas. For more effectiveness perennial grasses like *Lasiurus sindicus, Panicum turgidam, C. ciliaris* may be sown in between the trees.

Strategy of Achieving 1000 m Neem Trees

1. **Neem in forestry:** Every year huge forest area are planted with 800-1000-1250 plants/ha. If we plant 20% neem plants for 3 lakh ha annually and with 80% survival we can get 54 m trees of neem. Villagers do prefer neem as has been revealed in recent PRA exercises conducted for JFM in Rajasthan.
2. **Silvipastoral system for wasteland:** With 25% plants of neem in silvipastoral system @ 800 MPTS/ha and with 80% survival and covering annually 2 lakh ha, we get 36 m neem trees.
3. **Agrisilvicultural system:** Our farmers may be induced to grow 50-100 neem trees/ha in one lakh ha annually in their farm land with 90% survival we get about 10 m neem plants.
4. **Agri-horticultural:** In agri-hortisilvicultural and hortipastoral system growing 40-50 neem tree/ha with about 90% survival covering annually 50,000 ha we get another 2 m neem plants.

So in next ten years we achieve the above target of having about 1000 m more or 1020 m neem plant i.e. one neem plant/head. After 10 yeas from assuming that neem trees are properly protected and managed the annual value of only fruits and leaves have been calculated to be around Rs. 2,66,875 million.

REFERENCES

1. Anonymous. 1997. Annual Report, NRCAF, Jhansi.
2. Deb Roy, R. and Pathak, P. S. 1983. Ind. Rev. Life Sci., 3: 247-264.
3. Deb Roy, R.; Patil. B. D. and Pathak, P. S. 1980. Ind. J. Range Mgmt. 1 (2): 113-119.
4. Deb Roy, R.; Patil, B. D.; Pathak, P. S. and Gupta, S. K. 1980. Ind. J. Range Mgmt. 3 (122): 45-52.
5. Deb Roy, R. 1994. In: Natural resources management for sustainable agriculture and environment (ed. Deb B. L.) Angor Publ. Ltd., New Delhi.
6. Deb Roy, R.1995. In: Sustainable development of dryland agriculture in India (ed. Singh, R. P), Sci. Pub. Jodhpur.
7. Deb Roy, R. 1997. Neem the Kalpa Vriksha: Souvenir Sarita Vihar Sarvajanin Puja Sanity. New Delhi.
8. Dhanda, R. S. and Verma, R. K.1993. In: Agroforestry in 2000 A.D. for semi arid (Eds.: Gill, A. S., R. Deb Roy and AK Bisaria), NRCAF, Jhansi.
9. Gupta, J. P. and Venkatswrulu, J. 1993. In: Agroforestry in 2000 AD for semi arid (Eds.: Gill, A. S.; R. Deb Roy and AK Bisaria), NRCAF, Jhansi.
10. Pathak, R. K. 1997. In: Recent advances in Agroforestry and its role in conservation of soil, water and biodiversity (Eds.: Solanki, K. R.; AK Bisaria, V. K. Gupta, P. Rai and R. Newaj), NRCAF, Jhansi.
11. Pathak, P. S. and Khan, T. A.1990. In: Silvipastoral system in India (Eds.: Pathak, P. S. and Punjab Singh), Range Mgmt. Soc. India, IGFRI, Jhansi.
12. Parmar, B. S. and Ketkar, C. M.1993. Commercialization in Neem research and development. Soc. Pesticide India.
13. Rethinum, P. 1993. In: Agroforestry in 2000 AD for semi arid and arid tropics (Eds.: Gill, A. S. R. Deb Roy and AK Bisaria), NRCAF, Jhansi.
14. Solanki, K. R. 1997. In: Recent advances in agroforestry and its role in conservation of soil, water and biodiversity (Eds.: Solanki, K. R.; AK Bisaria, V. K. Gupta, P. Rai and R. Newaj), NRCAF, Jhansi.
15. Saxena, R. C. 1993. In: Souvenir World Neem Conference, Ind. Soc. Tobacco, Sci.
16. Singh Panjab 1990. In: Silvipastoral systems in India. Range Mgmt Soc. of India, IGFRI, Jhansi.
17. Vashisht, B. B. 1997. In: Recent Advances in agfoforestry and its role in conservation of soil, water and biodiversity (Eds.: Solanki, K. R.; AK Bisaria, V. K. Gupta, P. Rai and R. Newaj), NRCAF, Jhansi.
18. Vevekanandan, P. 1998. Agroforestry Today, Jan-March.
19. Tejwani, K. G. 1994. Agroforestry in India. Oxford & IBH Pub. Co. Ltd., New Delhi.

MPTs Inventory and use in Agroforestry: Current Scenario and Tasks Ahead

P. S. PATHAK
Indian Council of Agricultural Research, New Delhi

ABSTRACT

Trees have been the basic component of the traditional farming systems of India. Their uses were intimately linked with the traditional wisdom and knowledge. It is only in the recent times that the trees are being categorised as MPTS or SPTS and much emphasized, since large scale deforestation and harvest of trees on farm lands has created deficit of vital timber, firewood and fodder resources. Although, Indian vegetation in different Botanical regions (7) with 16 forest types and 5 grass covers has been extensively studied but there has been no serious attempts to inventoried the MPTS in the farming systems and also forest areas to highlight their use potential in different farming systems. MPTS research movement although is recent but much head way could be made due to goal identification and research prioritization by the MPTS Research Network supported by Winrock International during the period. Although much progress made on MPTS and fodder trees inventory in India has been based on more of Academic compilation from literature and some institutions through D&D exercise have given the eco-regional priority species but it lacks information regarding farming systems preferences, tree crop association, farm holding size suitability, edaphic preferences, socio-cultural and economic suitability etc. Paper presents the MPTS scenario in the country, species prioritization and the industry driven species adoption by the farmers. Differences between house hold use preferences and the high economic turnover through single species plantations are analysed for their adoption and use.

MPTS trials have been taken up and promising tree species are identified in the country but their potential for adoption in the farming systems and use are not well understood. Looking to the need of households for timber, firewood and fodder and the industry for pulpwood, potential species of MPTS are identified for their improvement and propagation strategy to maintain sustained high production. Research thrusts have been indicated in view of the launching of the IXth plan and meeting the future needs of the country.

INTRODUCTION

Tree plantation has been an age-old practice. Ever since launching of Van Mahotsava in 1954, tree planting continued as a regular practice in India. During the past 14 years there has been a national drive to afforest the degraded and wastelands in the country to check the process of desertification and land degradation. In the process of choosing species for afforestation different objectives are kept in mind alongwith matching of site potential and market demands.

With the initiation of agroforestry research, objective search for tree species to suite the farming systems and meet the farmer's demands started and thus multipurpose trees research became agenda of many national and International organizations. Note worthy is the contribution of Nitrogen Fixing Tree Association with its head quarters in Hawaii and the Winrock International's Forestry and Firewood Research and Development project, with head quarters at Bangkok with ascent on Multipurpose Tree Species research. ICRAF also initiated MPTS research programme during the eighties. In India, although tree research was on the agenda of many organizations, its organised research thrust was initiated during 1983 with the launch of All India Coordinated Research Project on Agroforestry. By now much has been achieved. This paper attempts to inventories the MPTS and indicate priorities of research and the potential species.

MPTS DEFINED

As the abbreviation denotes, some people call it Multipurpose Tree species while some as Multipurpose Trees and Shrubs. One can very conveniently call it Mix of Preferred Tree Species. Van Carlowitz and Burley (1984) defined MPTS as the trees grown deliberately or kept and managed for preferably more than one intended use, usually economically motivated major products and or services in any multipurpose land use system, specially agroforestry systems. Thus, it involves more than one purpose and also multipurpose land use.

The other important role assigned for trees is site improvement for which nitrogen fixing tree species are desired. The trees are known as Nitrogen Fixing Tree Species (NFTS). Such species fix nitrogen from the atmosphere. Mostly they are leguminous but some times non-leguminous trees with association of *Frankia* also fix nitrogen. There are more than 1000 species of such woody perennials, which fix nitrogen.

HISTORICAL PERSPECTIVES

India has a very old history of plantations. Survey of records shows:

1842 ☐ *Teak plantations of Nilambur*
1843 ☐ *Eucalyptus pycnantha in Tamil Nadu*
1866 ☐ *Fuelwood plantations of Changamanga*
1869 ☐ *Teak plantations of Makum*
1872 ☐ *Teak plantations of Kulsi*
1874 ☐ *Rubber plantations of Charuduar*
1890 ☐ *Dalbergia sissoo, Mesua and some exotic*
1912 ☐ *Dalbergia sissoo plantations in Bundelkhand*
1934 ☐ *Tectona grandis, Shorea robusta, Garcinia indica, Acacia catechu, Salmelia melabarica*
1948 ☐ *Commercial plantations and taungya*
1951 ☐ *Scheme on afforestation, rehabilitation planting and commercial plantations. Special emphasis given to industrial plantations. Species were Casuarina, Eucalyptus, cinchona, wattles and match wood species.*
1956 ☐ *Afforestation and improvement of poor areas, species of commercial and industrial value (Teak, wattle and blue gum).*
1961 ☐ *Commercial and industrial plantations. Plantations outside the forest areas.*
1969 ☐ *Replacement of slow growing species by fast growing ones. Largescale introduction of exotics like Eucalyptus, tropical pines and poplars.*
1974 ☐ *Forests for food, Forests for people and forests for wood. Emphasis on many species of MPTS.*
1976 ☐ *Man made forests and Social forestry.*
1985 ☐ *Organisation of National Wastelands Development Board and call for planting at least 5 million hectares every year.*
1992 ☐ *National Wastelands Development Board split into Department of Wastelands Development and National Afforestation and Eco-development Board to cater to the requirement of non forests and forest wastelands separately.*

This subject has attracted wide attention during the past two decades. Besides research papers regularly published on the MPTS in research journals, many compilations and books have been published. Notable among them are the proceedings of MPTS Research consultation meetings of 1987, 1989 and 1990. Six reports of MPTS Research committee meetings and also their steering committee meetings. Nitrogen Fixing Tree Association continued to publish an annual volume on Multipurpose Tree Research Reports regularly bringing out salient research achievements on the subject. Symposia proceedings of MPTS research by the countries of South and South-east Asia etc. were also compiled regularly.

In India a comprehensive book dealing with MPTS has been published by Gupta (1993). Pathak (1993) while discussing the MPTS for agroforestry systems gave the list of promising species for different climatic regions of the country.

INVENTORYING MPTS

MPTS identification for agroforestry is very complicated since it is expected to fulfill many requirements, which are often poorly defined, less well understood and lack scientific information about its holistic value in a landscape (Table 1).

TABLE 1: *Knowledge gap analysis of MPTS in agroforestry systems*

Attributes	Well understood	Poorly understood
Production	*Production of timber and firewood production*	*Coppicing cycle*
Above ground		*Lopping management and products viz; fibre, thatch, silk, fodder, medicine*
Wood	-	
Leaf	*Production*	*Production and role/use drink, medicines*
Flower	-	
Fruit	-	*Management techniques for improving production*
Bark	-	
Fibre	-	*Production and use.*
Litter	-	*Production*
Gum	-	*Production and nutrient cycling in agroforestry system*
Resin	-	
Below ground		*Tapping techniques and yield associated and microbial biomass*
Wood		*Production, structure*
Fine root		*Production, dynamics*

Attributes	Well understood	Poorly understood
Interactions	*Crop-production decline*	*Competition, shading, selection of ideal crops under trees.*
Above ground	-	
Below ground		*Symbiosis, microbial interactions, nutrient transfer, habitat enrichment*
Environmental	-	*Not well understood*
Trees and rains	*Role of shelterbelts*	*Tree mixtures and their planting geometry.*
Arresting wind erosion	*Vegetative barriers, alley cropping*	
Arresting water erosion		*Water uptake and hydrological cycle*
Microclimate moderation	*Some factors viz; light*	*Microclimate diurnal flux and impact on nutrient turnover*
Pest and Weed control		
Degraded land rehabilitation		
Bioremidiation wildlife habitat		
Socioeconomic	*Horticultural value*	*Economic value*
Amenity and Tourism	*Through planting and products*	*Indirect employment*
Employment generation		*Role through traditional knowledge*
Public education	-	*Quantification*
Risk reduction	*By product diversification*	*Proper understanding and valuation*
Counter seasonality		*Medicines*
Improved nutrition	-	
	Through fruits, wets	

Criteria for selection of trees in agroforestry involves many parameters viz; fast growth, tree farm, response to management practices, compatibility with associated crops, nitrogen fixing ability, products variability and their easy marketability. Some such criteria cannot be combined in any one species but one can attempt to search for many of them in one. Earlier approach proposed by wood (1989) was modified by Pathak and Roy (1992) and 15 attributes were highlighted (Table 2).

Burley (1987), highlighted in details the products and services from trees. He also gave a list of semi-arid species with their distribution and phenology. An important progress in this direction was preparation of MPTS database by ICRAF (Von Carlowitz 1986) and development tree seed directory and information system (Von Carlowitz 1986, 1987). Though the data base it was possible to match climate, soil, tree characteristics and products for any specific site.

TABLE 2: *General characteristics for selection of trees/shrubs in agroforestry*

Tree attribute	Relationship of attribute with performance in agroforestry
1. *Height*	*Ease of harvesting leaf, fruit, seed and branch wood* *Shading/wind effects*
2. *Stem form*	*Suitability for timber, poles, posts,* *Shading effects*
3. *Crown size, shape and density*	*Quantity of leaf, mulch and fruit production* *Shading / wind / allelopathic effects*
4. *Rooting pattern*	*Competitiveness with other components Suitability for soil conservation*
5. *Physical and chemical composition of leaves and pods*	*Fodder and mulch quality* *Soil nutritional aspects/allelopathic effects*
6. *Thorniness*	*Suitability for barriers*
7. *Wood quality*	*Acceptability for fuel and various other wood products*
8. *Phenology*	*Timing and labour demand for fruit, fodder and seed harvest*
9. *Di/monoeciousness*	*Sexual composition of individual species community* *Importance for seed production and pollen flow*
10. *Pest and disease resistance*	*Important regardless of function*
11. *Vigour*	*Biomass productivity/early establishment*
12. *Site adaptability and ecological range*	*Suitability for extreme sites or reclamation sites*
13. *Phenotypic or eco-morphological variability*	*Potential for genetic improvement* *Need for culling unwanted phenotypes*
14. *Response to pruning and cutting management practices*	*Use in alley farming or for lopping/ coppicing management systems*
15. *Possibility of nitrogen fixation*	*Use in alley farming, planted fallow and rotational systems*

Hegde (1990) while presenting inventory of MPTS gave their characteristics for agroforestry systems and listed tree species with potential for fodder cum fuelwood, fuelwood cum timber, pulp wood, fruits and vegetables, cold hilly regions, high rainfall areas, drylands and the irrigated areas.

TABLE 3: *List of important fodder trees in India having multiple uses.*

Species	Other uses
1. *Acacia albida Del.*	1, 2, 10, 7
2. *Acacia farnesiana (L.) Willd.*	1, 10, 11, 7
3. *Acacia ferruginea DC*	2, 11
4. *Acacia nilotica (L.) Willd ex. Del.*	1, 2, 10, 11
5. *Acacia planifrons Wight & Arn*	1, 2, 11
6. *Acacia senegal (L.) Willd.*	1, 2, 10, 11, 7, 6
7. *Acacia tortilis Hayne*	1, 2, 5, 7, 8
8. *Acacia cordifolia Hook, F.*	9, 2
9. *Ailanthus excelsa Roxb.*	3, 2, 1, 6
10. *Albizia amara (Roxb.) Boiv.*	1, 2
11. *Albizia lebbeck (L.) Benth.*	1, 2, 3, 6
12. *Albizia odoratissima Benth.*	2, 11, 4
13. *Albizia procera (Roxb.) Benth.*	1, 2
14. *Albizia stipulata Boiv.*	2, 11, 4
15. *Anogeissus pendula Edgew.*	1, 2
16. *Artocarpus hotcrophyllus Lamk.*	9, 1, 2, 6
17. *Artocarpus lakoocha Roxb.*	1, 2, 4, 9
18. *Azadirachta indica Adr. Juss.*	1, 2, 6, 3, 4
19. *Bauhinia purpurea Linn.*	1, 11
20. *Bauhinia retusa Roxb.*	9, 2, 10
21. *Bauhinia variegata Linn.*	1, 9, 6, 10
22. *Celtis australis Linn.*	2, 1, 9, 6
23. *Dalbergia sissoo Roxb. ex. DC.*	1, 2, 7
24. *Dendrocalamus strictus Nees*	2, 7
25. *Dichrostachys cinerea Wight & Arn*	1, 5, 7
26. *Ficus bengalensis Linn.*	4, 6, 14
27. *Ficus glomerata Roxb.*	9, 6, 4
28. *Ficus hookeri Miq.*	1, 15, 6
29. *Ficus religiosa Linn.*	4, 6, 15
30. *Gliricidia sepium (Jacq.) Walp.*	1, 3, 2, 9
31. *Grewia optiva Drummond ex. Burret*	6, 2, 12
32. *Hardwickia binata Roxb.*	2, 12, 1
33. *Holoptelia integrifolia Planch.*	1, 2
34. *Leucaena leucocephala (Lam) De Wit.*	1, 3, 2, 8, 4, 13, 9
35. *Mangifera indica Linn.*	9, 2, 1
36. *Moringa oleifera Lam.*	1, 13, 9

Species	Other uses
37. *Pithecellobium dulcee Benth.*	9, 1, 7, 10, 11
38. *Pongamia pinnata (L.) Plerre*	1, 2, 6
39. *Prosopis cineraria Druce*	2, 1, 9, 6, 7
40. *Prosopis juliflora DC.*	1, 2, 7, 5
41. *Pterocarpus marsupium Roxb.*	1, 2, 10
42. *Quercus semicarpifolia Sm.*	1, 2
43. *Quercus leucotrichophora Camas ex. Bahadur*	1, 2, 7, 10
44. *Sesbania grandiflora Pers.*	9, 7, 1
45. *Sesbania sesban Merrill*	1, 2, 7, 6
46. *Zizyphus jujuba Lamk.*	9, 1, 2, 6, 5
47. *Zizyphus nummularia (Brum. F.) W & A.*	9, 1, 2, 6, 5

Note: *1 = fuel wood/charcoal; 2 = timber/pole; 3 = mulch; 4 = shade; 5 = fence; 6 = medicine; 7 = soil conservation; 8 = windbreak; 9 = food/fruit; 10 = tannin/resin; 11 = gum; 12 = fibre; 13 = pulp; 14 = leaf plates; 15 = latex.*
*(**Source:** CSIR 1985; Mac Dicken and Mehl 1990; Tewari and Singhal 1991; FAO 1991).*

TABLE 4: *Potential multipurpose species for different agro-climatic situations and farming systems for small farmer in semi-arid India*

Species	Agro-climatic Preference		Farming System Preference
	Climatic	Edaphic	
1. *Acacia albida*	2,3	1,2,3	1,3,5,
2. *A. auriculiformis*	2,3	1,3,5	1,11
3. *A. leucophloea*	1,2	1,3	5,10
4. *A. nilotica*	1,2	1,3,4	1,3,4,5,9,10,11
5. *A. senegal*	1,2	2,4,1	7,10
6. *A. tortilis*	1,2,3	3,6,5	3,5,10,11
7. *Agave sisalana*	3	3,5	10,11
8. *Ailanthus excelsa*	1,2	3,4	3,5
9. *Albizia amara*	1,2	1,3,4	3,4,5
10. *A. lebbeck*	2,3	1,3,5,6	1,3,4,5,9,11
11. *A. procera*	2,3	3,4,1	1,3,4,5,9,8
12. *Annona squamosa*	1,2	1,2,3,4	2,4,6,9
13. *Anogeissus latifolia*	2,3,	1,3,4	1,3,4,5,9
14. *A. pendula*	2,3	1,3,4	3,4,5,9
15. *Artocarpus heterophyllus*	2,3	2,4	2,4,6,9
16. *Azadirachta indica*	1,2,3	1,5,3,2	1,2,4,5,9,11
17. *Bambusa aurundinacea*	2,3	1	9,11

Species	Agro-climatic Preference		Farming System Preference
	Climatic	Edaphic	
18. *Bauhinia variegata*	3	4,3	3,4,5,9,11
19. *Capparis decidua*	1,2	3,4	2,4,6,9
20. *Carica papaya*	2,3	2,4	2,4,6,9
21. *Cassia siamea*	2,3	3,4,2	1,11
22. *Casuarina equisetifolia*	2,3	3,5	1,11
23. *Cordia gharaf*	1,2	3,4	2,4,6,9
24. *Dalbergia sissoo*	3	1,3,2,5	1,3,4,5,9
25. *Dendrocalamus strictus*	2,3	3	1,3,4,9,11
26. *Desmenthus virgatus*	2,3	3,4,1	8
27. *Dichrostachys cinerea*	1,2	3,4,1,6	3,4,5,10
28. *Emblica officinalis*	2,3	2,5,3	2,4,6,9,11
29. *Eucalyptus camuldulensis*	2,3	1,3,4	1,2,7,11
30. *E. tereticornis*	2,3	1,3,4	1,2,7,11
31. *Ficus bengalensis*	3	2,4	9
32. *F. glomerata*	3	2,3,4	9
33. *F. religiosa*	3	2,3,4	9
34. *Gliricidia sepium*	3	1,2,3,4	1,8,9
35. *Grewia subinequalis*	2	3,4	2,6,4,9
36. *Hardwickia binata*	1,2,3	1,3	1,3,4,5,9
37. *Jatropha carcus*	3	3,5	10
38. *Leucaena leucocephala*	2,3	1,3,4	1,3,4,5,7,8,9,11
39. *Mangifera indica*	3	1,2,3,4	2,4,6,9
40. *Melia azaderach*	2,3	1,5,4	1,3,4,9,11
41. *Morus alba*	3	4,3	2,4,6,9,11
42. *Parkinsonia aculeata*	1,2	3,5,1,4	3,5,4
43. *Phoenix dactylifera*	1,2	3,4,5	2,4,6,9,11
44. *Pithecellobium dulce*	2	1,2,3,4	5,9,3
45. *Pongamia pinnata*	2,3	3,5,1	1,3,4
46. *Prosopis cineraria*	1,2	2,3,5	1,3,4,5,9
47. *P. juliflora*	1,2	3,5,1,6,2	
48. *Psidium guajava*	2,3	4,3,2	2,4,6,9
49. *Punica granatum*	1,2	2,5,3	2,4,6,9
50. *Salvadora oleoides*	1,2,3	3,5	3,4,5
51. *Sesbania grandiflora*	3	2,3,4	8,9,1
52. *S. sesban*	2,3	2,3,4	8,9,1
53. *Syzygium cumini*	2,3	2,4	2,4,6,9

Species	Agro-climatic Preference		Farming System Preference
	Climatic	Edaphic	
54. *Tamarindus indica*	2,3	2,3,5,1	2,4,6,9,11
55. *Zizyphus jujuba*	1,2,3	1,2,3,4	2,4,6,9
56. *Z. mauritiana*	1,2,3	3,5,4,1	2,4,6,9

*(**Note**: Climatic Preference: 1 = Rainfall 200-400 mm; 2 = Rainfall 400-800 mm; 3 = Rainfall 800-1000 mm. Edaphic Preference: 1 = Laterite soil; 2 = Black soil; 3 = Red soil; 4 = Alluvial soil; 5 = Saline-alkali soil; 6 = Calcareous soil. Farming System Preference: 1 = AS; 2 = AH; 3 = ASP; 4 = ASPH; 5 = SP; 6 = HP; 7 = EP; 8 = AC; 9 = HG; 10 = LF; 11 = FF).*

Identification of tree species for plantations was taken up seriously during 1976 when at the national level a list for different climatic regions was published by the Government. Patil and Pathak (1977) modified the list and included the grasses, legumes and the shrubs for the benefit of degraded lands rehabilitation. Identification of important fodder trees with multiple uses was made by Pathak and Roy (1992) (Table 3). Efforts were also made to identify the trees on the basis of climatic, edaphic and farming systems preferences (Table 4).

In a further attempt Pathak (1993) identified the tree species for different types of problem soils (Table 5). A detailed list of species for the rehabilitation of alkaline, saline and waterlogged soils has been presented by Pathak and Dagar (1998).

TABLE 5: *Multipurpose trees and shrubs suitable for different edaphic situations*

Edaphic situation	MPTS
1. *Acidic Soils*	*Albizia falcataria, A. procera, A. stipulata, Acacia auriculiformis, Alnus nepalensis, Gmelina arborea.*
2. *Sandy arid*	*Prosopis cineraria, P. juliflora, Acacia tortilis, Calligonum polygonicles, Tamarix aphylla, Zizyphus mauritiana, Z. nummularria*
3. *Coastal sandy*	*Prosopis juliflora, Casuarina equisetifolia, Anacardium occidentale*
4. *Poorly drained soils*	*Eucalyptus camaldulensis, Albizia procera, Terminalia arjuna, Acacia nilotica, Syzygium cumini, Casuarina equisetifolia.*
5. *Alluvial soils*	*Acacia nilotica, Melia azedarach, Azadirachta indica, Dalbergia sissoo, Eucalyptus tereticornis, Populus deltoides, Leucaena leucocephala, Mangifera indica, Bassia latifolia.*
6. *Calcareous soils*	*Acacia nilotica, Albizia lebbeck, Azadirachta indica, Eucalyptus camaldulensis, Melia azedarach, Tamarix aphylla.*
7. *Saline sodic soils*	*Acacia nilotica, A. tortilis, Pithecellobium dulce, Prosopis juliflora, P. cineraria, Sesbania sesban, S. aculeata, Tamarix aphylla, Azadirachta, Albizia procera, Terminalia arjuna.*

Edaphic situation	MPTS
8. Shallow gravelly soils	*Albizia amara, Zizyphus mauritiana, Hardwickia binata, Anogeissus pendula.*

Fodder Trees

Fodder trees have been enumerated, researched for their nutritive value and production through lopping management by many workers. Chaturvedi (1948), Ganguli *et al* (1964) Singh (1982) Everist (1969), IDRC (1990) are among the few noted publications specifically dealing with fodder trees. Deb Roy *et. al.* (1980) enumerated 74 species of trees for their fodder value and highlighted their role in dry lands and also flood affected areas. A comprehensive book of fodder trees (Singh 1982) has brought out the detailed description, growth attributes and nutritive value of species in India. Melkania (1991) described elite fodder tree species for mountainous regions giving their occurrence as per the altitude and also site preference. He also presented the nutritive characteristics of 27 species. He remarked that compared to the Himalayan fodder trees, the Nilgiris trees are least studied.

Pathak and Roy (1992) reviewed the fodder tree research and evolved the criteria for prioritizing species for the country. They highlighted the multiple uses of those species. Climatic, Edaphic, cultural and farming systems compatibility were the main attributes for selecting and prioritizing a species. Thus, based on utility index, adaptability index and productivity index, a superiority index was worked out (Table 6).

From the table it is apparent that species scoring more than 25 index value viz., *Acacia nilotica, Azadirachta indica, Leucaena leucocephala, Prosopis juliflora* and *Zizyphus jujuba* were most important while those with 20-24 were important (viz., *Sesbania sesban, Quercus leucotrichophora, Q. semicarpifolia, P. cineraria, Moringa oliefera, Mangifera indica, Grewia optiva, Gliricidia sepium, Dichrostachys cineraria, Dendrocalamus strictus, Dalbergia sissoo, Celtis australis, Artocarpus heterophyllus, Albizia procera, A. odoratissima, A. lebbeck, A. amara, Acacia tortilis, A. senegal* and *A. albida)*. They further remarked that a careful selection and management of fodder trees would improve the productivity and sustainability of agroforestry systems. Singh *et. al.* (1996) have given a comprehensive account of the fodder trees of eastern Himalayan region alongwith their nutritive value.

TABLE 6: *Superiority indices of some fodder trees in India*

Species	Productivity index (0-10)	Utility index (0-10)	Adaptability index (0-10)	Superiority index (0-30)
1. *Acacia albida*	6	7	8	21
2. *Acacia farnesiana*	6	5	4	15
3. *Acacia ferruginea*	5	5	5	15
4. *Acacia nilotica*	10	10	8	28
5. *Acacia planifrons*	5	6	5	16
6. *Acacia senegal*	7	9	6	22
7. *Acacia tortilis*	6	9	8	23
8. *Acacia cordifolia*	6	7	6	19
9. *Ailanthus excelsa*	5	6	7	18
10. *Albizia amara*	6	8	8	22
11. *Albizia lebbeck*	7	6	8	21
12. *Albizia odoratissima*	6	7	7	20
13. *Albizia procera*	6	7	8	21
14. *Albizia stipulata*	6	7	6	19
15. *Anogeissus pendula*	6	7	5	18
16. *Artocarpus heterophyllus*	9	7	8	24
17. *Artocarpus lakoocha*	6	7	6	19
18. *Azadirachta indica*	9	10	7	26
19. *Bauhinia purpurea*	5	7	7	19
20. *Bauhinia retusa*	6	6	6	18
21. *Bauhinia variegata*	6	6	7	19
22. *Celtis australis*	7	7	7	21
23. *Dalbergia sissoo*	8	7	7	22
24. *Dendrocalamus strictus*	8	6	7	21
25. *Dichrostachys cinerea*	6	9	5	20
26. *Ficus bengalensis*	5	8	6	19
27. *Ficus glomerata*	5	7	6	18
28. *Ficus hookeri*	6	7	6	19
29. *Ficus religiosa*	5	8	6	19
30. *Gliricidia sepium*	7	9	8	24
31. *Grewia optiva*	9	7	7	23
32. *Hardwickia binata*	6	6	6	18
33. *Holoptelia integrifolia*	5	6	5	16
34. *Leucaena leucocephala*	8	8	10	26

Species	Productivity index (0-10)	Utility index (0-10)	Adaptability index (0-10)	Superiority index (0-30)
35. *Mangifera indica*	8	6	7	21
36. *Moringa oleifera*	7	6	8	21
37. *Pithecellobium dulce*	6	6	7	19
38. *Pongamia pinnata*	5	5	5	15
39. *Prosopis cineraria*	10	9	5	24
40. *Prosopis juliflora*	6	10	10	26
41. *Pterocarpus marsupium*	6	6	6	18
42. *Quercus semicarpifolia*	7	7	7	21
43. *Quercus leucotrichophora*	8	7	7	22
44. *Sesbania grandiflora*	6	6	7	19
45. *Sesbania sesban*	6	8	8	22
46. *Zizyphus jujuba*	8	9	8	25
47. *Zizyphus nummularia*	5	7	6	18

Nitrogen Fixing Trees and Nitrogen Transfer

Nitrogen fixed by trees is transferred through fine root decay, nodule decay, excretion of nitrogen from nodules in the below ground parts or litter fall and pruning applied as mules above the ground.

Nitrogen fixing trees complete less with crop species than the non-NFTS for nitrogen in soil but in the nitrogen rich soils species rely on soil nitrogen and their fixation rates are low (Daniel 1993). Nitrogen addition to soil is very small through excretion. About 5 per cent of the nodule weight is nitrogen. Substantial amount of nitrogen is expected to add to soil through the nodule and root decay. A proper association of annual crops benefits from such processes depending upon the proximity with the scattered trees.

Benefits from above ground processes are more pronounced since many studies on the use of prunnings as mulch/green manure indicate the increase in soil fertility (Kang et al 1981, Pathak). Mostly NFTS are advocated for agroforestry since they can adopt to low nitrogen soils than the other species. ICRISAT (1989) have suggested the use of such prunnings as fodder than mulch in arid and semi arid climates due to their better utility compared to green manuring which is vapidly lost due to fast decomposition.

INDIGENOUS / EXOTIC WOODY SPECIES

Troup (1921) described over 500 species, 244 genera and 67 families of which 30 species and 22 genera are of outstanding value. There are about 270 commercial timber species listed from India (Pearson and Brown 1932). The Indian species have been worked on intensively for their silvicultural requirements and regeneration techniques. Luna (1989) opined that the species choice depends upon the purpose, demand and supply, cost of planting, growth rate, ease of establishment, coppicing power, vigour, growth habit, stem form, crown form and resistance to pests and diseases.

Indigenous species are preferred over exotic because they are well adapted to the local environmental, less vulnerable to pests and diseases, the end products are known, and their uses are known to people. Exotic introduction is recommended for increasing yields, improving wood quality and other special characteristics. When indigenous species can meet these requirements, it is felt that there is no need for introducing the exotic. Where fast growth is the major concern, poor sites sometimes do not suit exotic species.

Past experience in India has proved that exotic species demand good soil, assured supply of water and nutrients and ideal silvicultural practices for economic production. Hence the success of exotic species has been confined only to selected areas, while one third of the total geographical area, measuring over 125 million hectares, falls under the category of wastelands. We will have to look for different indigenous tree species for establishing green over.

Indigenous tree species have an added advantage of being accepted by the local population who are aware of making multiple uses of tree products, without depending on outside help for marketing. In plantation forestry normally three practices are followed viz; raising the same species, planting the area with non-indigenous species and planting of exotic. In many cases artificial restoration becomes important before planting the site. Luna (1989) has listed 67 successful exotic species in India.

PRIORITIZING MPTS

During the International congress of Genetics at New Delhi in 1983 a panel discussion was organized by SPWD on "*Choice of Tree Species*" for various types of degraded lands in different parts of India. The panelists included Prof. J. L. Brewbaker, Dr. J. Burley, Dr. K. G. Eldridge, Dr. Gene Namkoong, Dr. K. Ohba, Dr. M. R. Ahuja alongwith several senior foresters of the country and chaired by Dr. M. S. Swaminathan. They recommended that:

1. The problem of choice of tree species has to be seen in the wide variety of prevailing ecological conditions in different parts of India and keeping in view our expanding population and its essential requirements.
2. Discrimination between exotic and indigenous species in unfortunate. The tree species should be primarily chosen according to ecological requirements, and biotic limitations.
3. Special attention needs to be given to seed production and selection of seed bears for selected species in different agro-climatic zones, including import of appropriate seeds, and proper provenances from abroad.
4. Tree breeding can produce varieties of trees for fuelwood and timber production, which are faster growing than those usually planted at present. Tests of a large number of species and provenance are needed to evaluate the genctic resources of potentially suitable trees.
5. Collection, storage and distribution of germplasm of different species/provenances is necessary in cooperation with other international organisations.
6. It is necessary to identify and multiply multipurpose species for firewood, small timber, fibre and even food.
7. Special attention needs to be given in choosing nitrogen-fixing tree species in various agro-climatic zones according to their suitability, specially *Acacia, Casuarina, Dalbergia, Leucaena, Prosopis, Albizia, Calliandra, Parkinsonia, Samenea, Gliricidia, Erythrina, Sesbania, Alnus* and *Robinia*.
8. While much is known about the cylindrical bole, less branching of tree desired for timber production, specifications/plant types for fuelwood and fodder trees still need to be determined.

Internationally important nitrogen fixing tree species were identified in this meeting (Table 7). During the past few years through the experiences of the D & D exercise conducted by the centres most important species were identified and the priority species were identified for each centre to conduct the germplasm collection, evaluation and species improvement. The experiments conducted on the MPTS evaluation at different centres indicated best performing species for each centre. Based on the existing scenario of species research in the country an attempt has been made to identify the most important and important species for the country to undertake detailed research (Table 8).

TABLE 7: *Internationally important nitrogen fixing tree species*

Climatic region	Tree Species for			
	Fuelwood		Fodder	
	Priority 1	Priority 2	Priority 1	Priority 2
Arid and Semi-arid tropics	*Acacia albida, A. Senegal, A. tortilis, Cajanus cajan, Casuarina equisetifolia, Gliricidia sepium, Leucaena leucocephala, Prosopis juliflora complex.*	*Acacia auriculiformis, A. confusa, A. farnesiana, A. mangium, A. mearnsii, A. nilotica, A. saligna, Albizia falcataria, A. lebbeck, Calliandra calothyrsus, Casuarina cunnighamiana, C. glauca, D. sissoo, Derris indica, Parkinsonia aculeate, Pithecellobium dulce, Prosopis alba, P. tamarugo, Samanea saman, Sesbania grandiflora.*	*Acacia albida, L. leucocephala, G. sepium, Prosopis alba, P. cineraria, P. glandulosa, P. juliflora, P. tamarugo*	*Acacia farnesiana, A. nilotica, A. senegal, A. saligna, A. tortilis, Albizia falcataria, A. lebbeck, Derris indica, Inga vera, L. diversifolia, Parkia roxburghii, Parkinsonia aculeate, Pithecellobium dulce, Samanea saman, Sesbania grandiflora*
Humid Tropics	*Acacia auriculiformis, A.*	*Acacia confusa, A. farnesiana,*	*Gliricidia sepium,*	*Acacia farnesiana, A. mangium,*

Climatic region	Tree Species for			
	Fuelwood		Fodder	
	Priority 1	Priority 2	Priority 1	Priority 2
	mangium, Casuarina equisetifolia, Gliricidia sepium, Leucaena leucocephala	*A. mearnsii, Albizia falcataria, A. lebbeck, Cajanus cajan, C. glauca, D. sissoo, Derris indica, Erythrina berteroana, E. fusca, E. poeppigiana, Leucaena diversifolia, Nimosa soabrella, Parkia roxburghii, Parkinsonia aculeata, Pithecellobium dulce, Samanea saman, Sesbania grandiflora.*	*Leucaena leucocephala, Sesbania grandiflora*	*A. mearnsii, Albizia falcataria, A. lebbeck, Callindra calothyrsus, Derris indica, Erythrina berteroana, E. fusca, E. poeppigiana, Inga vera, Leucaena diversifolia, Pithecellobium dulce, Samanea saman.*
Highland Tropics	*Acacia mearnsii, Alnus nepalensis, Gliricidia sepium.*		*None*	*Acacia farnesiana, A. mearnsii, Alnus nepalensis, Leucaena diversifolia, Pithecellobium dulce*
Temperate	*Robinia pseudoacacia*	*Acacia farnesiana, Leucaena leucocephala, Prosopis glandulosa complex*	*Robinia pseudoacacia*	*Alnus acuminata, A. glutinosa, A. nepalensis, A. rubra, Casuarina cunninghamiana, Prosopis glandulosa complex.*

TABLE 8: *Priority MPTS for India*

Very Important Species	☐ ***Gmelina arborea, Dalbergia sissoo, Populus deltoides, Prosopis cineraria, Casuarina equisetifolia, Acacia mangium, Artocarpus heterophyllus***

Important Species	☐ *Acacia nilotica, A. auriculiformis, Alnus nepalensis, Grewia optiva, Morus alba, Calophylum inophyllum, Azadirachta indica, Melia azaderach, Ailanthus excelsa, Paraserianthes falcataria, Tamarindus indica, Michelia champaca, Thespesia populnea, Gliricidia sepium, Tecomella undulata, Leucaena leucocephala, Prosopis juliflora, Cieba pentandra*

RESEARCH THRUSTS

- ☐ Domestication of promising indigenous and exotic MPTS in agroforestry systems.
- ☐ Germplasm collection, evaluation and multiplication for optimizing production in specific habitats and niches.
- ☐ Establishment of seed orchards, seed multiplication and supply.
- ☐ Tree research in holistic perspectives (including Socio-economic) of the agroforestry systems to understand the complementarities.
- ☐ Post harvest technology for value addition of the tree related products.
- ☐ Creating ideal marketing infrastructure to ensure high returns.
- ☐ Establishing role of MPTS in bioremediation of polluted habitats and environments.

REFERENCES

1. Burley, J. P. and Von Carlowitz. 1983. Proc. planning workshop to discuss International Cooperation, ICRAF, Nairobi, pp. 298.
2. Burley, J. 1987. In: (Eds.: Steppler, H. A. and P. K. R. Nair) Agroforestry: a decade of development, 273-287.
3. Chaturvedi, M. D. 1948. U. P. Forest Deptt. Bulletin No. 19, 7 p.
4. CSIR 1985. The Wealth of India: Raw Materials (Vols. I - XI). Council of Scientific and Industrial Research, New Delhi.
5. Daniel, J. N. 1993. Range Management and Agroforestry 14 (2): 185-194.
6. Deb Roy, R.; Shankaranarayan, K. A. and Pathak, P. S. 1980. Indian Forester 106 (4): 306-311.
7. Enerist, S. L. 1969: Qld. Dept. of Primary Industries, Div. Plt. Ind. advisory Leaflet 1024, Queensland (Australia).
8. Ganguli, B. N.; Kaul, R. N. and Nambiar, K. T. N. 1964. Ann. Arid zone 3: 33-37.

9. Hangen, C.; Medema, L. and Lentican, C. B. 1990. Proc. of International Conference held Nov. 20-23, 1989, Jakarta, F/FRED and IDRC, Canada: 208.
10. Hanover, J. W. 1988. Feasibility study on small farm production of gums, resins, exudates and other non wood products. WINROCK International, Arlington (USA): 47.
11. Hegde, N. G. 1990. In: (Eds.: Pathak, P. S.; R. DebRoy and P. Singh) Multipurpose Tree Species fort Agroforestry Systems. Range Management Society of India, Jhansi, 15-27.
12. Hegde, N. G. 1993. Range Mgmt. and Agroforestry 14 (1): 73-77.
13. ICRISAT, 1989. Ann. Rep. Patancheru, AP. 5022324, India.
14. IDRC. 1990. Shrubs and tree fodder for farm animals, IDRC, Ottawa (Canada), 349.
15. Kang, B. T.; Wilson, G. F. and Sipkens, L. 1981. Plant and soil 85: 267-277.
16. Luna, R. K. 1989. Plantation forestry in India, International Book Distributors, Dehradun, India.
17. Melkania, N. P. 1991: Range Mgmt. and Agroforestry 12 (2): 179-194.
18. Pathak, P. S. 1993. In: (Eds.: Jha, L. K. and P. K. Sen Sarma) Agroforestry -Indian Perspective, 31-59, Ashish Publishing House, New Delhi.
19. Pathak, P. S. and Roy, M. M. 1992. Range Mgmt. and Agroforestry 13 (1): 63-87.
20. Patil, B. D. and Pathak, P. S. 1977. Invention Intelligence 12: 79-87.
21. Singh, K. A.; Yadav, B. P. S. and Goswami, S. N. 1996. Research Bulletin no. 39, ICAR Research Complex for NEH Regions, Barapani, Shillong (Meghalaya): 99.
22. Singh, M. 1982. (ed): Top-feed resources-their production, utilization and the constraints, Central Sheep and Wool Research Institute Avikanagar (India), 303.
23. Taylor, D. and MacDacken, K. G.1991. Proc. International Workshop held Nov. 19-23 Los Bonos, Philippines, Winrock International: 259.
24. Tewari, D. N. and Singhal, R. M. 1991. In: (ed. N. G. Hegde and J. N. Daniel), Selection of Multipurpose Tree Species for Small Farmers BAIF, Pune: 25-34.
25. Troup. 1921 Silviculture of Indian Trees, Oxford University Press, Oxford.
26. Von Carlowitz, P. 1986 a. In: (Adams, N. and R. K. Dixon eds.) Forestry Networks: Proceedings of the Forestry Network Workshop of F/FRED, 71-78, Morrilton, Arkansas, USA, Winrock International.
27. Von Carlowitz, P. 1986 b. Multipurpose Tree and Shrub Seed Directory, First edition, Nairobi, ICRAF, 265.

28. Von Carlowitz, P. 1987. Agroforestry Systems 5: 319-338

29. Withington, D.; MacDicken, K. G.; Sastry, C. B. and Adams, N. R. 1988: In: Proc. International Workshop held Nov. 2-5 1987, Pattaya, Winrock International Instt. for Agricultural Development, USA and IDRC, Canada: 281.

MPTs Inventory and Identification for Sustaining Agricultural Production in Eco-Friendly System

A. K. SRIVASTAVA AND S. A. KHAN
CS Azad University of Agriculture and Technology, Kanpur (UP)

ABSTRACT

As a result of green revolution, there was spectacular increase in gram production in Indo-Gangetic plains of India. The extensive use of land, chemical fertilizer and water resources for achieving high production potential caused various type of degradation of land and water resources. Keeping above facts in view a survey of districts Rai-bareli, Unnao and Kanpur have been conducted with the objectives- To identify the multipurpose tree species from different industrial areas and their adaptability in relation to pollution control. The survey was conducted at Sugar Mill (Munshi Ganj), Firoj Gandhi Thermal Power, (Uchahar), INCAN Fertilizer (Bacharawa) and Parag Dairy in the district Rai-Bareli, Panki Thermal Power Station (Panki), ICI Fertilizer (Panki) Parag Dairy (Nirla Nagar), and Jajmu Tannery area of Kanpur district and Distillery industry of Unnao district.

INTRODUCTION

The extensive use of land, chemical fertilizer and water resources for achieving high production potential caused various type of degradation of land and water resource. The major problems being faced in central plain zone of U.P. due to deforestation are development of alkalinity and sodicity, recession of ground water, loss of nutrient through different degradation forced. Similarly the exponential increase in industrialization and urbanisation has led to increasing level of several gaseous pollutants in the atmosphere affecting all life forms on earth. There is a substantial evidence that air pollution can harm people to a great extent as reported by Kar, 1995 and Srivastava et.al. 1996.

MATERIALS AND METHODS

The study was conducted in the vicinity of different industries responsible for air and water pollution in three districts namely Kanpur, Unnao and Raibarelly of Uttar Pradesh. The detailed study area is as follows:

Dist.	Industries	Type of pollution
Kanpur	*Thermal Power Station Site, Fertilizer Industry site, Jajmau Tannery*	*Water and Air*
Unnao	*Unnao Distillery*	*Water*
Rai Barelly	*Thermal Power Station Site, Sugar Mill Fertilizer Industry, Parag Dairy*	*Air and Water*

RESULTS AND DISCUSSION

On the basis of survey data presented in Table 1, it is revealed that the *Leucaena leucocephala*, *Acacia nilotica*, *Albizia lebbeck*, *Eucalyptus* hybrid, *Dalbergia sissoo*, *Pongamia pinnata*, *Pithecolobium dulce* were surviving successfully near the Thermal Power Station. The growth of *A. nilotica*, *Terminalia arjuna*, *Ficus religiousa*, *Cordia dichotma*, *P. pinnata* revealed that they were performing better near by the Tannery areas. *Eucalyptus* hybrid, *Acacia nilotica*, *Dalbergia sissoo*, *Azadirachta indica* and *Cassia siamea* also performed well near the Unnao Distillery areas. These findings are in line with results of Ahmed *et. al.*, (1989) and Srivastava *et. al.* (1995; 1996). The plant species like *D. sissoo, F. religiosa, Ailanthus excelsa, C. fistula, Calotropis procera, Moringa oleifera, F. bengalensis, A. indica, P. dulce, Mangifera indica, Tectona grandis* and *F. unfectoria* were resistant to the automobile exhaust pollutant, air pollutant (SO, SPM and Flyash) and worked as dust filtering agents (Table 2).

TABLE 1: *MPTS surviving near the industrial unit*

Name of Industries	Plant Species
Thermal Power Station	*L. leucocephala, A. nilotica, A. lebbeck, Eucalyptus hybrid, D. sissoo, P. pinnata, P. dulce*
Tannery Site	*F. religiosa T. arjuna, A. indica, A. nilotica, C. dichotoma, P. pinnata*
Distillery Site	*Eucalyptus hybrid, A. nilotica, D. sissoo, A. indica, C. siamea*
Parag Diary	*Eucalyptus hybrid, T. arjuna, A. nilotica, C. equisetifolia*

TABLE 2: *MPTS working as dust filtering agents*

Name of tree	Common name	Dust collected (g/sq.m. of leave surface)
T. arjuna	*Arjun*	*4.49*
M. indica	*Aam*	*4.05*
B. purpurea	*Kachnar*	*3.90*
B. Monosperma	*Plaw*	*3.05*
A. indica	*Neem*	*2.92*
C. fistula	*Amaltash*	*2.24*
T. indica	*Imli*	*2.08*

CONCLUSION

The tree plant like *A. nilotica, Eucalyptus hybrid, D. sissoo* and *A. indica* were survive very well in industrial area of all the three district. Tree species like *T. arjuna, M. indica* and *Bauhinia purpurea* were found better as dust filtering agent.

REFERENCES

1. Ahmed, K. J.; Yunus, M.; Singh, S. N. and Srivastava K. 1989 Study of plant in relation to air pollution - a report Environ, Bot. Lab. NBRI, Lucknow.
2. Kar, J. 1995 Van Anusandhan 12 (ii). 53-55.
3. Pandey, K. K. and Saleem M. 1994. Yojana 38 (10). 13-14.
4. Srivastava R. J.; Kumar A.; Rai, M. M. and Singh L. 1995. Indian Forester 11 (2). 81-87.
5. Srivastava R. J.; Dubey P.; Rai, M. M.; Mishra, C. M. and Dixit R. K. 1996. Paper presented in the Sixty Six Annual Session of the National Academy of Science, India held at Dr. B. A. Marathawada University, Aurangabad (M.S.) From Oct. 31 to Nov. 2, 1996.

Tree Breeding Status of Agroforestry Species in India

K. R. SOLANKI
National Research Centre for Agroforestry, Jhansi (UP)

ABSTRACT

In agroforestry and forestry sectors, few and sporadic efforts have been made for tree improvement research in India. Most of the agroforestry tree species have not been even gone through simple selection for superior types (plus trees). Exploration, collection and conservation of germplasm of woody agroforestry species, selection for fast growing trees, clonal propagation, seed orchards and other advances in tree breeding are still in fancy stage. At present, all the woody agroforestry species are propagated through collected seeds and/or propagules by unskilled labourers. It is well known that through tree breeding and establishment of seed orchards of improved genotypes, a significant achievement towards the improvement of productivity of economic tree products is possible as happened in agricultural crops in India and industrial trees in USA and Europe. Poplar, eucalyptus and *Leucaena* received popularity in India only due to availability of improved genotypes. The importance of quality seeds and/or propagules has been realized by all the concerned for increasing the productivity of planted species. Even by small efforts of tree improvement, productivity of economic tree products can be raised by 20-30 per cent (Chuntanaparb and Mac Dicken, 1991). In this paper breeding status and strategy of various agroforestry species has been discussed.

INTRODUCTION

In late 1970's agroforestry system, an alternative sustainable land management system started gaining importance in India. In the beginning, researchers and extension workers felt difficulty to identify suitable species of woody perennials matching to different agro-ecological zones under variegated situations, sites and needs of the farmers and users and also complementary to under storey annual crops. These problems have been solved to a great extent after obtaining the results of continuous exercises done in this field by various centres of All India Coordinated Research Project on Agroforestry, State Agricultural Universities and ICAR Institutes and tree species for various ecological zones and for different situations are identified. Still, the impact of agroforestry technology in Indian agriculture is not visible upto required level. One of the major reasons for this is non-availability of genetically Superior quality seedlings, thus losing large potential of returns on investment.

Brief review of the improvement work, breeding status and strategy for various agroforestry species has been discussed in this paper:

SUBABUL (*Leucaena leucocephala*)

Subabul is a self pollinated fast growing multipurpose leguminious tree species originated from Mexico and Central America (Brewbaker, 1987). Its potential as a fodder cum fuel tree in agroforestry systems in India is well known (Singh, 1986; Relwani, 1983; Deb Roy, 1988; Pathak, 1988) due to high protein and vitamin A content in the leaves, deep tap root system, erect growing habit and ability to improve soil fertility through fixation of aerial nitrogen and decomposition of protein rich leaves in the soil in a short duration. Average protein content varies from 22-30 per cent in leave (Brewbaker and Hutton, 1979). By simple selection in natural stands, Brewbaker and his team identified highly vigorous and straight growing trees which grew upto 20 m in southern Mexico and western Central America, By evaluating these selections, five outstanding varieties (K8, K132, K28, K29, K62) have been released. Out of these, K8 got maximum popularity. Forage yield of K8 variety was 19.5 t/ha/yr against 10.5 t/ha/yr of Hawaiian type (Brewbaker, 1979).

In CSIRO, Australia *Leucaena* breeding work started in 1950's for low mimosine content in the leaves and released the variety Cunningham by hybridization of Peru □ CPI 18228 followed by selection in 1976. In this variety, mimosine content is slightly less than cv Peru (Mackay, 1976). They further developed interspecific hybrids of *L. leucocephala* □ *L. pulverulenta* which were quite vigorous with mimosine content 14 per cent less as compared to var. Cunningham (Bray 1983) and increased forage yield by 42 per cent. These hybrids failed to gain popularity due to lack of simple and economic technology to multiply vegetatively and to retain the hybrid vigour.

In year 1984, *Leucaena* suddenly got setback due to wide spread of tiny homopterous insect called psyllids (*Heteropsylla cubana* craward) in whole of the *Leucaena* growing areas of world. Plants defoliated completely and growth hampered significantly. All the germplasm of *Leucaena* (more than 1000 varieties of 17 *Leucaena* species) were tested against this insect. Two varieties K636 and K584 of *L. leucocephala* were found tolerant and advocated for plantation in psyllid infested areas. Other species of genus *Leucaena* resistant to psyllids were *L. diversifolia*, *L. collinsii*, *L. pallida* (Brewbaker *et. al.*, 1989). Since all these species were not as vigorous as subabul, therefore, efforts were made to develop interspecific hybrids between these species rather than commercialising these varieties. Breeding efforts in this direction produced interspecific hybrids KX1 (*L. diversifolia* □ *L. pallida*), KX2 (*L. leucocephala* □ *L. pallida*) and KX3 (*L. diversifolia* □ *L. leucocephala*) by University of Hawaii which were quite vigorous and resistant to psyllids.

Subabul Breeding in India

Looking to the importance of subabul in agroforestry systems its breeding programme was initiated in 1975 in Indian Grassland and Fodder Research Institute, Jhansi. IGFRI collected 496 provenances of 14 *Leucaena* species. Detail of the species and number of accessions of each species are given in Table 1.

Biometrical studies like genetic variability between and within species, correlation and path coefficient analysis (Gupta and Patil, 1988) gene action and hybrid vigour studies (Gupta and Patil, 1986) have been done at IGFRI. Studies revealed that a good amount of variability including mimosine content was present in genus *Leucaena* (Table 2). Inter-specific hybrid vigour is quite significant but intervarietal vigour is absent (Table 3 and 4).

TABLE 1: *Germplasm collection of Leucaena species at IGFRI, Jhansi*

Species	Number of accessions
1. *L. leucocephala*	268
2. *L. diversifolia*	29
3. *L. shannoni*	55
4. *L. pulverulenta*	10
5. *L. macrophylla*	12
6. *L. esculenta*	10
7. *L. retusa*	12
8. *L. trichodes*	11
9. *L. lanceolata*	35
10. *L. collinsii*	11
11. *L. salvadorensis*	10
12. *L. gregii*	30
13. *L. pallida*	1
14. *L. multicapitulata*	2
Total	**496**

TABLE 2: *Range of variation in Leucaena germplasm at IGFRI, Jhansi, India*

Characters	Range
Leaf length (cm)	5.3-17.9
Pinna number	5.1-42.3
Leaflet length (cm)	0.4-3.9
Leaflet width (cm)	0.1-2.8
Seed number per pod	7-28
100-seed weight (g)	1.2-12.5
Plant height (cm) 3 yr. old	60-641
Collar dia. (cm) 3 yr. old	2.0-9.1
Mimosine in leaves (%DM)	1.2-5.4
Fodder yield/tree (DM kg) 3 yr old	0.07-4.1
Fuelwood yield/tree (DM kg) 3 yr. old	0.15-24.5
Days to first flowering	91-9820

Source: Gupta et. al., 1996

TABLE 3: *Hybrid vigour in Leucaena leucocephala for forage and fuelwood yields*

	No. of	Range of heterosis[a]	Average heterosis[a]	Mean performance

	F1s	Forage yield	Fuel yield	Forage yield	Fuel yield	Forage yield	Fuel yield
Trial-1	*15*	*-81.2 to 67.5*	*-60.9 to 86.8*	*-16.3*	*-6.0*	*-53*	*-45*
Trial-2	*10*	*-33.9 to 54.8*	*-32.5 to 70.2*	*-2.9*	*-0.5*	*-26*	*-27*
Trial-3	*24*	*-68.1 to 85.0*	*-69.1 to 71.0*	*-18.3*	*-31.1*	*-47*	*-60*

Source: Gupta and Patil, 1988 a= relative to better parent, b= relative to best parent.

TABLE 4: *Heterosis (%) over mid (MP) and best parent (BP) in inter-specific hybrids of Leucaena species*

Crosses	**Collar diameter**		**DBH**		**Plant height**	
	MP	**BP**	**MP**	**BP**	**MP**	**BP**
L. lanceolata □ L. leucocephala	*5.9*	*-20.6*	*24.1*	*-9.5*	*4.8*	*-8.3*
L. macrophylla □ L. leucocephala	*7.8*	*-23.5*	*0.0*	*25.2*	*3.8*	*-10.6*
L. macrophylla □ L. lanceolata	*13.4*	*13.4*	*20.0*	*15.6*	*8.1*	*-3.6*
L. shannoni □ L. leucocephala	*25.6*	*-10.0*	*13.3*	*-20.7*	*-21.1*	*-3.6*
L. leucocephala □ L. pulverulenta	*61.9**	*41.7**	*128.1**	*99.1**	*-4.1*	*-6.5*
L. leucocephala □ L. diversifolia	*94.3**	*51.5**	*86.1**	*59.5**	*23.8*	*8.3*

Source: *Gupta, 1990 *= significant at 5%, DBH= Dia. at breast ht.*

Several inter-specific hybrids were also developed at IGFRI, Jhansi. Hybrids of *L. leucocephala* □ *L. diversifolia* and *L. leucocephala* □ *L. pulverulenta* were promising (Table 4). One of the interspecific hybrids of *L. leucocephala* □ *L. pulverulenta* was superior over variety K8 by 88.9 % for fuelwood yield and 71.64% for forage yield (Gupta, 1988).

At IGFRI through simple selection and hybridization, seven promising selections were developed (Table 5) i.e. Sp1, S10, S14, S15, S22, S23 and S24 (Pathak and Gupta, 1994). Selection S24 was quite promising and better than K8 by 29% for bole yield and 65% for leaf yield in summer (May) and 36% for total biomass (Table 5). It also had 48% less pods than K8 and contributed 70% for bole yield and 10% for leaves against 64% and 8% by K8 to total biomass, respectively. This selection may be useful in northern India.

TABLE 5: *Growth and production traits of L. leucocephala provenances and selections*

Provenance	**Survival**	**Growth**	**Production (kg/tree)**

	(%)	CD (cm)	DBH (cm)	Ht (m)	Bole	Branch	leaf	Total
K8	55.90	11.60	9.10	12.80	26.40	11.70	3.24	41.34
K217	44.70	11.10	8.90	12.70	25.70	9.30	2.07	37.07
K601	75.00	12.10	9.20	12.20	30.60	8.80	1.87	41.27
K636	62.00	10.90	8.90	12.90	22.70	6.70	1.86	31.26
K397	54.10	10.90	8.30	12.50	21.80	8.10	1.40	31.30
K340	48.80	11.10	8.70	11.60	25.50	7.30	2.13	34.93
K29	61.90	10.70	8.90	11.70	25.50	13.70	2.78	41.98
IGFRI23-1	51.80	11.40	9.30	11.40	25.70	12.90	1.59	40.19
SP1	37.10	10.80	8.50	11.60	26.40	11.60	1.94	39.94
S10	67.90	13.30	10.10	12.50	36.00	12.70	1.46	50.16
S14	55.90	11.80	9.90	12.50	34.30	7.50	1.38	43.18
S15	59.20	11.00	9.00	11.70	25.50	9.60	1.48	36.58
S22	49.20	11.40	8.90	11.70	30.80	10.90	3.03	44.73
S23	43.10	10.10	7.70	10.70	15.60	6.20	2.11	23.91
S24	57.00	13.10	9.50	12.80	39.20	11.80	5.36	56.36
Mean	54.90	11.40	8.90	12.10	27.40	9.90	2.25	39.61
CD (5%)	19.60	1.30	1.10	0.97	7.30	NS	1.41	

Source: *Pathak and Gupta, 1994*

Vegetative Propagation

Grafting is a relatively simple and historic technique for *Leucaena*. The best results were obtained in cleft and whip and tongue grafting rather than bud or T-shield grafts, using four-months seedlings as stock and taping with grafting tape. Several types of inter-specific grafts were successful. Vegetative propagation through cuttings is reported by few workers. Hu (1978) obtained excellent rooting of cuttings under mist spray in Taipei greenhouse using 15 cm leafy twigs having terminal twigs intact in one year old trees. Leafless twigs from older trees failed to root. Micropropagation in *Leucaena* is successful but can not be considered as field applicable. Callus growth was achieved from germinating seeds and hormonal treatments were perfected leading to shoot differentiation and somatic embryos. Seed derived explants were carried out without callus formation through IBA and NAA media to vigrous growth shoots and with added ascorbic acid to healthy plantlets.

KHEJRI (*Prosopis cineraria*)

Prosopis cineraria is a better indigenous species than the exotics for survival, adaptation, growth, production of fodder and fuel, soil fertility improvement and to check desertification in Thar desert. Major limitation of this species is that it is extremely slow growing. It is not easy to meet ever increasing demand of fodder and fuelwood in Thar desert unless fast growing provenances/ genotypes of this species are identified. In this direction, little efforts have been made before 1981 to boost its growth and to increase the supply of fuel and fodder production.

With this background intensive genetic improvement work on *P. cineraria* was initiated in 1983 at Central Arid Zone Research Institute, Jodhpur to evolve fast growing genotypes by selection as well as to understand extent and pattern of genetic variability present in nature and its exploitation.

Survey

Eleven districts of Rajasthan were surveyed during 1984 and seeds from the best trees from each district were used to raise single tree progeny which were transferred to the field during August, 1984. During May and June, 1985, 7 districts of western Rajasthan, i.e. Jodhpur, Bikaner, Barmer, Jaisalmer, Churu, Sikar, and Nagour were surveyed. Observations were recorded on ten randomly selected trees from each district and pods from individual tree were collected. Observations were recorded on the locations of the trees, frequency of trees in each district in term of grading (abundant / frequent / occasional / rare) and tree characteristics like tree height, forking height, diameter at breast height, (DBH), canopy diameter, length of pod and number of seeds/pod. In addition to these, data on galls (profuse/moderate /poor), type of branching and spins habit were also recorded (Table 6). Study on 70 trees from natural stands spread over 7 provenances revealed that there was maximum variability for CD and DBH, coefficient of variation (CV) was approximately 65% followed by plant height (52.7%) and crown diameter (48.14%), similar findings were reported by Jatasara and Paroda (1983). Two types of angle of branches namely acute and horizontal observed during present survey apparently suggested the presence of two different plant types.

The present study also revealed the predominance of plant type with horizontal branches (69%). Jatasara (1982) during his survey of the Thar desert also reported two distinct type *P. cineraria i.e.* one spineless with dropping twigs and a compact crown, the other with spines, non-dropping branches and a comparatively wide open crown. He observed some intermediate ecotypes and suggested that these ecotypes can be crossed. The present study shows that wide variability exists in natural stands for all the characters studied. A wide range of variability was also observed in the percentage of crude protein, reducing sugars and total sugars of pods between trees within provenance and between provenances. It ranged from 8.05-15.05 % for crude protein, 0.83-4.67 % for reducing sugars and 6.04-15.83 % for total sugar. In juvenile progenies, peroxidase isoenzymes showed large variation at some loci both within and between progenies.

TABLE 6: *Mean and range of variation in P. cineraria in natural stands of Rajasthan.*

Trait	Range	Mean ± SE	SD	CV
Height (m)	*4.6-30.4*	*11.9 ± 0.75*	*6.28*	*52.7*
Forking height (m)	*1.2-04.6*	*3.12 ± 0.10*	*0.84*	*26.9*
DBH (cm)	*14.3-149.5*	*44.4 ± 3.4*	*28.8*	*64.9*
Collar dia. (cm)	*15.9-171.8*	*51.7 ± 4.0*	*33.8*	*65.7*
No. of branches	*2-9*	*4.5 ± 0.21*	*1.81*	*39.9*
Canopy dia. (m)	*1.20-9.72*	*4.6 ± 0.54*	*2.20*	*48.1*
Pod length (cm)	*6.4-18.0*	*12.2 ± 0.26*	*2.49*	*20.4*
No. of seeds/pod	*3-19*	*9.6 ± 0.39*	*3.33*	*34.8*

Progeny Trials

First progeny trial of 10 plus trees of *P. cineraria* was established during 1977 at the Central Research Farm, CAZRI, Jodhpur in randomised block design with three replications having 9 plants of each progeny spaced at 4 □ 4 m. These ten plus trees were identified at 2 locations i.e. Jodhpur (5 trees, JU-1 to JU-5) and Bikaner (5 trees, BK-1 to BK-5). The observations on fodder and other component characters were recorded during December, 1984 and December, 1986. The age of plants at the time of recording observations were 7.5, 8.5 and 9.5 years, respectively. There were significant differences among progenies and 2 locations during different years. Progeny number JU-5 has shown a consistently good performance over year (Table 7). Both heritability and genetic advance as a percentage of the mean have moderate implying that selection of progeny is likely to be moderate.

TABLE 7: *Genetic advance and variation studies in different progenies of P. cineraria.*

Progeny	Mean height (cm)		
	7.5 years	8.5 years	9.5 years
BK-1	1.83	2.02	2.10
BK-2	2.00	2.12	2.28
BK-3	1.86	2.06	2.18
BK-4	1.62	1.77	1.90
BK-5	2.34	2.58	2.92
JU-1	1.92	2.03	2.36
JU-2	2.70	3.12	3.43
JU-3	3.08	3.31	3.63
JU-4	3.50	3.71	3.81
JU-5	3.86	4.11	4.30
Mean	2.47	2.68	2.89
Sem.+	0.26	0.31	0.32
CD (5%)	0.73	0.85	0.89
CD (1%)	0.96	1.12	1.17
PVC	44.46	45.47	43.86
GCV	30.68	29.71	28.34
Heritability (%)	47.62	42.71	41.75
Genetic Advance	1.08	1.07	1.09
GA as % of mean	43.62	40.00	37.72

During May 1984, another survey of the arid region of Rajasthan was conducted and seeds of superior trees were collected. Of these, 11 half sib single plant progenies were transplanted into the field in July-August, 1984 in randomised block design with 3 replications. Results revealed that there were significant differences among 11 progenies for tree height for all the 5 years. Progeny number 264 from Jodhpur ranked first in all 5 growth years under study. Family heritability and single tree heritability exhibited consistently high values over the years. This stabilisation of components from the second year of growth onwards indicates that selection for height can be initiated at an early age (3 years). Even though heritability values for family selection and mass selection were more or less similar, the genetic gain in mass selection is almost double. This has happened because the difference between families means are much less than those among individual trees. As a result, selection differential is much less in family selection than in single tree selection. The high and uniform heritability values coupled with high genetic gain over different years show that effective improvement for tree height in *P. cineraria* can be achieved by following the simple procedure of selection the best trees among various families.

Inter-relationship of Height of Mother Trees with Juvenile Progeny

Selection of plus trees from mature stands is generally based on various morphological characters like height, DBH, collar diameter, crown diameter etc. and the presumption or rather than expectation is that the progeny from such trees will also have similar morphology and better growth. The correlation of 10 parents with their 7.5 year old progenies has demonstrated positive correlations between various characters of parents and progenies. The correlation of DBH is the maximum (0.763). It revealed that the approach of direct selection from natural stands appear to be effective.

Seed Orchards

Some elite selections of *P. cineraria* attained growth height of 7.3 m. against 6.5 m. attained by *Acacia tortilis*, (a fast growing species), in 6-years. Such identified elite trees were used to establish seed orchards. In the seedling seed orchards, there were 156 elite trees (mean height 3.54 m in 6-years). The best one among these are under further genetic exploitation.

Clonal seed orchard was also established by preparing seedlings through air layering of elite trees. Using the seeds collected from the seedling seed orchard and the clonal seed orchard, 7,000 saplings were raised in 1990-91 and supplied for multilocation trials.

Vegetative Propagation

There are several reports that *P. cineraria* have been propagated through root suckers (Rao, 19953) and by air layering (Rawat, *et. al.*, 1982; Solanki *et. al.*, 1984, 1986). There have been few reports that *P. cineraria* can be propagated through shoot cuttings (Arya and Tomar, 1989). The basic technique involved in air layering is removing the bark from 15 mm girdle and applying the hormones to the upper end of the girdle. The girdle is then covered with a layer of moss or clay and wrapped in polythene. Solanki *et. al.*, 1986 showed that with either 100 ppm IAA or with Seradix B3, rooting was induced in 10-15 mm branches but not 7 mm branches, the best results being obtained with Seradix and clay. They also observed a distinct seasonal pattern in the success of air layering with the best treatments giving 100% success in July and August, a little rooting in January and February and no rooting between October and December and between March and May. *P. cineraria* is successfully propagated through tissue culture (Arya and Shekhawat, 1986; Solanki *et. al.*, 1991)

NEEM (*Azadirachta indica* A. Juss)

In recent years, neem is getting much attention because of large number of liquid and powdered formulations of neem containing azadirachtin as active gradient are being marketed in world as a nontoxic biological insecticides or chemicals. Neem is also used to synthesis medicinal, cosmetic and animal health care products.

Realising its economic potential in world market and good adaptability in wide range of lands including degraded lands and climates, research work on tree improvement initiated in few research organisations i.e. National Research Centre for Agroforestry, Jhansi (Gupta *et. al.*, 1996), Central Research Institute for Dryland Agriculture, Hyderabad (Venkateswarula, 1996), Tamil Nadu Agricultural University, Coimbatore (Sivasamy and Karivaratharaju, 1993), Arid Forest Research Institute, Jodhpur (Mishra, 1996), CCCSHAU, Haryana and other organisations.

Survey and Studies on Geographical Variation

At NRCAF, Jhansi (1993), exploration studies were conducted in various parts of central India. Observations were recorded for morphological, growth and fruit characters. Seeds of 38 provenances and 28 plus trees were collected and sown in the nursery. During the exploration tour, data were recorded on DBH, clean bole height, canopy area, number of primary branches per plant and fruiting and ripening period of drupes. Genetic variation in fruiting was observed from plant to plant. Flowering and fruiting habits also varied from place to place. From Jaipur to Jhansi (longitude 74-78° E and latitude 26-28° N) trees were almost without fruits in 1993. On the other hand, heavy fruit crops were observed in the localities of Hamirpur, Bhognipur, Kanpur, Shivpuri and Guna (longitude 77-82° E and latitude 22-26 °N). Variation was also observed for plant height (7.2 - 23.47 m), clean bole height (0.84-7.92 m), canopy area (101.51-1400.12 m^2), DBH (53.5-214.6 cm), number of primary branches per plant (2-5), seed length (0.80-1.80 cm), seed diameter (0.40-1.00 cm) and weight per 100 seeds (11.5-22.6 g).

Seed size variation studies were conducted in TNAU, Coimbatore, they found seed length ranged from 1.25 - 1.57 cm. However, the breath of seed did not vary much. The 100-seed weight ranged between 17.0 - 22.7 g. In other collection of Karnataka, seed length ranged from 1.1-1.8 cm, seed width from 0.45-0.85 cm and 100 seed weight from10.0-53.0 g (Veerendra, 1995).

Variability for Azadirachtin Content

Jacobson (1989) reported that neem seed of Indian origin contained only 1-3% azadirachtin whereas seed of African origin had an average 5-6% azadirachtin with few lots exceeding even 9%. Rengasamy and Parmar (1995) evaluated azadirachtin of 11 of the 21 agroecological regions of India. Azadirachtin ranged from 0.14%-2.02% w/w of kernel with average of 0.68%. The trees growing in coastal, arid and semiarid ecosystem showed high (>0.75%) and those in sub-humid ecosystem showed very low (0.27%) average azadirachtin.

Provenance Trials

In NRCAF, Jhansi, twenty-five provenances were collected from central India and planted in a randomised block design with three replications on 15th July, 1994 in single row plot of 32 □ 4 m having 8 plants in each provenance spaced at 4 □ 4 m between and within rows. After 36 months, the collar diameter, height, length of dominant primary branch and canopy diameter was recorded for each plant. There was no variation among provenances for collar diameter, length of primary branches and canopy diameter. However plant height showed significant variation among provenances in 3 yr old trial (317.52, Charkha prov. to 425.83 cm Damoh prov.). Out of 25 provenances, 20 were superior over the Jhansi provenance for plant height. Provenances for Moth, Riva, Katani, Bhopal, Guna, Shivpuri and Karera were considered excellent. All these provenances belong to better soil with higher rainfall areas (Gupta, 1998). Monthly height increment was maximum during April to June when temperature remained more than 40 °C and soil moisture was quite low. Results of two year old trial of 37 provenances conducted at Arid Forest Research Institute, Jodhpur showed that provenances of Indore, Ujjain, Palanpur, Rajkot and Ravinagar are quite good. Central Research Institute for Dryland Agriculture, Hyderabad identified four superior clones, one each from Anantpur and Hyderabad in Andhra Pradesh, Sabarkantha in Gujarat and Philibani in Orissa.

Reproductive Biology

Intensive work has been done at National Research Centre for Agroforestry, Jhansi (Gupta *et. al.*, 1996). Studies revealed that bud initiation in neem take place in the second week of March and continued up to the first week of May. Anthesis started at around 9.00 pm and continued up to 4.00 am next day. Dehiscence of anther starts around 8.00 am just after sunrise, in closed flowers. Process remains continued in the afternoon (2.00 pm). Pollen germination starts around 5.00-6.00 pm on the same day and over by the night of the same day. Pollen viability, tested by staining pollens with 1% acetocarmine, ranged from 84.00 to 100.00% with an average pollen size of 53□. Pollen did not germinate in sucrose solutions (1, 5, 10, 15, 20, 25 and 30%) with or without boric acid (0.005 and 0.01%) in cavity slides. Fertilization time was around 2.00 am next day and continue upto 4.00 am Results of seed setting percentage in selfed and open-pollinated seed of same tree and absence of inbreeding depression suggested that neem may be a self-pollinated species (Table 8). Earlier workers reported that neem is a cross pollinated species.

TABLE 8: *Variation in growth parameters of self and open pollinated plants (12 months age).*

Character	Selfed	Open pollinated
Seedling height (cm)	55.20	54.25
Collar diameter (cm)	0.57	0.55
Leaf length (cm)	5.39	5.10
Leaf width (cm)	1.65	1.53

Source: Gupta et. al., 1996

Plus Trees Progeny Trial

Eighteen single plus tree progenies were planted in randomised block design in July, 1994 with two replications at a spacing of 4 □ 4 m having plot size of 4 □ 24 m at NRCAF, Jhansi. Each progeny has single row plot of 6 plants in each replication. Results of growth data of 36-months old trial indicated that maximum monthly height increment was recorded during April - June as compared to July - Sept. It indicated that water requirement for growth in neem is minimum. Progeny of plus tree no.19 of Mahuranipur showed excellent performance with annual increment of 1.34 m (Gupta, 1998).

Vegetative Propagation

Neem can be propagated through cuttings in mist chamber, air layering, grafting, budding and through tissue culture. Out of these methods, air layering is cheapest, simple and highly successful.

- **Air layering:** Rooting in air layered branch is 100 per cent successful by application of 800 ppm solution of IBA in rainy season (Gupta and Kumar, 1996). These air layered plants can easily be transferred in polybags without any mortality. Earlier Shanmugavelu (1967) also described vegetative propagation through air layering.
- **Branch cuttings:** IBA solutions of 100-200 ppm gave very good response in mist chamber under 24 hrs treatment.
- **Tissue culture:** Callus formation, somatic embroygenesis and direct multiple shoots induction from seedlings explant have been reported in neem (Rangaswamy and Promila, 1972; Sanyal *et. al.*, 1981; Narayan and Jaiswal, 1985, Gautam *et. al., 1993*).

POPLAR (*Populus* Spp.)

The history of poplar is quite old but it was largely confined to the Kashmir and the temperate arid zones. However, around 1950 with the import of large number of exotic clones and subsequent trials in north India, the usages pattern changed. Now these exotics (few clonal selections of *Populus deltoides)* are becoming an important part of agroforestry systems in some districts of Punjab, Haryana and Uttar Pradesh. These species are considered most successful and commercialised in agroforestry systems. *P. ciliata* is naturally distributed in Himalayan regions of India. In India, exotic poplar cultivars for plantation, mostly in foot hills, were introduce in 1956 (Chaturvedi, 1982). The programme of introductions and research on poplars was initiated by Forest Research Institute, Dehradun when 24 clones were introduced from England (Mathur and Sharma, 1983). The successful attempts were made by Kapoor (1980) to raise genetically fast growing clones of poplars. Initially 'IC' clones were considered to be the best performers, however, in 1978, these clones lost vigour and resistance to pests and diseases by the 10th cycle of clonal cuttings. In 1970, two *Populus deltoids* clones G3 and G48 were received from Australia along with D 100 and D 121 (obtained from Stoneville, USA) were found performing well in field conditions. By 1978, it was then realised that without proper breeding programme the large scale plantation being undertaken by introduction of new clones, may not remain viable (Chaturvedi and Rawat, 1994).

Seed Collections and Introduction

1. Seeds of 23 provenances of *P. maximowiczii* were imported from Oji Paper Co. of Japan during 1989 by YSPUHF, Solan.
2. Seeds of *P. deltoides* from the natural range of distribution sites of Texas, Louisiana and Mississippi states of the USA were collected. Out of the 103 open pollinated families collected, only 78 families could be retained after 1993 for improvement and further selection trials.
3. Seeds of the indigenous species *P. ciliata* commonly called the Himalayan poplar were collected from 22 plus trees and 72 open pollinated trees.
4. Seeds were obtained by interspecific hybridization *P. ciliata* □ *P. maximowiczii, P. euramericama* I-455 □ *P. deltoides and P. ciliata* □ *P. gunnanesis*

At TFRI, Jabalpur, 5 genotypes (G-3, 65/27, G-48, S7 Cl, D121 of *P. deltoides* were evaluated. Clone 65/27 has performed best in the field followed by D121.

SHISHAM (*Dalbergia sissoo*)

Sissoo is an important timber species belonging to family Leguminaceae and subfamily Papilionaceae. It provides fuelwood, fodder and enrich soil fertility. It provides important raw materials for a variety of wood based industries especially for furniture and building constructions. The wood is hard, heavy, strong, double elastic, seasons well and decay resistance.

Genetic Variability

A survey was conducted in northern India during February-March, 1990 in order to record pattern of variation in natural populations of *D. sissoo* by Bangrawa and group. It observed very high variability for various morphological and pod characters (Table 9).

TABLE 9: *Variation for morphological characters and pod characters*

Character	Mean	Range	Coefficient of variation
Stem straightness (score)	3.0	0.5 - 5.0	43.6
Tree height (m)	13.2	6.9 - 22.1	26.5
DBH (cm)	37.0	16.0 - 95.0	39.7
Clear bole height (m)	3.9	1.5 - 10.1	76.9
Crown diameter (m)	7.2	2.6 - 15.3	32.4
Seeds per pod	1.23	1.09 - 1.46	6.79
Pod length (cm)	4.90	3.2 - 6.60	12.57
Pod breath (mm)	8.10	5.20 - 11.00	16.37
100-seed pods wt. (g)	5.83	4.20 - 8.80	18.85

Provenance and Plus Tree Progeny Trials

Exploration and seed collection work in *Dalbergia sissoo* was done by several organisations out of these intensive work have been done by CCSHAU, Hissar, NRCAF, Jhansi, GBPUA&T, Pantnagar, TFRI, Jabalpur and FRL, Kanpur.

Reproductive Biology

Extensive work on reproductive biology of sissoo was done at CCSHAU, Haryana (Bangarwa, 1996). The studies revealed that flower buds appear in the first week of March in northern India and generally open (anthesis) between 10.30 and 13.30 hrs. The anthers dehisce earlier in the day before flowers open. Pollen viability varies from 56 per cent in March to 46 per cent in April. Stigma become receptive six hours before flower opening and continue upto six hours after flower opening. There was no difference in per cent fruit setting between isolated trees and trees in groups and fruit setting was not observed in emasculated flowers. The pod setting was around 40 per cent. It was observed that *D. sissoo* is predominantly self fertilizing species and maximum pod setting occurs from last week of March to first week of April.

Vegetative Propagation

It can be propagated easily in ordinary nursery by Cuttings (100 ppm. 150 ppm IBA), Layering (800 ppm IBA) and root suckers. It can also be propagated by tissue culture.

DESI BABUL (*Acacia nilotica*)

Acacia nilotica is a versatile indigenous multipurpose tree species of India. It is a good source of food, fodder, small timber and tannin. Its fast growth, good coppicing ability, fixing atmospheric nitrogen and wide range of adaptation has gained attraction of the people of south east Asia and Africa. Still mode of pollination of *Acacia nilotica* is not known clearly. It is considered to be highly cross pollinated species. Provenances trials were conducted in CCSHAU Hissar, FRI Dehradun, TFRI, Jabalpur and JNKVV, Jabalpur.

EUCALYPTUS (*Eucalyptus* Species)

In case of Eucalyptus two hybrids were developed in 1970 by FRI, Dehradun i.e. FRI 4 (*E. tereticornis* □ *E. camaldulensis)* and FRI 5 (*E. camaldulensis* □ *E. tereticornis*). These hybrids gave 109% more wood volume than Mysore Gum (Mysore hybrid) (Venkatesh and Sharma, 1977). However, these two hybrids could not become poplar for plantation due to lack of sufficient seedlings.

At present through tissue culture several plus trees of *Eucalyptus tereticornis* have been multiplied at TERI, New Delhi, NCL Pune, ITC Bhadrachalama Paper Board and FRI, Dehradun These clones are giving 40-60% higher wood volume than the seedlings raised through seeds. Farmers are purchasing these seedlings even on higher rate. In Institute of forest Genetics and Tree Breeding, Coimbatore, 17

Australian provenances received from Australia Tree Seed Centre, CSIRO, Australia were tested. It was concluded that the provenance Paluma, Ravenshoe, Gadgara and Herberton were superior in volume growth.

Tree improvement work has been done in small scale in *Albizia procera, Albizia lebbeck, Tecomella undulata, Casuarina equisetifolia* in isolated area. Major problem at present facing is lack of quality seeds of improved genotypes. Still all the forest nurseries are supplying poor quality of seedlings.

SIRIS (*Albizia lebbeck*)

Exploration team of NRCAF, Jhansi, made survey of 36 districts of UP, MP and Rajasthan. During exploration, we observed high variability for leaves, pods, tree canopy, diameter at breast height, tree height, number of seeds per pod, 100-seed weight, retention of leaves during dry period. There were few trees, which were entirely different than normal trees which are straight types. These tree attended the height more than 18 m against the normal tree of 10-12 m and very straight with black stems. These trees were named as giant tree. Seeds of these trees were collected. Progeny performance yet to be tested. In CCSHAU, Hissar, provenances trial of 12 provenances was conducted and found that provenance of Jammu was better than others for early growth parameters (Toky, 1995).

Methodology and Strategy for Improvement of MPTS for Agroforestry

The basic principles and practices of tree improvement are same as they apply for industrial trees. Superior phenotypes (plus trees) are selected from the best populations and their breeding potential evaluated in clonal or progeny tests on typical sites with typical managements. For immediate gain in 6-8 years (short cycle) following methodology may be successful.

1. Identify only 2-3 tree species for each zone according to the need of adaptability, accessibility and economic value of the species.
2. Identify areas where species in question are naturalised or native.
3. Identify areas where growth stands of a species to be improved are good. These area should be marked (seed producing zones) to

fulfill immediate needs of quality seeds of farmers. Trees raised by these quality seeds would be expected to be better than the trees raised by seeds of unknown genetic sources. Try to select the seed producing zones from area where plantation should be done. Expected gain would be 10-12 per cent for production traits and 3-5 per cent for quality traits.

4. Whole of the area where natural stands of a species present should be divided into 20 - 30 groups based on geographical, soil and climatic factors.
5. During exploration tour, based on eye judgment and measurements, best trees at particular site should be identified (Plus trees) among selected trees (candidate trees) of the same age groups within surrounding of 20 m. Number of plus trees should be more in the areas where performance of a species in question is good and minimum selection should be done from the poor areas.
6. Permanent spots should be marked in records so that same trees can be used to collect the seeds to raise nursery. Genetic advance is expected in the population of these trees would be 15-20%.
7. Elite trees will be identified based on the results of Elimination Trial (ET), Growth Trials (GT) and Provenances Trials (PT). Rest of the trees should be removed from the fields. This area can also be used as a seed orchard for supply of seeds. It can be said first generation seed orchard.
8. Clones or seed progenies of best trees should be used to establish 1.5 generation seed orchards to supply good quality of seeds. By using these seeds, improvement would be quite high.

Team of experts of All India coordinated Research Project on Agroforestry have identified following agroforestry woody species and organisations for intensive tree improvement programme.

FUTURE STRATEGIES

1. National network should be established for exploration, collection, multiplication, distribution and conservation of germplasm of important MPTS for agroforestry species.
2. In first phase, genetic improvement work should be started at national level in *Populus deltoides /P. ciliata*, *Alnus nepalensis,*

Ailanthus excelsa, Azadirachta indica, Dalbergia sissoo, Prosopis cineraria, Anthocephalus cadamba, Acacia nilotica, Tamarindus indica, Emblica officinalis.

3. All the germplasm should be conserved *in situ* as well *as ex situ.*
4. Mode of pollination and reproductive biology should be studied of important of tree species.
5. Simple and feasible techniques should be standardised for vegetative propagation.
6. Seed production zones should be identified to fulfill immediate need of the farmers.
7. Clonal seed orchards should be established.
8. Seed certification system should be started in tree species also.
9. Responsibility of multiplication of quality seeds/ propagules of MPTS should be given to NSC and SSC.
10. Data base of all the information related to tree improvement of agroforestry species should be prepared.

REFERENCES

1. Arya, H. C. and Shekhawat, N. S. 1986. Forest Ecology Management 16:201-208.
2. Arya, S. and Tomer, R.1989. Jeevanti 7: 65-68.
3. Bangarwa, S. K. 1996. In: Sissoo breeding (ed. S. K. Bangarwa), Systematic Printers, Hissar, Haryana, India.
4. Bangarwa, K. S.; Hooda, M. S. and Sunil Puri. 1990. In: MPTs for agroforestry system. (Eds.: P. S. Pathak, R. Deb Roy and Panjab Singh), Range Mgmt. Society, IGFRI, Jhansi. 111-114.
5. Bray, R. A. 1983. Leucaena Research Report 4:1-2.
6. Brewbaker, J. L. 1987. In: Agroforestry a decade of development (Eds.: H. A. Steppler and P. K. R. Nair), ICFRI, Nairobi Kenya: 289-323.
7. Brewbaker, J. L. and Hutton, E. M. 1979. In: New agriculture crop (ed. G. A. Ritchie), AAAS selection symposium Westview Press, Colordo, USA. pp 207-259.
8. Brewbaker, J. L.; Sorenss, C. T. and Whoeler, R. W. 1989. In: Leucaena psyllids problems and management (Eds.: B. Napombeth and K. G. MacDicken), Funny Publishing Limited Partnership, Bangkok, Thailand.

9. Burley, J. 1987. IUFRO Workshop on agroforestry for rural needs, Jan. 22-26, New Delhi, India.
10. Burley, J.1980. Forest Review 59 (2): 133-147.
11. Chaturvedi, A. N. 1982. Indian Forester 107: 661-664.
12. Chaturvedi, A. N. and Rawat, B. S.1994. Indian Forester 120:97-104.
13. Chuntanaparb, L. K. G. and MacDicken 1991. In: Biophysical research for Asian agroforestry (Eds.: M. E. Avery; M. G. R. Cannel and C. K. Ong), Winrock International, USA and South Asia Books, USA.
14. Deb Roy, R. 1988. In: Pasture and forage crop research A state of knowledge report, (ed. P. Singh), Range Mgmt. Society, Jhansi: 305-312.
15. Gautam, V. K.; Nada, K. and Gupta S. C. 1993. Plant Cell Tissue and Organculture. 34: 13-18
16. Gupta, A. K., Solanki, K. R., Kackar, N. L. 1984. Annals of Arid Zone. 23: 255-258
17. Gupta, C.; Dwivedi, AK; Shivkumar, P. and Banerjee, A. C. 1991. National seminar on provenance/species trials in arid and semi arid zones of India, Feb 25-27, FRI, Dehradun, India.
18. Gupta, V. K. and Patil, B. D. 1986. Indian J Tree Sci. 5: 67-73.
19. Gupta, V. K. and Kumar, R. V. 1996. Annual Report, NRCAF, Jhansi, India.
20. Gupta, V. K. 1988. In: Pasture and forage research: A state of knowledge report (ed. P. Singh), Range Mgmt. Society, Jhansi, India: 175-183.
21. Gupta, V. K. and Patil, B. D. 1988. Indian J. Agric. Sci. 58:483-484.
22. Gupta, V. K., Solanki, K. R., Gupta, R. and Dwivedi, P.1996. In: Tree improvement for sustainable tropical forestry (Eds.: M. J. Dieters; A. C. Matheson; D. G. Nickles and C. E. Hardwood), Proc. QFRI-IUFRO conference at Colaundra, Queensland, Australia, Oct. 27-Nov.2: 81-82.
23. Harm, H. 1940. In: Die NaturlichenPflanzenfamilien (Eds.: A. Engier and P. Prant): 1-172.
24. Hooker, J. P. 1973. Flora of British India. Vol. I Periodical Expert, New Delhi: 540-544.
25. Hu, Ta Wei. 1983. In: Leucaena research in the Asia-Pacific region. Ottawa: IDRC.
26. Jacobson, M. 1989. Focus on Phytochemistry of pesticides, 1:178 Florida, USA; CRC Press, Inc.
27. Jatasara, D. S. 1982. Forage Research 8:1-9
28. Jatasara, D. S. and Paroda, R. S. 1983. Transaction of the Indian Society of Desert Technology and University Centre of Desert Studies 8: 138-142
29. Kackar, N. L.; Solanki, K. R. and Jindal, S. K. 1986. Indian J. Forestry 9:113-115.

30. Kedarnath, S. 1982. In: Sym. on improvement of forest biomass held at Solan, India, (ed. P. K. Khosla), Indian Society of Tree Scientists: 13-19.
31. Khosla, P. K.1981. Advance in forest genetics. Ambica Publication, New Delhi, India.
32. MacDicken, K. G. and Mehl, C. B. 1990. In: Tree improvement of multipurpose species (ed. N. Glover): 45-54.
33. Mathur, R. S. and Sharma, K. K. 1983. Indian Forester 109:591-631.
34. Mishra, R. N. 1996. Indian Forester 121: 997-1002
35. Narayan P. and Jaiswal, V. S. 1985. J. Tree Sci. 4: 65-68
36. Pathak, P. S.1988. In: Pasture and forage crop research- A state of knowledge report (ed. P. Singh), Range Mgmt. Society, Jhansi.
37. Pathak, P. S. and Gupta, V. K. 1994. In: Agroforestry system degraded lands (Eds.: P. Singh; P. S. Pathak and M. M. Roy), Oxford & IBH Publishing Co. Pvt. Ltd. New Delhi, India.
38. Rao, H. S. 1953. Indian Forester 79: 176-183
39. Rawat, T. S.; Sharma, H. C. and Pundir, J. S. P. 1982. Udanika 5:43- 44.
40. Relwani, L. L. 1983. In: *Leucaena leucocephala* in India (Eds.: R. N. Kaul; M. G. Gogate and N. K. Mathur), Vangaurd Pree, Dehradun, India.
41. Sanyal, M.; Banergy, M. and Dutta, P. C. 1981. Indian J. Expt. Bio. 19: 1067 - 1068
42. Shanmugavelu, K. G. 1967. South Indian Hot. 15: 70-71
43. Shivasamy, M and Karivaratharaju, T. V. 1993. Abst. World Neem Conference, 24th - 28th February, 1993, Bangalore, India: 62
44. Solanki, K. R.; Kacker, N. L. and Jindal, S. K. 1984. Current Science 53: 1166 - 1167
45. Solanki, K. R.; Kacker, N. L. and. Jindal, S. K. 1986. Indian Forester 112: 202-207
46. Solanki, K. R. and Gupta, V. K. 1996. In: 2nd International crop science congress on crop productivity and sustainability-shaping the future from 17-24 Nov., 1996, New Delhi, India: 447.
47 Turnbull, J. W. 1983. In: Proceeding of ICRAF/IBPGR/CFI/NAS workshop on multipurpose tree germ plasm, June, 1983, Washington DC.
48. Venkateshwarlu, B. 1996. SAIC, News Letter, March, 1996:1 p.
49. Venkatesh, C. S. and Sharma, V. K. 1977. Indian Forester 103 (12): 795-802.

Growth Performance of Different Provenances of Neem (*Azadirachta indica* A. Juss) in Semi-Arid Region of Central India

V. K. GUPTA, R. V. KUMAR, A. DATTA AND K. R. SOLANKI
National Research Centre for Agroforestry Jhansi (UP)

ABSTRACT

An experiment was conducted to evaluate growth of 36 provenances of neem (*Azadirachta indica* A. Juss) collected from various parts of central India in randomised block design in three replications in July, 1994. At the age of 36-months after plantation, provenances were assessed for plant height, diameter at breast height, basal diameter and volume. From the results of growth attributes it appeared that provenances of Damoh, Bhopal, Jabalpur, Sagar, Kota, Vidisha, and Boira were performing well. Provenances belonging to higher rainfall areas had better growth than provenances of low rainfall areas. Maximum growth of neem took place during April - October and minimum during October - March. Tree growth was affected appreciable by temperature while soil moisture had little effect. Temperature more than 30 °C favours the growth of neem.

INTRODUCTION

A large number of multipurpose trees and shrubs with potential use in agroforestry through out the tropics and subtropics as well as arid and semi-arid regions are well documented in India (Hegde, 1990; Solanki *et. al.* 1991; Tiwari, 1994; Singh, 1996; Pathak, 1998). However, paucity of information on their natural variability, adaptability, growth rate and promising provenances/ genotypes for different ecological zones has restricted their utilization (Venkatesh, 1988). Therefore, most of the agroforestry programmes in tropics in general and India in particular use limited number of well known and widely introduced multipurpose trees. Some examples are *Leucaena leucocephala*, *Populus deltoides Eucalyptus* species etc. Recent evidences showed that *Leucaena* got setback due to wide spread of tiny insect called psyllids (*Heteropsylla cubana* Craword) in whole of the *Leucaena* growing areas of the world. It is largely a consequence of the narrow genetic base (spreading of single variety i.e. K8) of the *Leucaena* germplasm used (Brewbaker, 1985; Hughes, 1988; Gupta *et. al.* 1996a). Therefore, identification, introduction, collection and testing for adaptability and growth of various multipurpose tree species and provenances are necessary alternative to provide new MPTS and/or provenances so that agroforestry programmes may become popular among farmers. Provenance trials are also important to develop genotypes for specific characteristics that make the genotypes fit in agroforestry systems.

Neem (*Azadirachta indica* A. Juss), a native to dry forest area of India, Pakistan, Sri Lanka and Mayanmar, was considered to have potential in agroforestry programmes as a tree component (Randhawa and Parmar, 1993). Neem is widely grown in 70% area covering most of the states of India except temperate zones due to its ability to survive and grow in varied stress environments. Neem is considered as an economic tree because of growing use of neem kernels to formulate bio-pesticides at global level, which are safe for mankind and eco-friendly. Because of its wide natural distribution and the diversity of ecological conditions where the species occurs, neem is considered to have considerable genetic variation (Roche, 1969; Write, 1976). Despite its importance as an agroforestry tree species, the information on genetic variation in neem is still limited (Gupta, 1996b; Mishra, 1995). Provenance trial is, therefore, important in determining genetic variation within species and selection of promising provenances for plantation. This paper reports the performance of 26-provenances collected from the central India.

MATERIALS AND METHODS

Twenty-six provenances for this study were collected from varied geographical areas belong to four states i.e. Uttar Pradesh, Madhya Pradesh, Rajasthan and Haryana. The latitude, longitude and altitude of collected sites ranged from 23°12' N to 29°10' N, 75°46' E to 80°55' E and 130.40 m to 512.92 m from mean sea level, respectively (Table1). The study was conducted in the experimental farm of National Research Centre for Agroforestry, Jhansi at 78°8' E longitude, 24° 26' N latitude and 251 m altitude. The annual total mean rainfall is 900 mm which lasts from last week of June to September followed by long dry spell from October to 15th June with little rainfall in winter season (January). The soil is black containing organic carbon 0.29%, pH 7.07 and available NPK 82.3 kg/ha, 3.16 kg/ha and 83.66 kg/ha, respectively.

Twelve-month-old polybags seedlings (50-80 cm tall) were planted in the field in July, 1994 in a randomised block design with three replications in a single row plot of 32 □ 8 m having 8 plants in each provenance spaced at 4 □ 4 m between and within rows. Fertilization was done at the rate of 20 kgN/ha and 49 kg P_2O_5 in the pits along with BHC (10%) to protect the seedlings from the attack of termites. At 36-moths after planting, tree height, diameter at breast height (DBH) and basal diameter (BD) were measured. Expected wood volume was estimated by using equation $V=D^2h$ where V was wood volume, D was DBH and h was tree height from the ground level. Analysis of variance for randomised block design was performed for tree height, DBH and BD.

TABLE 1: *Seed source of neem provenances*

Provenance	Latitude °N	Longitude °E	Altitude (m)
Datia	*25°40'*	*78°27'*	*261*
Dabra	*25°56'*	*78°21'*	*206.33*
Murena	*26°30'*	*78°00'*	*174.31*
Jhansi	*25°27'*	*78°37'*	*321*
Moth	*25°44'*	*78°57'*	*174.64*
Kalpi	*26°80'*	*79°43'*	*129*
Bhognipur	*26°14'*	*79°48'*	*140*
Ghatampur	*26°12'*	*80°09'*	*130.4*
Gulganj	*25°44'*	*80°45'*	*280.08*
Tikkamgarh	*24°45'*	*78°53'*	*244.54*
Panna	*24°44'*	*80°14'*	*380.7*
Satana	*24°34'*	*80°55'*	*260.8*
Riva	-	-	-
Katani	*23°48'*	*80°27'*	*332*

Provenance	Latitude °N	Longitude °E	Altitude (m)
Jabalpur	*23°12'*	*79°59'*	*411.78*
Damoh	*23°56'*	*79°29'*	*420*
Sagar	*23°48'*	*78°50'*	*512.92*
Kota	*25°10'*	*75°52'*	*253.28*
Charkha	*25°25'*	*79°45'*	*240.38*
Vidisa	*23°31'*	*77°51'*	
Bhopal	*23°16'*	*77°36'*	-
Boira	-	-	-
Guna	*24°40'*	*77°00'*	*440.65*
Shivpuri	*25°40'*	*77°44'*	*410.3*
Karari	*25°30'*	*78°37'*	*320*
Hisar	*29010*	*75°46'*	*215*

RESULTS AND DISCUSSION

The analysis of variance indicated that differences among the provenances were significant for tree height, bd and DBH suggesting that a wide range of variability exist in the collected provenances. The overall mean height of all the provenances was 3.73 m with mean annual increment (MAI) of 1.24 m at 36-months after plantation (Table 2). The best provenance (Damoh, Madhya Pradesh) recorded mean tree height growth of 4.25 m while poorest provenance (Charkha, Uttar Pradesh) had a mean height growth of 3.18 m. It was observed that most of the fast growing provenances (Damoh, Bhopal, Shivpuri, Kota, Sagar, Guna and Jabalpur) belong to higher rainfall areas. MAI for best provenance was 1.42 m which was appreciably high compared to MAI for tree height growth in other trials conducted at different parts of India. MAI for height was 1.19 m at Nanodara, Gujrat; 1.30 m at Urlikanachan, Maharashtra; 0.99 m at Lakkihalli, Karnataka; 0.82 cm at Rahuri, Maharashtra, 0.95 m at Jodhpur, Rajasthan and 0.55 m at Jhansi, Uttar Pradesh (Hedge et.al., 1990; Mishra, 1995; Rai, 1995; Kasole et.al., 1997).

Results of the trial also showed that MAI for plant height did not vary much from year to year. It was almost similar in 1st year, 2nd and 3rd year. On the other hand, seasons had strong influence on plant height growth. MAI for tree height was minimum (<0.25 m) during October-March while it was maximum (>1.25 m) during April-September in all the three years of the study.

TABLE 2: *Mean plant height, diameter at breast height, basal diameter and expected volume of 26 provenances of neem (Azadirachta indica) at 36-months age*

Provenance	Pl.ht. (cm)	MAI (Pl.ht.)	DBH (cm)	MAI (DBH)	BD (cm)	MAI (BD)	Expt. volume $m^3 ha^{-1}$
Datia	325.67	108.56	4.08	1.36	6.12	2.04	4.11
Dabra	361.00	120.33	5.14	1.71	8.36	2.78	6.42
Murena	319.72	106.57	4.30	1.43	6.77	2.25	3.80
Jhansi	374.17	124.72	4.73	1.57	7.90	2.63	5.42
Moth	356.11	118.70	5.33	1.78	7.98	2.66	6.53
Kalpi	389.17	129.72	5.28	1.76	7.19	2.40	6.81
Bhognipur	369.05	123.02	4.30	1.43	7.14	2.38	4.33
Ghatampur	366.29	122.10	4.72	1.57	7.46	2.49	5.17
Guljanj	354.44	118.15	4.91	1.64	7.39	2.46	5.98
Tikkamgarh	360.22	120.07	5.21	1.74	7.97	2.66	6.13
Panna	386.11	128.70	5.78	1.93	8.24	2.75	8.94
Satana	341.33	113.78	4.87	1.62	7.84	2.61	5.13
Riva	394.17	131.39	5.56	1.85	9.80	3.27	7.73
Katani	369.17	123.06	5.63	1.88	8.22	2.74	7.33
Jabalpur	388.67	129.56	5.88	1.96	9.36	3.12	9.33
Damoh	425.83	141.94	6.31	2.10	9.23	3.08	10.86
Sagar	396.73	132.24	6.18	2.06	8.18	2.72	9.89
Kota	397.50	132.50	6.02	2.00	8.62	2.87	9.04
Charkha	317.52	105.84	4.24	1.41	7.35	2.45	9.91
Vidisa	385.83	128.61	6.03	2.01	8.99	3.00	8.77
Bhopal	412.00	137.33	6.24	2.08	9.03	3.10	10.44
Boira	377.92	125.97	6.36	2.12	8.49	2.83	9.72
Guna	389.17	129.72	4.12	1.37	6.62	2.20	4.18
Shivpuri	398.33	132.78	4.77	1.59	7.64	2.54	5.83
Karari	406.33	135.44	5.54	1.84	8.53	2.84	8.28
Hisar	345.00	115.00	3.47	1.16	5.08	1.69	3.67
Sem ±	23.86	NA	0.64	NA	0.82	NA	NA
CD (5%)	66.13	NA	1.76	NA	2.28	NA	NA

This observation suggested that temperature more than 30□ C favours the tree height growth while less than 30□ C slow down height growth. Soil moisture had not much influence on height growth because MAI during April-June (dry period) was almost same as MAI during July - September (rainy period). This suggested that water requirement of neem for tree height growth is low. Mishra (1995) also reported slow growth for tree height during October-January (MAI 0.14 m) compared to March-October (MAI 0.98 m) in neem provenance trial at Jodhpur. Provenances of Ujjain, Indore, Palampur, Jhansi and Ravinagar were superior than others in 40 provenances trial at Jodhpur (Mishra, 1995).

Diameter at breast height (DBH) at age of 36- months ranged from 3.47 cm (Hisar provenance) to 6.36 cm (Boira provenance). The provenances which showed excellent DBH growth (MAI > 2.00 cm) were Boira, Damoh, Bhopal, Sagar, Vidisha and Kota. All the provenances except Boira also had good MAI for tree height. MAI for DBH was maximum during April-October as compared to October-March as observed in case of tree height growth. Results of other trials conducted in different parts of India had MAI for dbh 1.32 cm at Nanodara, Gujarat; 1.88 cm at Urlikanchan, Maharashtra; 0.73 cm at Lakkalli, Karnataka and 1.23 cm at Rahuri Maharashtra (Hegde et.al., 1990; Kasole, et.al. 1997). In contrast to these results best provenance of this trial had MAI 2.12 cm.

Basal diameter at collar region at 36- months after plantation ranged from 5.08 cm (Hisar, Haryana) to 9.80 cm (Riva, Madhya Pradesh). The provenances exhibiting excellent growth for BD (MAI > 3.00 cm) were Riva, Jabalpur, Damoh, Vidisha, and Bhopal. All these outstanding provenances belong to better rainfall areas while Hisar provenance, which had poorest BD growth, belongs to poor rainfall area. MAI for bd growth in other MPTS trials conducted in other regions of India was 1.71 cm at Nanodara, Gujarat; 2.72 cm at Urlikanchan, Maharashtra and 1.45 cm at Lakkihall, Karnataka (Hegde et.al.1990). MAI for BD of best provenance of present study was 3.27 cm. Results of the trial also showed that MAI for BD was maximum in 1st year of plantation and minimum in 3rd year. MAI for BD during October - March (<0.25 cm) was minimum and maximum between July- September (>1.72 cm). In case of dry period (April-June), recorded MAI was 1.26, which was slightly less than rainy period MAI. This suggested that soil moisture had little influence on diameter growth but temperature had remarkable influence on diameter growth.

The overall expected volume of 26 provenances was 7.08 m^3 ha^{-1} at 36 months after plantation which suggested that neem is medium growing species. It ranged from 3.67 m^3 ha^{-1} (Hisar) to 10.86 m^3 ha^{-1} (Damoh). Outstanding provenances for this trial were Damoh, Bhopal,

Jabalpur, Sagar, Kota, Vidisha, and Boira suggested that these provenances had good potential as a source of wood.

CONCLUSIONS

Overall it was concluded that better rainfall areas had superior genotypes and poor rainfall produced low productive genotypes. Longitude, latitude and altitude did not have any relationship with growth attributes. From the results of growth attributes it appeared that some provenances i.e. Damoh, Bhopal, Jabalpur, Sagar, Kota, Vidisha and Boira, were promising based on the results of early age evaluation.

REFERENCES

1. Brewbaker, J. L.1985. Leucaena Res. Rep. 6: 81.
2. Gupta, V. K.; Pathak, P. S. and Solanki, K. R. 1996a. In: Tree Improvement for Sustainable Tropical Forestry (Eds.: M. J. Dieters, A. C. Matherson, D. G. Nickles and C. E. Hardwood). Proc. QFRI-IUFRO Conf. at Colaundra, Queensland, Australia: 455-58.
3. Gupta, V. K.; Solanki, K. R.; Gupta, R. and Dwivedi, P. 1996b. In: Tree Improvement for Sustainable Tropical Forestry (Eds.: M. J. Dieters, A. C. Matherson, D. G. Nickles and C. E. Hardwood). Proc. QFRI-IUFRO Conf. at Colaundra, Queensland, Australia: 81-82.
4. Hegde, N. G. 1990. In: Multipurpose Tree Species for Agroforestry Systems (Eds.: P. S. Pathak, R. Deb. Roy, and P. Singh). Proc. National Workshop on Multipurpose Tree Species for Agroforestry Systems. Jhansi, India, February, 1-3: 15-28.
5. Hegde, N. G, Relwani, Gandhe, A. M., S. R. Joglekar and Kumar, K. T. D.1990. In: Multipurpose Tree species for Agroforestry Systems (Eds.: P. S. Pathak, R. Deb Roy, and P. Singh). Proc. National Workshop on Multipurpose Tree Species for Agroforestry Systems. Jhansi, India. February 1-3, 1990: 53-60.
6. Hughes, C. H. 1988. New opportunities for *Leucaena* genetic improvement. IUFRO Conference on Breeding Tropical Trees. December, 1988, Pattaya, Thailand.
7. Kasole, K. E.; Patil, F. B.; Jhambhale, N. D.; Kunjir, N. T. and Deshmukh, R. B. 1997. Range Mgmt. and Agroforestry 18: 99-101
8. Mishra, R. N. 1995. Indian Forester 121: 997-1002.
9. Rai, P. 1995. Annual Report, National Research Centre for Agroforestry, Jhansi: 92-97.

10. Pathak, P. S. 1996. Abstract and Souvenir National Symposium on Multipurpose Tree species for Agroforestry, Jhansi, India. July 11-13: 31.
11. Randhawa, N. S. and Parmar, B. S. 1993. In: Neem Research and Development (Eds.: N. S. Randhawa and B. S. Parmar Society of Pesticides India: 1-5.
12. Roche, L. 1969. A gynecological study of the genus *Picea* in British Columbia. New Phytol. 68: 505-554.
13. Singh, K. A. 1996. Rang. Mgmt. and Agroforestry 17:93-97.
14. Tiwari, D. N. 1994. Proc. National Workshop on Multipurpose Tree species for Agroforestry in India. Pune, India. April 6-9: 5-14.
15. Venkatesh, C. S. 1998. Int. Tree Crop. 5: 109-124.
16. Write J. W. 1976. Introduction to Forest Genetics. Academic Press, NewYork: 463.

Genotypic Variation in Physiological Characteristics of *Leucaena leucocephala*

SEEMA, VANDANA, L. P. MISRA, R. K. BHATT AND P. S. PATHAK
Indian Grassland and Fodder Research Institute, Jhansi (UP)

ABSTRACT

The fifteen promising provenances of *Leucaena leucocephala* were evaluated for their comparative physiological characteristics at Indian Grassland and Fodder Research Institute Jhansi during 1995-96. Rate of photosynthesis (PN) and stomatal conductance (CS) ranged from 4.195 mol/m^2/s to 16.63 mol/m^2/s and 0.168 cm^2 s^{-1} to 0.637 cm^2 s^{-1}, respectively. The selection S24 showed maximum rate of photosynthesis and stomatal conductance followed by S10, K8, S22 and IGFRI 101 indicating their superiority over other genotypes. Similar to the rate of photosynthesis the selections also showed higher rate of transpiration exhibiting their fast assimilatory function contributing towards higher biomass production. The accumulation of chlorophyll content varied among the genotypes may be due to differential synthesis of chl a. In general the selection accumulated high chlorophyll content carotene and soluble protein in fresh leaves. The higher specific leaf mass production in various genotypes contributes towards this total above ground biomass production potential. Selection S24 produced maximum biomass followed by S10 and K8 as check whereas, the other selections were at par in their biomass production potential. On the basis of growth and assimilatory characters, selections S24, S10 and S14 were identified as superior genotypes for maximization of productivity under semi-arid tropical condition and may be taken for further improvement programme. The variation in physiological characters in the genotypes clearly indicate that variability exist among the genotypes of *Leucaena* and there is further scope of development of superior types by incorporating all these characters in the improvement programme.

INTRODUCTION

Leucaena leucocephala, the miracle multipurpose tree species provide fodder, fuel, fertilizer, green manure, shade and conservation to the environment. Leucaena is a tropical species requiring warm temperatures for optimum growth (Brewbaker *et. al.*, 1985). Several collections in the areas of its origin have brought out the information about the existence of large variability in this genus with many races. During the past 24 years, research has been conducted on Leucaena at IGFRI Jhansi on its germplasm collection, selection and breeding. Fourteen species and 524 collections were made and many interesting material were generated. In this study 15 promising provenances of *L. leucocephala* were evaluated for their physiological characteristics on the degraded land for their suitability and sustainability.

MATERIALS AND METHODS

Physiological characteristics and biomass production of selected 15 promising accessions (8 promising provenances and 7 selections (viz., K8, K217, K601, K636, K397, K340, K29, SP1, IGFRI 23-1, S10, S14, S15, S22, S23 and S24) of *L. leucocephala* has been studied at the Military Padav area of IGFRI experimental farm (25□ N, 78□ E and 275 msl) with average annual rainfall 800 to 1000 mm, most of which received during July to September. The annual mean maximum and minimum temperatures were 33□ C and 17.4□ C respectively. Three distinct seasons i.e., rainy (hot wet), winter (cold dry) and summer (hot dry) occur during a year which affects growth, physiology and the production of the *Leucaena*. The soil of the site was calcareous and degraded with very poor fertility and water retention capacity. The plantation was raised during 1989 and coppiced in 1994 were used for this study. Rate of photosynthesis (PN), stomatal conductance (CS), intercellular CO_2 concentration (CINT) and transpiration rate (TR) were recorded in the leaves using LI 6200 portable Photosynthesis System (LI-COR, USA). The ratio of PN/TR, PN/CINT was also calculated. All the observations were recorded at the full vegetative growth in a clear sky day between 11.00 hr to 12.00 hr under natural environment. Three plants of each variety were uprooted for recording the biomass production during Oct-Nov. 1997. Nitrate reductase activity in fresh leaves was determined using method of Bar-Akiva and Sterbaum (1965) and modified by Kleeper *et. al.*, (1971).

RESULTS AND DISCUSSION

Rate of photosynthesis (PN) and stomatal conductance (CS) ranged from 6.91 (mole $m^{-2}s^{-1}$ to 16.63 (mole $m^{-2}s^{-1}$ and 0.210 cm^2s^{-1} to 0.637 cm^2s^{-1}, respectively. The selection S24 showed maximum rate of photosynthesis and stomatal conductance followed by K601 and S10, which were at par, and S22 indicating their superiority with respect to assimilatory performance over the other genotypes. Similar to the rate of photosynthesis the selection S24 and S22 maintained higher stomatal conductance for CO_2 assimilation. Lowest rate of photosynthesis and stomatal conductance was recorded in provenance K636. Among selection S24, S10 and S22 showed higher rate of transpiration indicating their fast assimilatory function contributing towards higher biomass production. The ratio of PN/CINT, which represents the carboxylation efficiency (Farquhar and Sharkey, 1982) was highest in S24 followed by S22, S10 and K601 which were at par confirming their highest productivity performance. No definite trend was seen in the PN/TR ratio, which may be due to differential rate of transpiration and photosynthesis. However the ratio of PN/TR was observed highest (water use efficiency) in S14 and S24 (Table 1).

The ecophysiological studies have provided functional aspect of the growth and production of *Leucaena leucocephala* provenances / genotypes under semi-arid environmental condition. The measurements of photosynthesis were important component of growth (Ledig and Botkin, 1974). Because of the potential utility of growth components in plant improvement, it is desirable to quantify genotypes and genotype-environment variation in photosynthesis. The process of transpiration (TR) is important in controlling the leaf temperature and it was promising in all genotypes. The Nitrate reductase activity (NRA) was observed highest in selection S10, S22 and S14. The higher specific leaf mass (SLW) in various genotypes contributes towards their total above ground biomass production potential. The higher SLW was observed in selection S24 predicting its superiority over other selections and provenances. The total above ground biomass (dry matter) was recorded in selection S24 followed by S10 and S22. These selections showed their superiority over the promising provenances for all the characters.

Rate of photosynthesis is positively associated with stomatal conductance and transpiration which shows interdependence of these characters. A weak correlation was analyzed between photosynthetic rate and dry weight as also reported by Ledig and Perry (1967). PN/CINT was positively correlated with stomatal conductance (CS) indicating that these traits are important for selection of a genotype for fast growth and biomass production.

The variations in physiological characters in these genotypes clearly indicate that variability exists among the genotypes of *Leucaena* and their further scope of development of superior types by incorporating all these characters in the improvement programme.

Table 1: *Total dry weight and other physiological characters in L. leucocephala*

Genotypes	Total Dry Wt. g/p	NRA µmol/g/h	CS cm^2/s	PN/TR µmol H_2O / mmol CO_2	PN/CINT
K8	41.34	221	0.40	2.14	0.054
K217	37.07	192	0.30	1.63	0.046
K601	41.27	282	0.47	1.89	0.165
K636	31.26	210	0.21	2.04	0.031
K397	31.30	294	0.47	2.17	0.053
K340	34.93	165	0.36	2.33	0.054
K29	41.98	152	0.35	2.37	0.043
SP1	39.94	181	0.28	2.27	0.042
IGFRI 23-1	40.19	205	0.27	2.16	0.045
S10	50.16	319	0.46	1.82	0.062
S14	43.18	308	0.45	2.45	0.043
S15	36.58	216	0.31	1.97	0.041
S22	44.73	323	0.48	2.11	0.066
S23	23.91	214	0.47	2.03	0.043
S24	56.36	223	0.64	2.19	0.081

On the basis of growth and assimilatory characters, selection S24, S10 and S14 were identified as superior genotypes for maximization of productivity under semi-arid tropical condition and may be taken for plantation and improvement programme of *L. leucocephala*.

ACKNOWLEDGEMENTS

We are grateful to Director, Indian Grassland and Fodder Research Institute, Jhansi for providing facilities and encouragement. We are thankful to Shri A. K. Panday, Technical Assistant for data recording.

REFERENCES

1. Bar Akiva, A. and Sternbaum, J. 1965. Plant and Soil 23: 141-144.
2. Brewbaker, J. L.; Hegde, N.; Hutton, E. M.; Jones, R. J.; Lowry, J. B.; Moog, F. and Van den Beldt, R. J. 1985. Leucaena- Forage production and use, NFTA, Hawaii: 39.
3. Farquhar, G. D. and Sharkey, T. D. 1982. Ann. Rev. Plant Physiol. 33: 317-345.
4. Ledig, F. T. and Perry, T. O. 1967. In: Proceedings of the Ninth Southern Conference on Forest Tree improvement, Knowville, Tennessee: 120-128.
5. Ledig, F. T. and Botkin, D. B. 1974. Can. J. Bot. 47: 519-527.

Provenance Studies in Indian Blackberry

S. L. MADIWALAR, S. I. HANAMASHETTI, S. M. MUTANAL AND B. S. NADAGOUDAR
University of Agricultural Sciences, Dharwad (Karnataka)

ABSTRACT

Indian blackberry (*Syzigium cumini*) is highly valued for its fruits, medicinal value and as a fodder tree, and produced a strong, heavy timber. It grows best in moist locations and tolerates water logging, but also survives and is productive in semi-arid conditions. Provenance studies were initiated on black clayey soils at the main Research Station, University of Agricultural Sciences, Dharwad during 1988. The experiment consisted of ten sources viz. Krishnagiri families, 1, 2, 3, 4, 5, Devaragudihal families 26, 28, 30, Badami and local. The seedlings were planted at 6 □ 6 m spacing with three replications in randomised block design. At the end of 10 years, Krishnagiri families (No. 2, 3 and 4) were significantly superior in growth attributes like stem girth, crown spread and plant height compared to other provenances.

INTRODUCTION

Indian blackberry is native of India, Burma, Sri Lanka and Philippines but widely planted and naturalized in many tropical countries (Gupta, 1993). It is valued for its fruits, medicinal values (in *Aurveda*) timber and fodder. Although it grows in moist locations, it tolerates water logging and is productive in semi arid conditions. A great variation in growth rate, fruit production and fruit size is observed between plants naturally growing at different locations in Karnataka (Anon, 1989). Presence of large variation between provenances indicate the potential of achieving genetic gain through selection (Oduol, 1998). Hence, provenance studies were initiated to know the extent of genetic variation and there by to select productive trees for forestry / agroforestry.

MATERIALS AND METHODS

The superior trees were selected at Kolar (eastern dry zone) Badami (northern dry zone), Kalaghatagi (hill zone) and Dharwad (northern transition zone), Forest Ranges. The selection was based on stem form, crown size, fruit quality, fruit size and yield. The selection involved preliminary listing, verification, deleting inferior ones and collecting lot of details of each tree. Five sources from Krishnagiri (Kolar Range), 3 sources from Devaragudihal (Kalaghatagi Range) one source each from Badami and Dharwad (local) were selected for the studies. The seeds from Dharwad (local) were selected for the studies. The seeds from each source were collected during 1987. The seedlings were raised in nursery and one year old seedlings of uniform size were used for planting. The seedlings were planted on black soils at main Research Station, Dharwad during 1988 adopting a spacing of 6 □ 6 m. Each treatment was replicated 3 times in Randomized Block Design The provenances were assessed for growth attributes like stem diameter, crown spread and plant height.

RESULTS AND DISCUSSION

The data on assessed variable among different provenances is given in Table 1. Maximum stem diameter at ground level was recorded in Krishnagiri-3 (24.5 cm) followed by Krishnagiri 2, 1 and 5 which were at par but significantly superior to local. The lowest stem diameter was recorded in local check (12.93 cm).

The plant height was significantly high in Krishnagiri-3 (6.24 m) followed by Krishnagiri-2 (5.68 m), Krishnagiri-4 (5.42 m) and Krishnagiri-1 (5.28 m), which were at par with each other. Krishnagiri-5 and Devaragudihal-9 were at par but significantly superior to local check (4.06 m). Maximum crown spread both in North-South and East-West was recorded in Krishnagiri-3 (4.15 m and 3.96 m) followed by Krishnagiri-2, and 4 which were at par but significantly superior to local check. Krishnagiri-5 and Devaragudihal series were at par but significantly superior to local check.

TABLE 1: *Growth parameters as influenced by different provenances*

Treatements	Collar diameter (cm)	Plant hight (m)	Crown Spread (m)	
Krishnagiri-1	21.83	5.28	3.95	3.52
Krishnagiri-2	22.13	5.68	3.90	3.88
Krishnagiri-3	24.50	6.24	4.15	3.96
Krishnagiri-4	20.70	5.42	3.67	3.73
Krishnagiri-5	21.33	5.13	3.33	3.66
Devaragudihal-6	17.23	4.90	3.00	2.66
Devaragudihal-7	18.15	4.00	2.46	2.67
Devaragudihal-8	15.42	4.10	2.91	2.80
Devaragudihal-9	18.13	5.05	3.00	3.06
Local -10	12.93	4.06	2.35	2.28
Sem	2.38	0.31	0.38	0.35
CD @ 5%	7.00	0.91	1.12	1.03

CONCLUSIONS

Thus at the end of 10 years, Krishnagiri sources (No. 2, 3 and 4) were significantly superior in growth attributes like stem diameter, crown spread and plant height compared to other provenances. These significant differences between provenances in assessed characters show that *S. cumini* has an ecotypic variation making selection at provenance level worthwhile.

ACKNOWLEDGEMENT

The help rendered by Silviculturist (Northern Zone), Karnataka Forest Department, in providing seeds/seedlings of elite trees of *S. cumini* is gratefully acknowledged.

REFERENCES

1. Anonymous, 1989. Plus trees. Karnataka Forest Department, Silviculturist Northern Zone, Dharwad.
2. Gupta, R. K. 1993. Multipurpose trees for agroforestry and Wasteland Utilization, Oxford & IBH, New Delhi.
3. Oduol, P. A. 1998. In: Sunil Puri ed. Tree Improvement. Oxford & IBH Pub., New Delhi 169-187.

Establishment and Growth of Tissue Culture Raised Clones of Plus Trees of *Eucalyptus tereticornis* under Semi-Arid Conditions

AJIT, V. K. GUPTA, K. R. SOLANKI, A. DATTA AND R. V. KUMAR
National Research Centre for Agroforestry, Jhansi (UP)

ABSTRACT

An adaptability cum growth trail of tissue culture raised clones of ten plus trees of *Eucalyptus tereticornis* conducted at National Research Centre for Agroforestry, Jhansi revealed that clones ET-6R, ET-12, ET-362 and Et-2116 were quite promising based on growth attributes as well as total biomass. The mean annual increment (MAI) for plant height was 3.02 m and for DBH was 2.38 cm. At the age of 3.5 years the mean above ground biomass (AGB) was 51.75 kg/tree, although the variations in AGB amongst the clones were not significant. The correlation coefficients amongst growth parameters and biomass components revealed that circular growth attributes (DBH, CD) plays dominant role for determination of biomass as compared to polar growth (height).

INTRODUCTION

A large number of Indian population living in villages depend upon fuelwood for cooking purpose, as a result estimated fuelwood requirement increased by many folds (250 millions m^3). Similarly, requirement of round wood for construction, agricultural implements, cattle sheds, fencing materials and for other purposes is estimated to be 65 millions m^3 by 2000 AD as against the present production of about 6 millions m^3. In contrast to these constraints, the productivity of forests in India is much below (0.7 m^3/ ha/ year) against the world average (21 m^3/ ha/ year) which further exaggerate the problem of fuelwood requirement. Under these situations, plantation of fast growing tree species under agroforestry and social forestry programmes must be encouraged to meet these challenges.

In tropics, *Eucalyptus tereticornis* is one of the fastest growing species with 20-40 m^3 /ha/yr. wood production (Eldridge, 1978; Chaturvedi, 1973). Fast growing habit, large seed production, easy multiplication, good establishment and varied uses makes *Eucalyptus tereticornis* popular in India particularly in northern India which occupies more than 1 million hectare area (Sundararaju *et. al.*, 1995). Present populations of *E. tereticornis* used for wood production are derived from very few original trees (Pryor, 1966; Chaturvedi, 1976). As a consequence, many trees of eucalyptus have poor growth and unsatisfactory stem straightness. Inbreeding depression is considered to be major attribute of poor growth and stem crookiness (Eldridge, 1978).

The major objectives of many programmes of *Eucalyptus* improvement throughout the world are to increase the yield and quality of wood (Eldrige, 1978). Improvement in *Eucalyptus* is possible one way by avoiding selfing and making controlled pollination for further selection. Appreciable hybrid vigour exist in hybrid populations (Pryor, 1976). F_1 hybrids of the *E. tereticornis* □ *E. camaldulensis* (FRI-4) and *E. camaldulensis* □ *E. tereticornis* (FRI-5) showed rapid height and diameter growth increment in India (Venkatesh and Sharma, 1977). The large-scale use of eucalyptus hybrids for plantation is handicapped for want of methods of mass production. Mass production of F_1 or their hybrids might be achieved either by a cheap means of controlled pollination, which is not yet available or by vegetative reproduction of hybrids. Looking to the needs to supply productive propagules in mass, 10 plus trees selected in various parts of Haryana were multiplied through tissue culture at TERI, New Delhi to retain the hybrid vigour. These were tested at NRCAF, Jhansi to explore the potential of these plus trees as compare to seedlings raised at forest nurseries in agroforestry programmes.

MATERIALS AND METHODS

An adaptability cum growth trial of tissue culture raised clones of 10 plus trees of *Eucalyptus tereticornis* was conducted at National Research Centre for Agroforestry, Jhansi. Jhansi is situated between latitude 24°11' – 26°27'N and longitude 78°17' -81°34' E with mean annual precipitation of 936 mm of which 80-90% is received during July-September. The climatic pattern of Jhansi is true representative of semi-arid region with mean maximum temperature ranges from 22°C-47°C and minimum from 2.5°C to 29.9°C. The soil of the experimental site is black containing organic carbon 0.29%, pH 7.07 and available NPK 82.3 kg/ha, 3.16 kg/ha, 83.66 kg/ha, respectively.

Tissue culture raised clones of 10 plus trees (ET-12, ET16, ET-107, ET-236, ET-330, ET-348, ET-360, ET-362, ET-6R and ET-2116) collected from TERI, New Delhi with local check were transplanted in first week of September 1994 in randomized block design with three replications. Each plot consisted of 25 plants in five rows spaced at 3 □ 2 m. Whole of the experimental area was surrounded by guard rows. Fertilizer was given in the pits at the rate of 60 kg N/ha and 40 kg P_2O_5/ha and 25 kg BHC (10%)/ ha as a basal dose at the time of plantation. Data on basic growth attributes i.e. tree height (H), collar diameter (CD) were initially taken at an interval of 3 months whereas the diameter at breast height (DBH) were initiated to record at the age of 1.5 years after plantation. Leaf length and breadth were taken at the age of 1.5 years. Felling of 5 trees per clone per replication was done at the age of 3.5 years after plantation to assess leaf, branch, main bole weight and total biomass. SPSS (Statistical Package for Social Sciences) has been used for computing the descriptive statistics, correlation matrix and analysis of variance.

RESULTS AND DISCUSSION

All the clones under study survived well (more than 80%) except local check and ET-236. At 3.5 years after plantation, survival ranged from 58.66% for check to 92.00% for clone ET-107 and ET-348 (Table 1). Relatively lower survival for check and ET-236 might be due to comparatively more attack of termites on these clones. High survival of all these clones suggested that establishment of tissue culture raised clones were very good and commercialization of tissue culture raised seedlings is feasible.

TABLE 1: *Growth attributes of various clones of Eucalyptus tereticornis*

Clone	Survival (%)	Leaf length (cm)	Leaf breadth (cm)	Height (m)	DBH (cm)	CD (cm)
ET-16	*90.66*	*11.60*	*2.98*	*11.00*	*7.97*	*10.70*
ET-360	*90.66*	*13.74*	*2.80*	*11.28*	*7.63*	*10.48*
ET-2116	*89.33*	*13.25*	*3.94*	*11.00*	*8.33*	*12.06*
ET-236	*62.66*	*13.45*	*4.02*	*10.89*	*8.65*	*11.99*
ET-330	*90.66*	*14.60*	*2.91*	*10.30*	*8.31*	*11.68*
ET-348	*92.00*	*14.54*	*2.45*	*10.32*	*8.03*	*11.31*
ET-6R	*84.00*	*13.53*	*3.63*	*10.74*	*8.55*	*12.60*
ET-12	*84.00*	*14.75*	*4.41*	*10.29*	*9.16*	*12.44*
ET-107	*92.00*	*14.17*	*2.94*	*10.41*	*7.55*	*10.71*
ET-362	*82.66*	*12.35*	*3.16*	*9.93*	*8.90*	*13.26*
Local Check	*58.66*	*11.89*	*3.70*	*10.21*	*8.75*	*12.20*
Sem ±	*NA*	*NS*	*0.13*	*NS*	*NS*	*0.78*
CD (5%)			*0.36*			*1.53*
Grand Mean	*83.39*	*13.44*	*3.35*	*10.58*	*8.35*	*11.77*

Statistical analysis did not show significant differences in tree height. All the clones grew very fast in vertical direction with mean annual increment (MAI) of 3.02 m in rainfed conditions. The ET-360 clone gave the highest mean height (11.28 m) at 3.5 years age of assessment followed by ET-2116 and ET-16. All these clones were superior to local for plant height growth. Chandra *et.al.* (1998) reported average MAI for plant height as 2.21 m in high rainfall area (Rudrapur, Nainital). Similarly, in other trials, conducted at Kurukshetra, Haryana; Marakkanam, Tamil Nadu; Ludhiana, Punjab; Ferojpur, Punjab; MAI for plant height was less than 3.00 m (Tandon, et.al. 1993; Sundararaju, et.al. 1995; Sidhu, 1995; Aulokh and Sandhu, 1990). Best hybrids developed at FRI, Dehradun, also had MAI of 2.06 m at Dehradun (Venkatesh and Sharma, 1977).

Analysis of variance for DBH suggested that variation among clones for this trial was not much appreciable. It ranged from 7.55 cm (ET-107) to 9.16 cm (ET-12). Only two clones were superior than check which were ET-12 and ET-362. MAI for DBH (2.38 cm) reported in this trial was quite good when compared to the results of other trials conducted at various places in India (Tandon, et.al. 1993; Aulokh and Sandhu, 1990). Chandra et.al., 1998 reported average MAI of 2.27 cm of best provenance in 25 provenances trial of *Eucalyptus* species at

Rudrapur, Nainital which is high rainfall area. Promising hybrids FRI-4 and FRI-5 had MAI of 1.67 cm at Dehra Dun (Venkatesh and Sharma, 1977). Similarly, the overall mean of the clones for CD was 11.77 cm at the age of 3.5 years with MAI for CD was 3.36 cm. The best clone (ET-362) had MAI 3.79 cm followed by ET-12 (3.55 cm) and ET-6 R (3.60 cm).

The above ground biomass (AGB) and its components for all the clones were shown in Table 2. At the age of 3.5 years, the average biomass was 51.75 kg/tree, although the variation in AGB amongst the clones were not significant. ET-362 (63.51 kg/tree), ET-6R (56.37 kg/tree, ET-12 (55.90 kg/tree) and ET-2116 (55.78 kg/tree) showed higher values than the check (53.00 kg/tree). ET-6R (43.68 kg/tree), ET-12 (42.41 kg/tree) and ET-362 (42.08 kg/tree) were found superior than check (40.19 kg/tree) for bole weight. ET-362 and ET-2116 were also found better than check for both branch weight as well as leaf weight.

TABLE 2: *Biomass production (kg/tree) of different tree components for various clones of Eucalyptus tereticornis at the age of 3.5 years*

Clone	Leaf	Branch	Bole	Total Biomass
ET-16	5.077	4.827	35.923	45.827
ET-360	6.483	5.150	36.557	48.190
ET-2116	7.839	8.096	39.850	55.786
ET-236	5.945	6.210	40.230	52.385
ET-330	5.554	7.500	35.799	48.853
ET-348	5.293	7.080	35.200	47.573
ET-6R	5.841	6.855	43.683	56.378
ET-12	4.954	8.533	42.413	55.900
ET-107	5.007	5.450	31.369	41.826
ET-362	7.104	14.33	42.082	63.518
Local Check	5.921	6.936	40.193	53.050
Sem ±	NS	2.48	NS	NS
CD (5%)		4.86		
Grand Mean	5.91	7.360	38.482	51.753

The correlation coefficients of all the attributes amongst growth parameters and biomass components were found significant (1 tailed significance at p=.001) and the results were presented in Table 3.

TABLE 3: *Correlation matrix among the growth attributes and biomass components in Eucalyptus tereticornis at the age of 3.5 years*

	Plant height	DBH	Collar diameter
Total biomass	*0.633***	*0.930***	*0.877***
Bole weight	*0.725***	*0.932***	*0.848***
Leaf weight	*0.487***	*0.686***	*0.666***
Branch weight	*0.229***	*0.668***	*0.700***

Fairly strong linear dependence of total biomass on DBH and CD was observed ($r_{\text{biomass, dbh}} = 0.9295$, $r_{\text{biomass, cd}}=0.8768$), as compared to the linear association of biomass with height ($r_{\text{biomass, height}} =0.6333$). It revealed that circular growth attributes plays more dominant role for determination of biomass as compared to polar growth. Patil et.al. (1997) have worked out correlation among growth attributes and reported high positive correlation between girth at breast height and total height.

Overall, clones ET-6R, ET-12, ET-362, and ET-2116 were quite promising based on growth attributes as well as total biomass, suggesting that plus tree selection in *Eucalyptus tereticornis* followed by mass multiplication through micropropagation (tissue culture) is beneficial breeding procedure to increase the productivity of eucalyptus by way of capturing hybrid vigour generated by free pollination among genetically dissimilar trees.

REFERENCES

1. Aulokh, K. S. and Sandhu, S. S.1990. Indian Forester 116: 442-445
2. Chandra, J. P.; Yadav, M. P. S. and Gandhi, J. N. 1998. Indian Forester 124: 232-236
3. Chatuvedi, A. N. 1976. Indian Forester. 102: 63-67
4. Chaturvedi, A. N. 1973. IFR 12 (14)
5. Eldrige, K. G. 1978. Silvae Genetica 27: 205-209
6. Patil, J. V.; Desmukh, R. B.; Jambhale, N. D.; Patil, S. C. and Kunjir, N. T. 1997. Indian Journal of Forestry 20: 133-135
7. Pryor, L. D. 1966. Indian Forester 92: 615-22
8. Pryor, L. D. 1976. The biology of *Eucalyptus*. Arnold, London.
9. Sidhu, D. S. 1995. Indian Forester 121: 126-130
10. Sundararaju, R.; Bharathi, R. K. and Chinnathurai, A. K. 1995. Indian Forester 121: 96-102
11. Tandon, V. N.; Rawat, J. K.; Singh R. 1993. Indian Forester. 119: 232-237
12. Venkatesh, G. S. and Sharma, V. K. 1977. Indian Forester 103: 795-802

Preliminary Evaluation of Growth Performance of Different Plus Tree Progenies of *Dalbergia sissoo*

R. V. KUMAR, V. K. GUPTA, A. DATTA AND K. R. SOLANKI
National Research Centre for Agroforestry, Jhansi (UP)

ABSTRACT

Dalbergia sissoo Roxb (sisham) is known internationally due to its multifaceted use and now-a-days this tree is getting much importance in agroforestry system owing to its ideotype needed for the system. A study on preliminary evaluation of growth performance of nineteen different plus tree progenies of sisham of 14 months age was conducted on the NRCAF Farm, Jhansi at two different sites in completely rainfed situation. The analysis of variance for different characters under study was found to be highly significant indicating the presence of sufficient variation among plus tree progenies. For the character plant height, progenies PT-19 (250.67 cm) and PT-2 (239.50 cm) were found significantly superior over check (205.00 cm) at first site whereas progeny PT-3 (201.80 cm) registered highest plant height at second site. Similarly, for the character collar diameter, progenies PT-19 (4.21 cm) and Prov-4 (4.19 cm) at first site and progenies PT-9 (2.61 cm) and PT-6 (2.56 cm) at second site was adjudged best performer. Highest expected volume was recorded for the progeny PT-19 (4442.90 cm^3) followed by prov-4 (4088.82 cm^3) at first site and for the progeny PT-3 (1312.20 cm^3) at second site. After comparing the relative performance of all the progenies for all the characters under study at two different sites, progenies PT-3, PT-9, PT-19 and Prov-4 were found to be promising ones at this particular stage of the growth.

INTRODUCTION

Sisham is known internationally as a premier timber species of rosewood genus. Early perceptions of natural distribution of sisham perhaps points to its long-term use as a cultivated tree which has considerably expanded its habitat (White, 1990). Sisham is native to the foothills of the Himalayas of India, Pakistan, Nepal and Indo-Gangetic basin. It is used extensively for shelter belts and fuel wood planting in the sub-humid and drier areas of Ghana and Nigeria (Ladipo, 1994). Friend of farmer's no less than that of foresters, sisham is a tree that pays rich dividends. It is a very important raw material for a variety of wood based industries, especially furniture and building construction. Its pruned branches are the cheap source of firewood with calorific value ranging from 4900-5200 K cal/kg. Being protein rich (containing up to 24.1% crude protein and 26.1% crude fiber on dry weight basis), the leaves of sisham can serve as a suitable alternative of cattle fodder during scarcity period. Young leaves appear at the end of February and leafing is complete by early April, making April to May the best time of year for the production of high quality fodder (Jackson, 1987). Now-a-days this tree is getting important place in agroforestry system due to its ideotype which is very much suitable as per requirements of agroforestry systems. It has very light crown and growth of the main stem is quite straight thus interfering least with the under story crop. It enriches soil through atmospheric nitrogen fixation with the help of nitrogen fixing *Rhizobium* spp. and rich leaf fall. For the success of this tree species in agroforestry system, we need fast growing plus tree progenies with clean bole height of this species. In recent past, several workers have tried to study the growth performance of different *Dalbergia sissoo* progenies (Haque and Kamaluddin, 1995; Singh and Bhargawa, 1995, Gera et al, 1996). Keeping in view the information available and research needed for future tree improvement programme, a study on preliminary evaluation of growth performance of different plus tree progenies of sisham was conducted on the farm of NRCAF, Jhansi for which materials were planted in August, 1996. The results of the study and their long term effect in future tree improvement programme have been discussed in brief in this paper.

MATERIALS AND METHODS

For conducting this experiment, a total of nineteen plus tree progenies including check were collected from different parts of Bundelkhand region. All the plus tree progenies were collected based on certain criteria viz., straightness, clean bole height, close canopy and fast growth. From all these materials two identical sets were prepared for planting the material at two different sites. All these materials were planted in August, 1996 in degraded land in completely rainfed situations on the farm of NRCAF, Jhansi. Experiment was laid out in randomized block design with two replications in each set of experiment. Each progeny had single row plot of 24 □ 5 m spaced at 5 □ 4 m between and within the rows. Evaluation was done under agroforestry system. Data were recorded on survival percentage, plant height, collar diameter and expected volume. Data were recorded at every three months interval. After taking the observations, data were put for analysis and comparative performance of each plus tree progenies were adjudged in relation to check.

RESULTS AND DISCUSSION

The analysis of variance for different characters viz. plant height, collar diameter and expected volume of plus tree progeny at the age of 14 months after planting were found to be highly significant indicating the presence of sufficient variation among plus tree progenies. This variation is very much needed for improvement of the character in question. Mean performance of twenty plus tree progenies including check for different characters have been presented along with values of critical differences in Table 1.

Survival of the material, which is the prime important thing for establishing the material, was found to be almost 100 per cent for all the plus tree progenies. This indicates that survival of the sisham plus tree progeny in degraded land of Bundelkhand region is not a problem of alarming situation. The plant height ranged from a minimum of 131.70 cm to a maximum of 250.67 cm in the 14 months old trial at first location. At this location, out of eighteen progenies, ten progenies exhibited their superiority over check. Out of these ten progenies, two plus tree progenies i.e. PT-19 (250.67 cm) and PT-2 (239.50 cm) were found significantly superior over check (205.00 cm) for plant height. On the other hand, at second location, all the eighteen progenies performed their superiority over check for the character plant height with minimum of 146.20 cm to a maximum of 201.80 cm. Six plus tree progenies were observed for exhibiting significant superiority over check for plant height. The highest average plant height was observed for progeny PT-3 (201.80 cm) which was followed by progenies PT-33 (187.73 cm) and PT-30 (182.73 cm).

TABLE 1: *Mean values for characters plant height, collar diameter and expected volume at two different sites*

Progeny	Plant Height (cm)		Collar Diameter (cm)		Expected Volume (cm³)	
	Site-I	Site-II	Site-I	Site-II	Site-I	Site-II
PT-2	239.50	157.50	3.67	2.24	3225.80	790.27
PT-3	219.42	201.80	3.31	2.55	2403.98	1312.20
PT-6	224.59	149.13	3.86	2.56	3346.30	977.34
PT-8	216.34	158.50	3.77	1.99	3074.82	627.27
PT-9	218.84	175.10	4.01	2.61	3518.97	1192.80
PT-11	204.67	148.70	3.53	2.10	2550.37	655.67
PT-19	250.67	163.50	4.21	2.36	4442.90	910.63
PT-22	167.25	164.70	2.94	2.53	1445.64	1054.23
PT-27	199.84	154.80	3.92	2.01	3070.82	625.41
PT-29	154.00	175.78	2.64	2.49	1073.32	1089.85
PT-30	201.33	182.73	3.92	2.40	3093.72	1052.52
PT-31	200.10	150.00	3.57	1.96	2550.25	676.24
PT-32	131.75	169.20	2.15	2.55	909.01	1100.22
PT-33	218.25	187.73	3.86	2.34	3251.84	1027.93
Prov-1	208.75	165.5	2.86	2.35	1707.49	913.97
Prov-2	208.13	173.54	3.44	2.56	2462.93	1137.31
Prov-4	232.9	167.9	4.19	2.38	4088.82	951.05
Hisar	180	167.87	2.99	2.45	1609.22	1007.64
Check	205	146.2	3.76	2.18	2898.21	694.8
S.E (M) ±	11.6	8.94	0.29	0.21	213.64	77.24
C.D.	32.17	24.78	0.82	0.59	628.34	227.18
C.V. (%)	8.33	7.6	12.15	12.76	11.4	11.61

Similarly, for the character collar diameter, progenies were ranged from a minimum of 2.15 cm to a maximum of 4.21 cm at location one. For this character, a total of eight progenies excelled their performance in comparison to check. Although, none of the progenies were found to be significantly better than check, progenies PT-19 (4.21 cm), Prov-4 (4.19 cm) and PT-10 (4.01 cm) registered their superiority over check (3.76 cm). On the other hand, at the second location, a total of fourteen progenies excelled their performance over check for the character collar diameter. The highest average collar diameter was observed for progeny PT-9 (2.61 cm) followed by progenies PT-6 (2.56 cm) and Prov-2 (2.56 cm) in location two. Expected volume was also calculated which is very important parameter for screening the better genotypes even at early stage of tree growth. In location one, a total of nine progenies proved their superiority over check for the character expected volume. Highest expected volume was recorded for the progeny PT-19 (4442.90 cm^3) followed by Prov-4 (4088.82 cm^3), PT-9 (3518.97 cm^3) and PT-6 (3346.30 cm^3) whereas for the check, expected volume was 2898.21 cm^3. Similarly, at the second location, a total of fourteen plus tree progenies were found to be superior than check for the character expected volume. The highest expected volume was observed for the progeny PT-3 (1312.20 cm^3) and it was followed by PT-9 (1192.80 cm^3), Prov-2 (1137.31 cm^3) and PT-32 (1100.22 cm^3) whereas for the check it was only 694.80 cm^3, which was almost half of the value of expected volume in comparison to best progeny PT-3 for this character at this location.

CONCLUSIONS

Comparison of the performance of all the progenies under study at two different sites revealed that the progenies PT-3, PT-9, PT-19 and Prov-2 are the promising ones at this particular age of tree growth. This preliminary evaluation is very helpful in predicting the performance of genotypes at an early stage thus reducing the time period required for any tree improvement programme.

REFERENCES

1. Gera, M.; Ginwal, H. S. and Srivastva, R. L.1996. Indian Forester 122 (3): 250-257.
2. Haque, M. N. and Kamaluddhin. M.1995. Science 19 (1):121-126.
3. Jackson, J. K.1987. Manual of Afforestation in Nepal. Department of Forest, Kathmandu, Nepal.
4. Ladipo, D. O.1994. In: *Dalbergia*; proceedings of an International Workshop, (Eds.: Sidney, B. Westley and James M. Roshetko). Morrilton (ARE) USA: NFTA pp. 27-37.
5. Singh, J. and Bangarwa, C.1995. Annals of Biology. 11 (½): 111-114.
6. White, K. J.1990. *Dalbergia sissoo*: an annotated bibliography. Winrock International F/FRED: Bangkok, Thailand, XIX+120 pp.

Multivariate Analysis in Clonal Progenies of Tamarind (*Tamarindus indica* L.)

S. I. HANAMASHETTI, G. S. SULIKERI AND P. M. SALIMATH
K. R. C. College of Horticulture, Arbhavi (Karnataka)

ABSTRACT

Multivariate analysis helps in the identification of plus trees for breeding objectives. The present study was undertaken to assess the genetic diversity in tamarind through multivariate analysis. The first canonical vector, pulp weight accounted 99.231 per cent towards divergence. Forty genotypes of tamarind were grouped into 10 clusters. Cluster VII, VIII and IX were the largest with 5 genotypes each, while cluster V was the smallest with two genotypes. Cluster mean for eight quantitative traits of forty genotypes were calculated. The present study indicates that pulp weight was the most important factor, which was responsible for discrimination in the genotypes studies. It is necessary to use only the best trees for multiplication and for incorporating them into the breeding population.

INTRODUCTION

Multivariate analysis is a suitable statistical methods, for analyzing various causative factors, and their inter relationships operating within the tree populations. It also helps in the identification of plus trees for specific breeding objectives in order to improve the genetic characters of tamarind and thereby economic values of future generations. The present study was undertaken to assess the genetic diversity in tamarind through multivariate analysis.

MATERIALS AND METHODS

The material for the present study consisted of forty plus trees identified in the northern Karnataka and maintained at Forest Research Centre, Gungaratti, Dahrwad. The data were collected in different genotypes during the fruiting season 1993, 1994 and 1995. Mahalanobis D^2 and canonical variate were done to estimate the dissimilarity coefficients and cluster the genotypes by using dendrogram.

RESULTS AND DISCUSSION

The data on canonical variate analysis, clustering pattern and intra cluster distance are presented in Table 1, 2 and 3, respectively.

Canonical Variate Analysis

In the present study 40 genotypes of tamarind belonging to five districts were included to assess the nature of genetic diversity by multi variate analysis. As revealed by the first vector, pulp weight alone contributed maximum in discriminating the genotypes (Table 1).

TABLE 1: *Elements of the first canonical vector in tamarind*

S. No.	Character	Vector-I
1.	*Pulp weight (g)*	1.000
2.	*Shell weight (g)*	0.007
3.	*Pod width (cm)*	0.007
4.	*Seed weight (g)*	0.004
5.	*Number of seeds*	0.004
6.	*Pod weight (g)*	0.004
7.	*Pod length (cm)*	0.003
8.	*Pod thickness (cm)*	0.002

Shell weight and pod weight are the other two important characters which contributed to divergence. The first canonical vector accounted 99.231 per cent towards divergence (Table 2).

The contribution of the remaining vectors, towards divergence was meager. In the canonical variate analysis, the number of variable were reduced to a linear expression called principal component and this accounts for most of the variation produced by these characters.

Similarly, a second succeeding principal component was also formed to account for the residual variability. The first principal component was significant enough to use for the purpose of discriminating the genotypes and to cluster on the canonical configuration. The contribution of the remaining eight vectors was very low and hence, they were not considered.

TABLE 2: *Canonical roots and variability accounted by them in tamarind*

Canonical roots	Variability accounted	Percentage variance	Cumulative variance
1.	14287.182	99.231	99.231
2.	62.036	0.431	99.661
3.	40.036	0.278	99.940
4.	6.400	0.044	99.984
5.	1.779	0.012	99.996
6.	0.316	0.002	99.999
7.	0.167	0.001	100.000
8.	0.044	0.000	100.000

Clustering pattern, intra and inter cluster distance of 40 genotypes of tamarind.

The horizontal line drawn at a dissimilar coefficient of 1.63 resulted into 10 clusters. Clusters VII, VIII and IX were the large with 5 genotypes each (Table 3) followed by clusters I, II, III, IV and □ with four genotypes I each case. Cluster VI had three genotypes, while cluster V was the smallest with two genotypes (Table 3).

Cluster-II recorded the highest intra cluster distance (17.13), followed by cluster VII (15.63) and cluster-I (14.27), while cluster-X recorded the lowest intra cluster distance (3.00) Table 3.

Cluster I was close to cluster III and IV, while it was moderately distance from clusters II, V and VII. It however, showed high inter cluster D2 value, with clusters VI, VIII, IX and X. Cluster II and III were also nearer to each other, and both of them showed greater diversity from cluster VI and cluster X, followed by cluster IX. Cluster IV was closer to cluster V. It showed greater diversity, with cluster distant from cluster VII, VII, IX and X. While it was nearer to cluster I, II, III and VI. Cluster IV also appears to be placed further away from clusters VII, VIII, IX, X, II and III. Likewise cluster VIII was very near to cluster □ and cluster VIII. Both cluster IX and □ showed greater diversity from cluster I, IV, V and VI, while they were moderately distant from cluster II, III and VII.

TABLE 3: *Clustering pattern and intra cluster distance (D^2) among 40 genotypes of Tamarind*

Cluster	No. of genotypes	Genotypes	D2 value	Maximum inter cluster distance and cluster number
I	4	*BGL 1, BEL 61, 75, 85*	14.27	1403.42 (X)
II	4	*BEL-2, UK-31, 73, 78*	17.13	1616.68 (VI)
III	4	*BEL-7, BEL-16, 79, 83*	10.07	975.31 (X)
IV	4	*BGL-5, BEL-14 BEL-15, 77*	6.55	2034.26 (X)
V	2	*BEL-57, BEL-62*	3.28	2774.91 (X)
VI	3	*DWD-19, DWD-71, S-4*	8.97	4271.85 (X)
VII	5	*BEL-6, DWD-70, BEL-60, S-13, 82*	15.63	2252.90 (VII)
VIII	5	*DWD-55, BEL-59, S-7, 74, 84*	11.47	2784.11 (VI)
IX	5	*BEL- 17, UK-32 S-54, S-3, 80*	12.47	36.9.50 (IX)
X	4	*BGL-21, BEL-56, BEL-58, 76*	3.00	4271.85 (VI)

The clustering pattern of different genotypes is presented in Table 3. The data clearly indicates that no parallelism can be drawn between geographic and genetic diversity. The genotypes collected from Belgaum are relatively more diverse among these selves than the genotypes from other region.

CONCLUSION

The information obtained from the present study indicates that pulp weight was the most important factor, which was responsible for discrimination in the genotypes studies. It is also note worthy that the present findings are in agreement with those of earlier workers (Illoh and olorode, 1990; Ganga Prasad, 1993 and Samiullah *et. al.*, 1993).

REFERENCES

1. Ganga Prasad, S. 1993. Ph.D. thesis submitted to UAS, Bangalore.
2. Illoh, H. C. and Olorode, O. 1990. Euphytica, 51 (3):197-205.
3. Samiullah, R.; Sheriff, R. A. and Chickadevaiah, C. 1993. In: Golden Jubilee Symposium on Horticulture Research- Changing Scenario. Horticulture Society of India, New Delhi: 473.

Multipurpose Browse Shrubs of India: Diversity in Tropical and Temperate Species

VINOD SHANKAR AND J. P. SINGH
Indian Grassland and Fodder Research Institute, Jhansi (UP)

ABSTRACT

There is a great potential for shrubs still untapped in vast rangelands of the world (Riveros 1988). Use and misuse of shrubs and trees as fodder-the Imperial Agricultural Bureaux Publication No.10 (IAB, 1947) is probably the first worldwide compendium on fodder trees and shrubs. Browse shrubs also find place in the relevant Indian literature but the information is diffused and it needs to be carefully picked up (Watt 1889-96, Gamble 1902, Haines 1910, 1921-1925, Duthie 1903-29, Kanjilal 1928, Laurie 1945, Chaturvedi 1949, Anonymous 1948-76, Negi, 1977). Shrubs not only show taxonomic diversity in respect of a large number of genera, species and varieties, but also show diversity in terms of range of distribution, growth habits, forms, nutritional value and palatability etc. Diversity is also noticed within and between populations of a species within its natural distribution range. Genetic variations within the species are associated with races or genotypes adapted to specific situations (Dogra 1986). Intra-specific variations in forest trees viz., *Terminalia, Pterocarpus, Bridelia, Castonopsis* etc. have been reported by Brandis (1906).

INTRODUCTION

Browse shrubs exhibit considerable diversity of form and distributional pattern due to India's diverse agroecology, which results from variations in rainfall, altitude, slope and soils. The Indian woody flora is represented by about 2417 dicot trees, 21 gymnosperms and 69 monocot trees (Pokhriyal *et. al.*, 1992). About 1420 shrub species are distributed in different parts of India. Tree and shrub species of fodder value and grown over wide or restricted areas have been identified (Pokhriyal *et. al.*, 1992, Singh 1982, Singh and Yadav 1995, Purohit and Samant 1995). But a comprehensive and systematic inventory of browse shrubs in India is yet to take off.

In the Himalayan region the polymorphic leguminous shrub viz., *Lespedza juncea* shows distinct genetic diversity (Rao and Hussain 1993 and Hussain and Kapur 1990). Some species of *Lespedza (L. gerardiana, L. eriocarpa)* exhibit a broad range of distribution in the western and the eastern Himalaya, and *L. sericea* also occurs in Nilgiris and Palni hills in southern India (Arora and Chandel, 1972). The other leguminous taxa viz., *Astra galus* and *Caragna* are distributed to 5000 m in the Himalaya. Genus *Astragalus* shows strong development in the dry western Himalayan region with 75% endemics (Chatterjee, 1939). The genus *Caragna* is endemic to Euro-Siberian region though a few species extend beyond the territory (Good 1964)., its five species occur in alpine subalpine and Ladakh region in India.

Tropical species of *Acacia, Carissa, Capparis, Flacourtia, Securinega, Woodfordia, Vitex, Ziziphus* are widely distributed up to the subtropical Himalayan region. In *Ziziphus nummularia*, which is an important browse species of arid and semi-arid regions, both tetraploid and hexaploid plants are reported (Khoshoo and Singh, 1963). Both, *Z. nummularia* and *Z. xylopyrus* show a considerable diversity of form and leaf fodder yield in the range of occurrence covering a number of habitats in the arid and semi-arid regions (Kaul and Ganguli 1963).

The main focus in this paper is on browse shrub diversity in tropical and temperate regions and their exploitation for multipurpose uses. Efforts have been made to scan the relevant information, which is diffused in a variety of literature sources, and highlight these in a systematic, classified manner. Also presented at the end is an overview alongwith future needs and research opportunities.

ENVIRONMENTAL ADAPTATIONS

Several shrubs show adaptation to harsh environments viz., *Acacia jacquemontit, Capparis decidua, Calligonium polygonides, Crotalaria burhia* are able to survive in shifting sand dunes due to their leathery, low transpiration, xerophytic foliage and deep rooting habits. Other arid shrubs viz., *Salvadora oleoides* and *Clerodendrum phlomidis* orient their leaves in such a manner that they absorb minimum amount of solar energy, which enables them to remain cool and hence green in the face of the most scorching sun (Bhandari 1988). *Tecomella undulata, Balanites aegyptiaca* and many other arid shrubs/trees also develop a large amount of cutin on their leaves as well as on their seeds which protect them from high transpiration (Bhandari 1988). In the cold desert species viz., *Caragna versicolor, Ephedra gerardiana* and *Hippophae rhamnoides* form dense bushy habit with woody branches to overcome harsh conditions of the cold desert (Rao, 1994).

PALATABILITY

Palatability is a complex phenomenon determined by both, the livestock and the herbage on offer (Ivins 1955). Seasonal variations in the browse choice are also reported Singh *et. al.* (1996). Cattle in general show low preference to browse compared to sheep and goat (Singh *et. al.*, 1996). Goat showed highest browse preference and sheep equally preferred grasses and browse. In case of goats and wildlife, browse usually provides 25% to 50% of the forage during growth season and up to 75% or more during the dry periods (Ibrahim 1988). Species having thorns, oily or pungent leaves are not preferred by cattle (Chippendale 1963). Goats browse all type of shrubs. Camel mostly prefers browse species of the family Chenopodiaceae, Asclepiadaceae and Salvadoraceae. The sheep and goat commonly browse species belonging to family Rhamnaceae, Rosaceae, Oleaceae, Salicaceae, Acanthaceae, and Asteraceae. Elephants browse plants of family Moraceae, Leguminosae and Anacardiaceae. Locations and habitats also affect palatability of the species, as edaphic conditions influence the growth characteristics of plants which indirectly influence their nutritive value (Cook and Harris, 1950). The leaves of *Butea monosperma* is lopped and fed to buffaloes in parts of Gujarat and Rajasthan but it is not so in the Bundelkhand region of UP and MP. Certain species are browsed regularly, whereas others at a particular growth stage or under certain seasonal conditions. Heady (1964) mentions that climate is one of the major factors influencing relative preference, because of its influence on plant chemical composition, turgidity and harshness of foliage. Dried leaves of *Calatropis procera* are eaten by cattle (Walt, 1889-1896). The leaves of *Grewia optiva* are not relished by cattle till they are fully mature. The chief factor influencing palatability of a particular plant is succulence of herbage which varies with the time of year and degree of grazing (Davies 1925).

NUTRITIVE VALUE

Rangeland browse species add a variety of diets to livestock and wildlife. Nutritionally browse species are generally rich in protein content and minerals in the lean period, when mature grass show poor protein content. Nutritive value is related to intake, chemical composition, digestibility and the presence or absence of anti-nutritional factors (Gutteridge, 1995). About 30 per cent of the fodder tree leaves so far evaluated in India contain less than 50 per cent total digestive nutrients (TDN) on dry matter basis (Upadhyay, 1997). Leaves of *Grewia optiva* which is extensively used as fodder tree in the lower Himalayan regions are fairly rich in protein (21.68.1) and other nutrients (Khosla *et. al.*, 1982). The high digestibility of its crude protein (72.4%) is comparable to that of cultivated leguminous fodder. Ganguli *et. al.* (1964) reported that in general the protein, nutrient and mineral contents of most of the top feed species were higher than critical levels of stock maintenance in western Rajasthan. A multipurpose shrub *Helicteres isora* is rich in essential nutrients and vitamin A and is a good browse species during the lean period. *Sauropus androgynus* occuring in Sikkim, Khasi hills, Kerala and Western Ghats at 1200 m and grown as a border and shade plant in vegetable gardens, is reported for its high vitamin content and is called multivitamin green (Anonymous, 1972).

ANTINUTRITIONAL FACTORS IN BROWSE

Anti-nutritional factors such as tannins, saponins, alkaloids, hydrocyanic glucosides, coumarine, oxalates and nitrates etc. cause toxicity to the animals. It is observed that high levels of tannin reduce palatability of species, but its level shows seasonal fluctuations. Goats appear to be more tolerant of tannins than sheep but reasons for this is not fully understood (Upadhyay 1997). It has, however, been suggested that the increased salivery function produces mucin that binds tannins and spares the protein for digestion (Hoffman, 1987). In the high altitudes of the Himalaya, the shrubby species of *Astragalus* are browsed by sheep and goat but some of the species are toxic/poisonous as they contain high concentrations of selinium picked up naturally from soil (Anonymous, 1985). Animal species show differential tolerance to toxicities as some species not toxic to a particular animal are found toxic to other animals. For example, *Sophora mollis* is browsed by goats but it is toxic to other animals. Some species of *Crotalaria* and *Indigofera* have good forage value but are toxic to livestock.

A plant rated toxic in one part of the country may be less toxic in other part because the nature of soil and the amount of moisture also influence toxicity in plants. In this context, collection of diverse genepool of a species from its whole distributional range is important because toxicity level may vary within species. For example, the leaves of *Acacia leucophloea* are commonly used as fodder in southern India but are reported toxic to the animals in northern parts of India.

MULTIPURPOSE BROWSE SHRUBS

Minor Fruits

Several species of *Ziziphus, Grewia, Carissa, Flacourtia, Cordia, Capparis* have a great potential for leaf fodder and edible fruits throughout India. The edible fruit producing shrub *Flacourtia jangomas* shows high diversity and its several races varying in edibility of fruits existed in India but not enough attention has been given to it (Kanjilal, 1933; Haines, 1921-25). High diversity in habit, size and shape of leaves is shown by *Flacourtia indica* which is extensively used as forage. Wild species of *Grewia* viz., *G. flavescens, G. hirsuta, G. tenax* produce edible fruits besides browse. In the Himalayan region browse species viz., *Elaeganus angustifolia, E. umbellata, Hippophae rhamnoides* have edible fruits rich in vitamin C. Leaves of several browse shrubs are used as vegetables by tribal and rural populations in India.

Fibre

The common fibre-cum-browse species belong to the plant family Leguminosae, Malvaceae, Tiliaceae, Sterculiaceae in the tropical region and Urticacea in the sub-tropical Himalayan region. In the species of *Crotalaria* particularly in the sunhemp *(C. juncea)* a great diversity in terms of maturation, resistance to pest and diseases existed in India and this is cultivated even now as forage-cum-fibre crop (Littlewood 1936, Mann 1916, Nnarayan and Dabadghao 1972).

This has a great potential for introduction in wet rangelands. The other *Crotalcaria* species viz., *C. burhia* and *C. bialata* are also dual purpose undershrubs. The left over after harvest of species of *Corchorus* viz., *C. capsular is* and *C. olitorius* is browsed by animals. Variability in this species and also in species of *Grewia (G. optiva), Helecteris, Leptadenia, Sida, Malvastrum* and *Trema (T. orientalis)* can be exploited for selection of species for silvipastoral system of forage-cum-fibre production. In the subtropical Himalaya traditional dual-purpose browse shrubs are: *Boehmeria platyphylla, B. platyphylla* var. *scabrella, B. regulosa, Debregeasia hypoleuca, Maoutia puya* and *Urtica parviflora* - all belonging to the family Urticacea. Ramie (*Boehmeria nivea)* produces one the largest and strongest fibre in the world. The inhabitants grow some of the *Boehmeria* species. Diverse germplasm of this species is yet to be exploited.

Hedge Rows

A number of browse shrubs traditionally used as hedge rows or live fences throughout India show diversity of form, growth and habitat affinities. Among these the genus *Vitex* shows a great diversity as both, wild and domesticated types occur in India. Animal preference for this varies from placed to place; in N.E. region it is used for forage and in northern India for its other uses. Fodder shrubs are also planted throughout the subtropical Himalayan region on agricultural field bunds, farm boundaries, terrace risers and as hedge rows. In Garhwal Himalaya, *Boehmeria regulosa* is planted in the villages and lopped for fodder (Todaria and Bhatt, 1992). *Debregeasia sacifolia* is usually planted near moist agricultural fields.

Fuelwood

Traditionally farmers protect or encourage growth of wild shrubs on their farmland for both, fodder and fuelwood needs and the choice of such species varies across agroecological regions. In the temperate region *Salix* species are used for fodder and fuelwood. The wild and cultivated *S. wallichiana* is found growing upto 3750 m in the Himalaya. Variability of form and growth habit is found in *Maytenus emarginata, Ziziphus nummularia, Capparis decidua, Calligonum polygonoides* in arid Rajasthan. Particularly *Maytenus emarginata* and *Z. nummularia* are important for fuel and fodder and are very hardy species with a range of variability.

Medicinal

Forage shrubs combining medicinal values are raw materials for pharmaceutical industries and their incorporation in agroforestry systems will enable farmers to earn additional income. This will help in their conservation as most of them figure in the endangered list of species in the arid, semi-arid and the Himalayan regions. The important Himalayan species are: *Astragalus multiceps, Berberis aristata, Ephedra gerardiana, Potentilla fruticosa, Myricaria brachiata, M. elegans, Sauropus androgyanus, Ziziphus vulgaris, Colebrookea oppositifolia* etc. In the range of distribution all these species show considerable variability which awaits exploitation. Similarly in the tropical and subtropical regions there are browse shrubs of medicinal value as *Gymnema sylvestre* (Gudmar) which is a well known herbal ingredient of diabetic medicines.

Economic Value

Forage shrubs have economic values too. For example, several shrubs are employed for woodworks viz., agricultural implements, toys, household articles; basketry, walking sticks, brooms, cooling mats and thatching. These are in use traditionally but their diversity can be exploited for selection of superior types. Forage shrubs can be put to multiple uses viz., bee keeping, tanning materials, gums and resins, vegetable dyes, washing materials, halophytes as source of sodium bicarbonate and also as shade and ornamental plants.

Agricultural Uses

Many forage shrubs are traditionally used as green manure, shelter belts and wind breaks and cover crops.

For Revegetation

Indigenous shrubs have an important yet unrealized role in rehabilitation of degraded rangelands. Most of the multipurpose browse shrubs are adapted to a wide range of altitude, soil type and environment. In a degraded rangeland situation planting of trees may not be possible and planting of native shrubs may be technically more desirable in the initial stage (Chaturvedi, 1985).

HIMALAYAN REGION

Leguminous Shrub

A rich leguminous shrub diversity exists in the tropical and subtropical regions and the temperate Himalaya. The genus *Lespedza* and *Caragna* have great potential as soil cover in the temperate and subalpine Himalaya (viz., *L. cuneata, L. stenocarpa, L. juncea* var. *juncea*). The species of *Desmodium* specially, *D. elegans* are much used on the Himalayan region for fodder and soil conservation. *D. gyroides* is adapted to acid infertile soils with high soluble aluminum content and it will grow on very difficult sites where very few leguminous shrubs can grow (Gutterdige, 1994). This species is also resistant to fire and more tolerant to waterlogging. The shrubby species of *Indigofera* are also important as they are widely distributed on the Himalaya viz. *I. hebepetala* which is found from Kashmir to Sikkim upto sub alpine region between 1800-4000 m. *Flemingia macrophylla* which occurs in the sub-tropical Himalaya from Kumaon to Sikkim, grows well on acid soils with high level of soluble aluminum (Gutterdige, 1994). This evergreen shrub can resist long dry periods and some degree of waterlogging and may be a potential dry season browse. The shrubby species of *Astragalus* are also promising as they naturally occur in the alpine, temperate Himalaya and in cool arid region. The shrubby species of *Calliandra* are useful for North-Eastern Region.

Non-leguminous Shrubs

The species of *Salix* and *Ribes* are adapted to acid soils. For multipurpose uses species of *Boehmeria, Debregeasia, Grewia, Ziziphus* etc. deserve much attention in the subtropical Himalaya. The species viz., *Boehmeria rugulosa, Vitex negundo, Woodfordia fruticosa* etc. are utilized to control land slide in the Himalaya (Mathur, 1975). *Abelia trifolia* which occurs from Kashmir to Nepal between 1200-3245 m is planted on limestone ridges. The species viz., *Parrotopsis jacquemontiana* and *Plectranthus mollis* are also used for soil conservation.

TROPICAL REGION

Leguminous Shrubs

The widely distributed species viz., *Crotalaria, Acacia, Desmodium, Indigofera, Flemingia* etc. are important for tropical rangelands. The native species of *Stylosanthes* i.e. *S. fruticosa* has a great potential for revegetating degraded rangelands. The potential of this has not been exploited so far. A great diversity in this species in respect to habitat, habit, flowering, fruiting etc. is reported from its natural occurrence in southern India. The species of *Crotalaria* viz., *C. burhia* and in *Indigofera* viz., *I. oblongifolia* are important browse shrubs for sand dune areas in the Indian Thar desert. The species of *Desmodium* viz., *D. heterocarpon, D. gyroides* are useful for soil conservation.

Non-leguminous Species

In this group the species of *Carissa, Clerodendrum, Flacourtia, Grewia, Helicteres, Securinega, Vitex, Ziziphus* etc. are important for different situations. The species of *Carissa* have great potential to revegetate the degraded/barren lands in semi-arid region and also in the lower hills of the Himalaya. This is a common species of degraded lands and also resistant to browsing. In arid regions species viz., *Calligonum polygonoides, Maytenus emarginata, Ziziphus nummularia* etc. are important for degraded lands. *Z. nummularia* also acts a as wind-break in arid regions. *Dodonea viscosa* is utilized for reclaiming denuded tracts, as it is a good sand binder and is also used for reclaiming marshy lands. The species viz. *Clerodendron phlomidis, Vitex negundo* etc. are also good for sodic soils. *Flacourtia indica* has a great potential, as it is very hardy and is extensively lopped for fodder. It is distributed throughout India also in the subtropical Himalaya. *Adhatoda zeylanica* is also used for reclamation of degraded lands, as this shows good adaptability.

Halophytes

For rehabilitation and reclamation of *usar* lands, halophytic shrubs have a prime place. A number of halphytic shrubs having forage value grow naturally in highly saline soils and are adapted to extreme situations. The species of *Atriplex, Haloxylon, Salsola, Salvadora, Salicornia, Suaeda* are adapted to high saline environment. Only few species of *Atriplex* viz., *A. nummularia, A. amnicola* etc. are introduced in India for fodder and other purposes. The promising adapted genotypes of these shrubs can be evaluated for salinity tolerant forage species. *Avicennia* spp. are important for sea coasts and marshy areas and can be much utilized for fodder and fuel purposes by local population.

FUTURE RESEARCH

Compared to trees, shrubs in general and browse shrubs in particular, have not received much needed research attention. Most browse shrubs have multipurpose uses and will be useful introductions in agroforestry systems particularly when the systems are now in greater needed of diversification towards side income generating components described in the text. Browse shrubs show high diversity of forms and habitat affinities across agroecological zones of India. Hence it will not be possible for a single agency to tackle a large issue thrown up by browse shrubs. It would be desirable to go for a Browse Shrub Network Programme running across the country. Therefore a detailed proposal on this aspect may be mooted and funding sources explored. The tasks that the proposed network would like to address are outlined below:

- ☐ Based on extensive literature sources identification of important browse shrubs.
- ☐ Extensive surveys to map the distribution pattern of selected shrubs and also to collect diverse materials (PGR).
- ☐ Introduction, evaluation and documentation of important browse shrubs.
- ☐ Measures for *ex situ* and *in situ* conservation of browse shrub PGR.
- ☐ Emphasis on collecting ethnobotanical information on browse shrubs of multiple uses.
- ☐ Chemical screening for antiquality factors in browse shrubs and standardization of chemical ameliorates for browse as feeds.
- ☐ Particular emphasis on halophytes and also on browse shrubs having drought and frost tolerance and high nutritional value.

REFERENCES

1. Anonymous. 1948-1976. The Wealth of India: A dictionary of Indian raw materials and natural products. Vol. 1 -11 CSIR, New Delhi.
2. Anonymous. 1972. The Wealth of India. Vol. IX: Rh-S0, CSIR, New Delhi.
3. Anonymous. 1985. The wealth of India. Vol. I: A (Revised) CSIR, New Delhi.
4. Arora, R. K. and Chandel, K. P. S.1972. Tropical Grassland 6: 213-221.
5. Bhandari, M. M. 1988. In: (Eds.: R. S. Paroda, P. Kapoor, R. K. Arora and Bhag Mal) Life Support Plant Species: Diversity and Conservation, NBPGR, New Delhi.
6. Brandis, D. 1906. Indian Trees (Reprinted, 1971). Bishan Singh Mahendra Pal Singh, Dehradun.
7. Chaturvedi, M. D. 1949. UP Forest Dept. Bull. No. 19.
8. Chaturvedi, A. N. 1985. In: (ed. J. S. Singh) Environmental Regeneration in Himalaya. CHEA and Gyanodya Prakashan, Nainital.
9. Chatterjee, D. 1939. Journ. Royal Asiat. Soc. Bengal. Sci. 5: 69 p.
10. Chippendale, G. M. 1963. Aust J. Exp. Agric. Anim. Hus. 8:303-304
11. Cook, C. W. and Harris, L. E.1950. Utah Agric. Exp. Stan. Bull. 344.
12. Davies, W. 1925. J. Min. Agri. 32: 106-116.
13. Dogra, P. D. 1986. Indian Forester 112 (7): 596-607.
14. Duthie, J. F. 1903-29. Flora of Upper Gangetic Plains and of the adjacent Siwalik and Sub-Himalayan Tracts. 3 Vols. Calcutta (Repted ed. 1960).
15. Gamble, J. S. 1902. A Manual of Indian Timbers. London
16. Ganguli, B. N., Kaul, R. N. and Nambiar, K. T. N. 1964. Ann. Arid Zone 3; 33-37.
17. Good, R. 1964. The geography of flowering plants. Longmans. Great Britain.
18. Gutteridge, R. C. 1994. In: (Eds.: R. C. Gutteridge and M. M. Shelton) Forage Tree Legumes in Tropical Agriculture CAB International UK.
19. Gutteridge, R. C. 1995. Paper presented in International Workshop on Nitrogen Fixing Trees for Fodder held in Pune, India, March 20-25, 1995.
20. Heady, H. F. 1964. J. Range Mgmt. 17: 76-82.
21. Haines, H. H. 1910. A Forest Flora of Chota Nagpur. Delhi.
22. Haines, H. W. 1921-25. The Botany of Bihar and Orissa. 3 Vols. London.
23. Hoffman, R. R. 1987. In: (ed. A Dobson) Aspects of digestive physiology in ruminants. Cornell Univ. Press, New York, USA.
24. Hussain, T. and Kapoor, S. L. 1990. Enumeration of legumes in India. NBRI, Lucknow.

25. Ibrahim, K. M. 1988. In: (ed. M. A. Habit) The current state of knowledge in *Prosopis juliflora*. FAO.
26. Jones, R. J. and Lawry, J. B. 1984. Experimentation 40: 1435-1436.
27. IAB, 1947. Joint Pub. no.10. Imp. Bur. Pasture and Field Crops, Aberystwyth, UK.
28. Ivins, J. D. 1955. Herb. Abs. 25; 75-79.
29. Kaul, R. N. and Ganguli, B. N. 1963. Indian Forester 89:623-630.
30. Kanjilal, UK. 1928. Forest Flora of Chakrota, Dehradun and Saharanpur forest Divisions, United Provinces, Calcutta.
31. Kanjilal, P. C. 1933. A. Forest flora for Pilibhit, Oudh, Gorakhpur and Bundelkhand, Allahabad.
32. Khoshoo, T. N. and Singh, N. 1963. Silvae genetica 12; 158-174.
33. Khosla, P. K., Pal, R. N., Negi, S. S. and Kaushal, P. S. 1982. In: (ed. P. K. Khosla) Improvement of Forest Biomass. ISTS, Solan
34. Laurie, M. V. 1945. Indian For. Leaflet No. 82 (Silv.), FRI, Dehradun.
35. Littlewood, R. W. 1936. Livestock of Southern India. Government Press, Madras.
36. Mann, H. H. 1916. Bull. No. 77. Bombay Dept. Agriculture.
37. Mathur, H. N. 1975. Indian Farming 25: 17-19.
38. Narayanan, T. R. and Dabadghao, P. M. 1972. Forage Crops of India. ICAR, New Delhi.
39. Negi, S. S. 1977. Indian Forester 103 (9): 616-622.
40. Pokhriyal, T. C.; Kumar, A.; Nautiyal, S.; Naithani, H. B. and M. Mishra. 1992. Fodder from Forests. ICFRI, Dehradun.
41. Purohit, K. and Samant, S. S. 1995. Fodder Trees and shrubs of Central Himalaya. Gyanodaya Prakashan, Nainital.
42. Riveros, F. 1988. In: (ed. M. A. Habit) The current state of knowledge in *Prosopis juliflora*. FAO.
43. Rao, R. R. 1994. Biodiversity in India: Florical Aspects. Bishan Singh Mahendra Pal Singh, Dehradun.
44. Rao, R. R. and Hussain, T. 1993. In: (ed. U. Dhar) Himalayas Biodiversity: Conservation strategies. Gyanodayan Prakashan, Nainital, India.
45. Singh, R. V. 1982. Fodder Trees of India. Oxford & IBH Pub. Co., New Delhi.
46. Singh, J. P.; Vinod Shankar and Upadhyay, V. S.1996. IGFRI, Annual Report, Jhansi. 33-35.
47. Singh, K. A. and Yadav, B. P. S. 1995. In: (Eds.: C. R. Hazra and B. Misri) New Vistas in Forage Production. AICRP (FC), IGFRI, Jhansi.

48. Todaria, N. P. and Bhatt, B. P. 1992. In: (Eds.: P. S. Pathak, R. Debroy and P. Singh). Multipurpose tree species for Agroforestry Systems. RMSI, IGFRI, Jhansi.
49. Upadhyay, V. S. 1997. FAO Training Course on Management and Utilization of fodder trees and shrubs in subtropical and temperate Himalaya, IGFRI, Jhansi. 62-69.
50. Watt, G. 1889-1896. Dictionary of Economic Products of India. Vol. I-VII. Govt. Press, Calcutta.

Status of Genetic Diversity and Biotechnological Research in Arid Zone Trees

MANJIT SINGH, N. L. KACKAR AND S. K. JINDAL
Central Arid Zone Research Institute, Jodhpur (Rajasthan)

ABSTRACT

Start of the term *Biological Diversity* in the current sense can be traced to three publications (Lovejoy, 1980a, b; Norse and McManus, 1980) which appeared in 1980's. Norse *et. al.* (1986) expanded the usage to refer unequivocally to biological diversity at three levels: genetic (within species), species (species number) and ecological (community) diversity. At the most fundamental evolutionary level, the genetic diversity within species is the raw stuff upon which evolutionary processes act. On shorter time scale, such genetic variability enables a species to cope with old and new pathogens, environmental fluctuations, and so on. We still lack a clear understanding of how the long-term survival of many species is likely to be affected by recent and severe reduction in the sizes of their populations (May, 1995).

INTRODUCTION

Trees and shrubs are extremely important in arid ecosystems as both human and livestock population depend on these, especially in drought years. In Indian arid zone there is a sharp decrease in forest cover and reduction in the number of trees and shrubs on cultivated fields in the last few decades because of farm mechanisation and over exploitation of vegetal resources for meeting the ever increasing needs of fuel, food and fodder. Since mid-1950s, tree breeders have demonstrated that simple selection-type breeding operations can substantially increase yield and modify many traits and adaptabilities (Namkoong *et. al.*, 1988). Forest genetics and breeding has not made the desired impact on less productive lands. The importance of trees in arid zone was realised very early in Indian arid zone but the actual programme on exploring genetic diversity was initiated in late seventies. Since then work on arid zone forestry has also been initiated in other institutes of the region which include Haryana Agricultural University, Hissar; Gujarat Agricultural University, Ahmedabad; and Punjab Agricultural University, Ludhiana. Understanding of breeding system can have an important bearing on conservation aspects and proper utilisation of genetic diversity. For rapid gains in improvement programmes, it will be imperative to collect and multiply superior genotypes using biotechnological methods. The information available on genetic diversity, breeding systems and micropropagation of arid zone trees is summarised here.

GENETIC DIVERSITY

Morphological Variation *in situ*

Over a century ago, Darwin (1859) noted the universality of natural variation and this formed the inspiration for his great work on the evolution of species by natural selection. Genetic diversity found in tree species is a part of nature's strategy for defence and survival against all types of risks encountered in the long life span of forest trees (Heybrock, 1978). The use of diversity of wild species for genetic gain is the basis of tree improvement work.

Till the end of seventies no planned explorations were undertaken for collection of germplasm of arid zone trees in Indian arid zone and its use for genetic improvement. Brief account of variability observed in the important indigenous tree species is given below.

Prosopis cineraria

Although wide distribution of *Prosopis cineraria* suggests that it is likely to be a variable species, the variation within *P. cineraria* was not investigated scientifically till 1980's (Leakey and Last, 1980). Jatasra and Paroda (1981) further added that though no published report is available on genetic diversity in this species, travelling through the vast Thar desert, one could easily conclude that wide range of variability is available particularly for growth habit, canopy area, fodder yield, spiny habit and so on. They were of the opinion that large genetic variability available in this species offers a vast scope for selection of fast growing, spineless and high yielding genotypes.

Jatasra (1982) collected two hundred fifty strains of *P. cineraria* by the 'random' sampling from Thar desert during 1981. The quantitative data recorded on 223 trees indicated a wide range of genetic variability for tree height, DBH, stem height, canopy height and canopy diameter. Ranking for leaf fodder and seed yield revealed vast exploitable variations. He further selected 15 best trees for various quantitative traits. Maximum number of plus trees were selected from Sikar, Salasar and Laxmangarh areas. He emphasised the necessity of recurrent germplasm collection along with various precautions and suggestions for undertaking such surveys.

Continuing this study Jatasra and Paroda (1983) reported that in natural stands tree height varied from 4 to 22 m with mean value of 9.19 m, stem height and canopy height ranged from 1 to 8 m and 1 to 20 m respectively. Mean stem/canopy ratio value (0.56) indicated that canopy grew almost double the height of stem. Branches per tree ranged from 4 to 43. Very high range for canopy volume indicated the presence of wide variability for fodder yield in *P. cineraria*. Maximum variation was observed for canopy volume and least variation was observed for stem height. The strains from Sikar, Salasar and Laxmangarh areas were comparatively less spiny. It was observed that high forage yielding tree was not necessarily high seed producer; rather there appeared to be negative association between these two attributes. The comparison of mean of selected strains and the population mean revealed that number of branches, canopy height and canopy diameter of the selected genotypes were almost double in magnitude. The canopy volume of the selected tree was more than thrice that of the base population. Hence these four traits appeared to be most important contributing traits of leaf fodder yield in *P. cineraria*. The stem height of the two populations was almost same and so stem height appeared to have exerted little influence on forage yield. They concluded that there existed a vast exploitable genetic variability for leaf fodder and its component traits.

A regular programme on genetic improvement of *P. cineraria* was initiated at Central Arid zone Research Institute, Jodhpur in 1983. An exploratory survey was undertaken in 1984 and germplasm collection from 140 trees was made from various places of Indian Thar desert viz., Ajmer, Barmer, Bikaner, Churu, Jaipur, Jaisalmer, Jalor, Jhunjhunu, Jodhpur, Nagaur, Sikar and Tonk districts. Analysis of three characters i.e., seed weight, pod length and number of seeds per pod revealed enough variability. Maximum range of variability for pod length was observed in Jodhpur provenance (8.6-26.0 cm with mean 13.46±1.12), whereas for the number of seeds per pod, range was maximum (2.0-17.0 with mean 9.60±1.48) in Nagaur provenance and for seed weight maximum range was observed in Churu provenance (4.0-7.4 g with mean 5.34±0.25 g). Coefficient of variation was highest for number of seeds per pod and lowest for seed weight. Barmer region represented more than 30 per cent C.V. for all the three characters studied (Kackar *et. al.*, 1986a). Subsequently detailed surveys and germplasm collection from arid parts of Rajasthan and Gujarat were undertaken during 1986 and 1991 during which seed from 70 and 79 plus trees was collected and utilised for nursery studies and establishing progeny trials and germplasm (Kackar, 1988; Singh *et. al.*, 1993). Provenance variation in seed and pod characteristics has also been studied by Arya *et. al.*, (1992).

Kackar *et. al.*, (1986b) reported variation for other morphological characters of *P. cineraria* in natural stands from various edaphic sites and rainfall zones of western Rajasthan. The height of the trees was found to be variable in all the eleven provenances. The maximum and minimum mean height of 16.3 and 8.8 m were observed in Barmer provenance and Tonk provenance, respectively. The mean values of forking height varied from 2.04 to 4.28 m in Tonk and Barmer provenances (over all mean, 2.79±0.28 m). Average forking height was lower than three metre in all the provenances except Barmer and Nagaur.

Tecomella undulata

Tecomella undulata (Rohida) belongs to family Bignoniaceae. Its seeds are very light and papery and easily get wind dispersed. In Indian arid zone it is locally called as "*Marwar Teak*" or Desert Teak because of its wood quality. Presently it is one of the most important desert timber tree species. Last three decades have witnessed large scale export of its carved furniture. This factor has caused a heavy demand for quality wood. Thus indiscriminate felling continued and at present the desert tract has almost exhausted large bold and healthy plants. Consequently the population of this species decreased drastically and it is now listed as threatened species (Saxena, 1993).

In western Rajasthan, rohida (*T. undulata*) is generally recorded on sandy plains, undulating hummocky plains and undulating interdunal plains with deep to very deep sandy soils. There are two distinct concentration zones of high density of rohida: (i) district Barmer and (ii) Sikar. In Barmer district, tehsil Chohtan, Barmer and north Bhakasar show high density of rohida (20-60 plants ha^{-1}) on flat and hummocky interdunal areas. This is a low rainfall tract (150-250 mm) but the soil moisture in lower depth is always present round the year and perhaps this factor helps in providing high tree density of rohida and khejri. Rohida is always a common associate of khejri (*P. cineraria*) community in these areas. Low dunes and longitudinal ridges of these three tehsils of Barmer district are also occupied by *P. cineraria* with 10-20 plants $ha^{-1.}$

Jindal *et. al.*, (1985a, 1987) reported wide variation for tree height, basal diameter and DBH Maximum frequency of this species was observed in interdunal areas of Barmer district. Coefficient of variability for tree height, DBH, basal diameter and canopy diameter was also maximum in this district.

Mean, range, standard deviation for various quantitative characters of *T. undulata* have been studied (Jindal *et. al.*, 1985a). Tree height ranged from 2.0 m to 12.8 m with mean value 5.78 m. Forking height 0 to 4.2 m with 1.63 m mean. Branches per tree ranged between 1 to 15. Maximum coefficient of variability was observed for basal diameter (61.3%) followed by DBH (59.2%) and minimum for leaf size (30.6%).

Acacia spp.

There are reports available on natural variability in other important arid zone trees. Solanki *et. al.*, (1985) reported phenotypic variations in pod and seed size in natural stand of *A. senegal* in western Rajasthan. From their studies of 52 individual trees they reported enough variability for various pod characters like pod length, pod breadth, seeds per pod and 100-seed weight. In an earlier study, Solanki *et. al.*, (1984) compared tree height, basal diameter, DBH, canopy diameter and estimated volume per tree of 22 year old *Acacia senegal* and *P. cineraria* grown in Jodhpur, Rajasthan. Coefficient of variation was more in *P. cineraria* for all the characters except volume.

A. senegal is an important fuel and gum yielding tree on rocky habitats. An exploratory survey was undertaken during 1984. Maximum variability was found in Barmer district. In 52 trees enough variability could be observed in various pod characters like pod length (4.2-10.3 cm), seeds per pod (2.8-7.0) and 100 seed weight (7.40-18.88 g). Variation in different traits has also been observed in progenies of 13 plus trees established in 1985 at Jodhpur, Bhopalgarh and Jaisalmer. At present there are about 700 trees in field representing 18 single plant progenies. Gum production has been observed in these plants. The gum yield was, however, more in some of the exotic plants (200 g tree^{-1}) in a provenance trial representing African provenances.

Variation in Nodulation in *P. cineraria*

Variation in 70 single juvenile plant progenies in nursery was investigated for number and size of nodules. Nodules were observed only in 14 progenies after six months. There were differences in number and size of nodules in plants within and between progenies (Kackar *et. al.*, 1990).

BIOCHEMICAL VARIATION

Solbrig *et. al.*, (1977) reported that the numbers and kinds of flavonoids found in different species and provenances of *Prosopis* differed appreciably, but their occurrence did not seem to be related to, or selected by habitat characteristics, when examined along east/west transect through Arizona and Texas. Besides variation in morphological traits there are number of reports available on biochemical variation in phenolics, isoenzymes and protein of *Prosopis* species. Studies at the Institute on six *Prosopis* species revealed that each species had a characteristic phenolic pattern and the degree of similarity as indicated by number of common spots among different species was in close agreement with their classification based on morphological criteria and geographical distribution (Pancholy *et. al.*, 1989).

The first report of genetic variation of crude protein content in *P. cineraria* was that of Gupta *et. al.* (1984). They reported wide range of variability in protein contents of pods (without seeds) of *P. cineraria* from different provenances of Rajasthan. Maximum range was observed in Tonk provenance (8.92-15.05%, mean = 12.4±0.74) and minimum in Sikar provenance (8.92-9.97%, mean = 9.62±0.15). The range of crude protein in different trees from whole population studied was 8.05-15.05 per cent. Highest coefficient of variation was observed in Jodhpur provenance (44.82%) and the minimum was in Sikar provenance (4.44%). In general in semi-arid region the variability was low as compared to that in arid regions. Variability for quality traits in leaves of *P. cineraria* and *A. senegal* was reported by Gupta and Jindal (1985). Genetic variation in minerals, crude protein and structural carbohydrates of foliage of young plants of *P. cineraria* has also been reported by Arya *et. al.* (1995).

Gupta *et. al.*, (1986) also reported variability for crude protein in leaves of nineteen lines of *Leucaena leucocephala* and it varied from 25.4 to 39.6 per cent. The phenotypic and genotypic coefficients of variation were 11.9 and 11.8 per cent, respectively.

BREEDING SYSTEMS

Prosopis cineraria

The information on mating system and detailed reproductive biology is scanty. Parihar and Vir (1993) reported pollinating insects of this species. The degree of outcrossing rates can have an important bearing on conservation strategy both *in situ* and *ex situ* as well as genetic improvement of the species under consideration. Reproductive biology of this species has been studied by Singh (1997).

Tecomella undulata

In *T. undulata* flowering like that in *P. cineraria* is asynchronous and occurs during November to February. In natural population flowers of red, yellow and intermediate colours are found. In the progeny trial comprising of eleven half sib plant progenies established at CAZRI, Jodhpur during 1984, besides variation in height, diameter at breast height (DBH), there was segregation for flower colour in ten progenies, suggesting the outcrossing nature of this species. Our observations showed that birds like purple sun bird (*Nectarinia asiatica*) visit the flowers for nectar and may be acting as pollinators.

Peak flowering period is from end of February to middle of March. The flowering duration varies from 59 to 103 days for individual tree and upto 135 days for the whole population. It has been observed that almost all the flower buds appearing in November, December and January fall before or after opening. Only a few of the flowers developed in February set fruits. Fruit set varied from 0.64 per cent for selfing to 3.94 per cent in cross pollination indicating the presence of self compatibility. Apomixis, an important mechanism for fixing heterotic advantage, was found to be absent (Jindal *et. al.*, 1985b).

Azadirachta indica

Apomixis is absent and self-pollination is dominant in this species. About 2 per cent of the flowers set into fruits, rest drop without fertilisation. Flowering occurs during March to May and flowers appearing in March and April drop and do not set fruits (Jindal and Vir 1994).

Salvadora persica

This species is also self-compatible and apomixis is absent. Flowers appearing in November to February fall without fertilisation. Very few flowers appearing in March develop into fruits and take about two months for ripening (Jindal *et. al.*, 1997b).

BIOTECHNOLOGICAL RESEARCH

Prosopis cineraria

Rapid multiplication through tissue culture has been attempted in number of arid zone trees. Goyal and Arya (1981) reported differentiation in *Prosopis cineraria* using hypocotyl segments. These workers in 1984 achieved the same by bud culture. The detailed account of clonal propagation of *Prosopis cineraria* as well as other important arid zone tree species like *Ziziphus mauritiana, Tecomella undulata, Aegle marmelos* and *Eucalyptus* was reported by Arya and Shekhawat (1986). Puri *et. al.*, (1992) achieved *in vitro* regeneration of *Prosopis cineraria* using epicotyl as explant. *In vitro* plantlets of *Prosopis cineraria* from juvenile and mature explants have been reported by various workers (Nandwani and Ramawat, 1993). Shekhawat *et. al.* (1993) studied the effect of various factors like genotype, morphology, location, season of explant collection, media composition, etc. during *in vitro* clonal propagation of *Prosopis cineraria*.

Work on micropropagation was also initiated at this institute in 1985. Success has been obtained on *in vitro* culture of *Prosopis cineraria* using both shoot and root segments as explants (Kackar *et. al.*, 1991, 1992), *Simmondsia chinensis* (Jojoba) using coppice shoots as explant (Kackar *et. al.*, 1993), *Salvadora oleoides* (Jal) using immature embryos as explants (Kackar *et. al.*, 1997) and *Phoenix dactylifera* (Date palm) using callus produced from apical meristem of offshoots (Kackar *et. al.*, 1989). Nodal segments from actively growing branches of five year old elite tree of *P. cineraria* when cultured on modified Murashige and Skoog, 1962 medium containing 3 mg/1 each of napthoxy acetic acid and napthyl acetic acid produced axillary shoots within 7 to 10 days. These shoots when cultured on modified MS medium containing 3 mg/1 of NAA developed tap roots in 25-30 days. The shoots from *in vitro* cultured plantlets could be further cut into 3-5 nodal segments, each of which produced complete plantlet on modified medium containing 3 mg/1 NOA (Kackar *et. al.*, 1991).

Root segments from 2 to 4 month old seedlings and also from 5 year old elite tree could also be used successfully for micropropagation. 1.0 -1.5 cm long root segment when cultured on modified MS medium containing 3 mg/1 NAA produced adventitious shoots, their number being more in explants from seedlings (5 to 20 shoots) than from adult tree (3-8 shoots). These shoots when cultured on the same medium produced complete plantlets (Kackar *et. al.*, 1992). Profuse callus could be obtained on MS medium containing 0.25 mg/l NAA and 4.5 mg/l Kinetin using cotyledonary leaves as explants. Callus development initiated after a week and the development was better in dark.

There are numerous reports on *in vitro* regeneration and multiplication of various tree species, but the number of reports on species hardened and successfully transferred to the field are very scanty. This task becomes more difficult in arid zones where climatic conditions are extreme and only limited period of monsoon can be considered favourable for transplantation of *in vitro* raised and hardened plantlets. Vitrification of shoots, impaired stomatal development and functioning, poor development of wax, inadequate photosynthetic efficiency, undesirable root system are among the frequently mentioned causes leading to death of plantlets during transfer from test tube to the soil. Success has been achieved at this Institute in field transfer of *Prosopis cineraria.*

Tecomella undulata

Various workers have also achieved Micropropagation in *Tecomella undulata*. Arya and Shekhawat (1986) described the procedure of clonal multiplication of this species. Rathore *et. al.*, (1991) propagated it by using nodal shoot segments as explant; Nandwani *et. al.,* (1995) described the method for *in vitro* propagation through cotyledonary nodes; and Bhansali (1993) micropropagated this species using axillary and terminal shoot buds from 10-15 year old trees.

Salvadora spp.

Micropropagation of *Salvadora*, an important non-edible oil yielding tree found growing on saline and sodic soils in Rajasthan and coastal area of Gujarat, has also been achieved. *In vitro* micropropagation of *S. oleoides* has also been achieved at Central Arid zone Research Institute, Jodhpur using immature embryos as explant (Kackar *et. al.*, 1997). A programme on tissue culture of *S. persica* was initiated at National Chemical Laboratory, Poona to multiply high yielding trees. Initially a technique for micropropagation mature trees was developed which was later on utilised to propagate elite trees (Rao, 1987). In a non-replicated small scale trial comprising of 15 plants each from tissue culture raised plantlets and seedlings, first flowering was observed in control plants after 20 months which was eleven months earlier than in tissue culture raised plants (Mascarenhas *et. al.*, 1987). These workers reported greater uniformity and greater crown diameter in plants raised through tissue culture. *Salvadora* trees are known to grow for over 100 years. Consequently, the initial cost of plants is not important since the end product is the seed and the annual gains will cover the initial cost (Mascarenhas *et. al.*, 1987). Protocols for micropropagation of *Salvadora* are available, but no commercial level propagation has been attempted possibly because raising plants through tissue culture involves additional cost, the consumers are sceptical about the cost benefit ratio of using such plants. Besides this, micropropagation in trees is not easy. Thirdly, field establishment of tissue culture raised plants in saline and sodic soils under arid conditions may not be very easy. The present scenario of no commercial level exploitation of useful scientific information is the outcome of such limitations on the side of producer as well as consumer.

Biotechnological methods can be successfully exploited for large scale production of superior trees derived through genetic improvement - all being exact duplicates of the original. This is bound to increase the economic yields of the forests. It may not be possible to achieve the same results using conventional methods in the short time. One objection which is often raised is that clonal forestry will lead to annihilation in the event of a disease outbreak. This problem can be overcome by using number of unrelated superior genotypes.

There is not only need to collect more germplasm but also to find ways and means to conserve it at low temperature or by adopting techniques like pollen storage and *in vitro* conservation.

REFERENCES

1. Arya, H. C. and Shekhawat, N. S. 1986. Forest Ecology and Management 16 (1-4): 201-208.
2. Arya, S.; Toky, O. P.; Tomar, R.; Bisht, R. P. and Harris, P. J. C. 1992. Journal of Tree Sciences 11: 86-94.
3. Arya, S.; Bisht, R. P.; Tomar, R.; Toky, O. P. and Harris, P. J. C. 1995. Agroforestry Systems 29 (1): 1-7.
4. Bhansali, R. R. 1993. Tropical Science 33 (1): 1-8.
5. Darwin, C. 1859. The origin of species. First edition, London.
6. Goyal, Y. and Arya, H. C. 1981. Current Science 50: 446-448.
7. Gupta, A. K.; Solanki, K. R. and Kackar, N. L. 1984. Annals of Arid zone 23 (3): 255-258.
8. Gupta, A. K. and Jindal, S. K. 1985. Nitrogen Fixing Tree Research Reports 3: 15-16.
9. Gupta, A. K.; Jindal, S. K.; Solanki, K. R. and Kackar, N. L. 1986. Annals of Arid zone 25 (4): 277-282.
10. Heybrock, H. M. 1978. Unasylva 30 (119-120): 27-33.
11. Jatasra, D. S. 1982. Forage Research 8: 1-10.
12. Jatasra, D. S. and Paroda, R. S. 1981. Forage Research 7: 1-12.
13. Jatasra, D. S. and Paroda, R. S. 1983. Transactions of Indian Society of Desert Technology and University Centre for Desert Studies 8 (1): 138-142.
14. Jindal, S. K. and Satyavir, S. 1994. Van Anusandhan 9-11 (1): 17-21.

15. Jindal, S. K.; Kackar, N. L. and Solanki, K. R. 1987. Indian Journal of Forestry 10 (1): 52-55.
16. Jindal, S. K.; Kackar, N. L. and Solanki, K. R. 1985a. Transactions of Indian Society of Desert Technology and University Centre for Desert Studies 10 (2): 33-37.
17. Jindal, S. K.; Bhansali, R. and Satyavir, S. 1997b. Journal of Tropical Forestry 13 (1): 30-34.
18. Jindal, S. K.; Satyavir, S. and Pancholy, P. 1997a. Journal of Tropical Forest Science, Malaysia (In press).
19. Jindal, S. K.; Solanki, K. R. and Kackar, N. L. 1985b. Indian Journal of Forestry 8 (4): 317-320.
20. Kackar N. L.; Joshi S. P, Singh M. and Solanki, K. R. 1993. Annals of Arid zone 32 (3): 175-177.
21. Kackar, N. L.; Singh, M. and Solanki, K. R. 1997. Agro Botanical Publishers India, Bikaner, pp. 277-283.
22. Kackar, N. L.; Jindal, S. K.; Solanki, K. R., and Singh, M. 1990. Nitrogen Fixing Tree Research Reports (USA) 8: 152-153.
23. Kackar, N. L.; Solanki, K. R. and Jindal, S. K. 1986a. My Forest 22 (2): 129-134.
24. Kackar, N. L.; Solanki, K. R. and Jindal, S. K. 1986b. Indian Journal of Forestry 9 (2): 113-115.
25. Kackar, N. L.; Solanki, K. R. and Joshi, S. P. 1989. Annals of Arid zone 28: 137-141.
26. Kackar, N. L.; Solanki, K. R.; Singh, M. and Vyas, S. C. 1991. Indian Journal of Experimental Biology 29: 65-67.
27. Kackar, N. L.; Vyas, S. C.; Singh, M. and Solanki, K. R. 1992. Indian Journal of Experimental Biology 30:429-430.
28. Kackar, N. L.1988. *Silvae Genetica* 36 (2): 68-71.
29. Leakey, R. B. B. and Last, T. T. 1980. Journal of Arid Environments 3: 9-24.
30. Lovejoy, T. E. 1980b. Changes in biological diversity. In 'The global 2000 report to the President, Vol. 2 (ed. G. O. Barney): 327-332. Harmondsworth: Penguin Books.
31. Lovejoy, T. E. 1980a. Foreword. In 'Conservation biology: an evolutionary-ecological perspective' (ed. M. E. Soule and B. A. Wilcox), Sunderland, Massachusetts: Sinauer Associates.
32. Mascarenhas, A. F.; Khuspe, S. S.; Nadganda, R. S.; Gupta, P. K. and Khan, B. M. 1987. In: Genetic Manipulation of Woody Plants (Eds.: J. W. Hanover, D. E. Keathley), Plenum Press, New York.
33. May, R. M. 1995. In:Biodiversity measurement and estimation (Eds.: D. L. Hawksworth. Chapman and Hal) l, London p 13-20.

34. Namkoong, G.; Kang, H. C. and Brouard, J. S. 1988. Tree Breeding: Principles and Strategies. Springer -Verlag, New York, 180 p.
35. Nandwani, D. and Ramawat, K. G. 1993. Indian Journal of Experimental Biology 31 (2):156-160.
36. Nandwani, D.; Mathur, N. Ramawat, K. G. and Mathur, N. 1995. Gartenbauwissenschaft 60 (2): 65-68.
37. Norse, E. A. and Mcmanus, R. E. 1980. In: Environmental quality 1980: The eleventh annual report of the Council on environmental Quality, pp 31-80. Washington D. C. Council on Environmental Quality.
38. Norse, E. A.; Rosenbaum, K. L.; Wilcove, D. S.; Wilcox, B. A.; Romme, W. H.; Johnston, D. W. and Stout, M. L. 1986. Conserving biological diversity in our national forests. Washington D. C. The Wilderness Society.
39. Pancholy, Anjly; Jindal, S. K.; Singh, M.; Kackar, N. L. and Solanki, K. R. 1989. Annals of Arid zone 28 (3-4): 299-303.
40. Parihar, D. R. and Vir, S. 1993. Annals of Arid zone 32 (4): 267.
41. Puri, S.; Jain, M. and Sharma, P. 1992. Nitrogen Fixing Tree Research Reports 10: 189.
42. Rao, S. M. 1987. M.Sc. Thesis, University of Poona, Poona, India.
43. Rathore, T. S.; Singh, R. P. and Shekhawat, N. S. 1991. Plant Science Limerick 79 (2): 217-222.
44. Saxena, V. S. 1993. Proceedings of Workshop on Tree Improvement and Provenance Research, Department of Forest, Govt. of Rajasthan, Jaipur, pp. 135-145.
45. Shekhawat, N. S.; Rathore, T. S.; Singh, R. P.; Deora, N. S. and Rao, S. R. 1993. Plant Growth Regulation 12 (3): 273-280.
46. Singh, M. 1997. In: Eds. M. S. Yadav, Manjit Singh, S. K. Sharma, J. C. Tewari and U. Burman. UNESCO-ICAR, Central Arid Zone Research Institute, Jodhpur, India, pp. 223-230.
47. Singh, M.; Kackar, N. L.; Jindal, S. K. and Solanki, K. R. 1993. Annals of Arid Zone 32 (2): 140.
48. Solanki, K. R.; Jindal, S. K. and Kackar, N. L. 1985. Transactions of Indian Society of Desert Technology and University Centre for Desert Studies 10 (2): 30-32.
49. Solanki, K. R.; Jindal, S. K.; Muthana, K. D. and Kackar, N. L. 1984. Nitrogen Fixing Tree Research Reports 2:13-14.
50. Solbrig, O. T.; Bawa, K. S.; Carman, N. J.; Hunziker, J. H.; Naranjo, C. A.; Palacios, R. A.; Poggio, L. and Simpson, B. B. 1977. In: Mesquite: Its Biology in Two Desert Ecosystems (Ed. B. B. Simpson), U.S./IBP, Synthesis Series 4, Dowden Hutchinson and Ross, 44-60.

Selection of Neem Ecotypes with High Azadirachtin and its Multiplication through Micropropagation

B. VENKATESWARLU, K. MUKHOPADHYAY, J. MUKHOPADHYAY AND J. C. KATYAL
Central Research Institute for Dryland Agriculture, Hyderabad (AP)

ABSTRACT

With the increasing use of neem products in integrated pest management (IPM), there is a renewed interest on the neem tree. A comprehensive survey of neem ecotypes in different agro ecological regions of the country was carried out and 6 plus trees of neem with relatively high azadirachtin content and other desirable traits have been selected. Study on the influence of a number variables on azadirachtin content revealed that none of the external factors like rainfall, soil type and relative humidity or plant dependent variables like tree age, oil content and seed weight, showed any definite relationship with the azadirachtin per cent in seeds. A micropropagation protocol was standardised for mass multiplication of the selected plus trees in order to produce true to type saplings. Preliminary results on the field performance of these saplings are promising and the plants have showed uniform growth.

INTRODUCTION

With the growing use of neem based products in integrated pest management (IPM), neem tree has been receiving increasing attention in recent times. In addition to pest management, neem based products are being used in medicines, cosmetics and livestock health (Parmar and Ketkar, 1993). Though a large number of bioactive compounds were reported in neem, azadirachtin, a limonoid received the maximum attention in view of its excellent antifeedant, repellent and growth regulatory properties on several insect species (Schumutterer, 1995a). Farmers in many parts of the country have started using neem cake, oil and kernel extracts in pest management. Several private manufacturers also started producing neem based commercial formulations. There is, thus, likely to be a good demand for quality seed in the immediate future. India has nearly 14 million neem trees producing an estimated 5 lakh tonnes of seed annually (Ketkar, 1976). Currently, only 20-30% of this seed is collected but with the expansion and diversification of the neem processing industry, there is likely to be a shortfall of quality seed in years to come.

In view of the importance of azadirachtin (aza) content in determining the quality of neem seed, it is essential that superior trees with high aza content are selected and adequate planting material produced from such trees to facilitate planting neem on a big scale. Neem trees show significant variation for azadirachtin content. Though the causes for this variability have been reported as high rainfall, extreme drought and high humidity in some earlier papers (Ermel *et. al.*, 1984; Rangaswamy and Parmar, 1995; Venkateswarlu *et. al.*, 1997), the results have not been conclusive in view of the small sample size and the influence of multiple variables.

Accordingly, an effort was made in the present study to survey and characterise neem ecotypes from major agro-ecological regions of the country, analyse the factors responsible for the variability and identify plus trees. It is not known as to how much of this variability in natural populations is due to outcrossing. There are conflicting reports on the pollination biology of the neem. Some earlier reports suggested that neem is cross-pollinated (Mohan Ram and Nair, 1993; Schumutterer, 1995b) which might explain the high variability, but recent studies at NRCAF, Jhansi (Gupta *et. al.*, 1998) indicated that neem is essentially self-pollinated. However, in view of the advantages of vegetative propagation in terms of producing "*true to type*" plants (particularly with respect to aza content), advancement of flowering and for establishment of seed orchards, the present study also attempts to standardise a micropropagation protocol for one of the plus trees of neem and evaluation of the field transferred progeny.

MATERIALS AND METHODS

Survey of Neem Ecotypes for Azadirachtin Content

A comprehensive survey of neem ecotypes from different rainfed areas of the country was conducted during 1995-97. The rainfall, soil type and the growing season vary widely in these locations. Seed samples were collected during two consecutive years following a uniform protocol with the help of collaborating scientists at the AICRPDA centres. A model data sheet was used, wherein rainfall, soil type, growing season length, girth of the tree, bole length, crown pattern, initiation of flowering, initiation and completion of fruit drop, total fruit yield (on a scale from 1-5) etc. were recorded. Most of the trees sampled were naturally growing and the site characteristics were also recorded.

In order to eliminate variables like rainfall and soil type on the aza content, about 30 seed samples were collected from neem trees growing at the Hayatnagar Research Farm, Hyderabad under conditions of identical rainfall (775 mm) and soil type (loamy sand). The trees had however varying GBH, which was recorded for each tree.

Collection and Processing of Seed Samples

About 500 gm seed samples were collected during each year from the identified trees. The trees were first labelled at the time of flowering and its GBH was recorded. One kg fruits at the yellow ripe stage (YRS) were hand collected from branches of each tree, just at the beginning of the fruit fall. The fruits were depulped by washing under running tap water and shade dried for 7 days. The dried seeds were then transported to Hyderabad from different locations in plastic bags and stored at 4°C before they were analysed. Samples which developed fungal infection due to high moisture were not included in the analysis particularly if the kernels became rancid.

Extraction of Azadirachtin

The depulped seeds were first dehusked and the moisture content in the kernels was brought to uniform level (5%) by slow drying in an incubator at 45°C. Twenty gm clean kernels were ground in a mixer grinder. Three replications were included for each treatment.

Extraction of azadirachtin was done following the method described by Govindachari *et. al.* (1990). Ground kernel sample was taken in a custom made thimble and placed in a 250 ml soxhlet apparatus. Two hundred ml n-hexane was added to each flask and subjected to heating for 4-5 hours for defatting. The n-hexane fraction was then discarded. 200 ml AR grade methanol was added to each flask and heated for 6-7 hours. With this step, the total azadirachtin was extracted with methanol from the kernel powder. The volume was concentrated upto 5 ml by using vacuum concentrator (Hetovac Model VR-1), then volume made up to 100 ml using by HPLC grade methanol.

HPLC Analysis

The azadirachtin-A content in the extracts was determined by HPLC following the method of Kearney et al (1994). The analysis was done on an isocratic Shimadzu HPLC model 1080 with UV-VIS variable detector coupled with class CR-10 software. An aliquot of 20 μl sample was injected by a Hamilton syringe into the 20 μl Rheodyne injector, from where sample passed through a 250□4.6 mm ODS 5μ C-18 column. Acetonitrile-water (40:60) was used as the mobile phase at a flow rate of 1 ml/min. Azadirachtin was determined by UV-VIS detector at 217 nm. The retention time was 11.5 minutes. All the calculations were done using the software mentioned above. A standard azadirachtin samples from Trifolio GMBH (99%) was used as reference.

Micropropagation of Plus Trees

Micropropagation protocol was evolved from identified plus trees based on axillary shoot multiplication method. Nodal explants were collected from juvenile shoots of a 22-year old plus tree of neem at HRF, Hyderabad (HY-33/94) during June-July, 1995. Surface sterilised nodal explants were inoculated on modified MS and WP media containing varying levels of Kinetin and BAP. Following bud induction, the cultures were transferred to the elongation medium containing Kinetin. A multiplication ratio of 1:4 could be achieved in 6 weeks. Micro shoots of 3 to 5 cm length were transferred following auxin treatment to vermiculite for *ex vitro* rooting and also to ½ MS medium with different IAA levels for *in vitro* rooting. Both *ex vitro* rooting and primary hardening were done in the green house. Rooted plants were transferred to poly bags (6" □ 4") filled with soil and kept in the green house under 70% humidity and 30□ C for 20 days. When new leaves started coming the plants were shifted to shade house for secondary hardening. About two months old hardened plants were field planted for further evaluation of survival and growth. A detailed description of the protocol was given by the authors in a related publication (Venkateswarlu *et. al.*, 1998).

RESULTS AND DISCUSSION

Variability for Azadirachtin Content

There was a significant variation in the soil and climatic features of the sampling locations. They fall in arid, semi-arid and sub-humid zones. They represent diverse agro-eco regions with good natural distribution of neem and cover red, black and laterite soils. The mean annual rainfall ranges from 300-1500 mm. A total of 110 seed samples were analysed for aza content and other variables from 13 locations. Data on azadirachtin, oil per cent, test weight and kernel to seed ratio of these samples are presented in Table 1. The aza content ranged from as low as 0.1 to as high as 0.95%. The variation was high in samples both from within a location and between locations.

TABLE 1: *Range and means of azadirachtin content and related characters in neem seed samples collected from different locations*

Locations	100 seed wt. (gm)		Kernel in seed (%)		Oil in seed (%)		Azadirachtin in kernel (%)	
	Range	*Mean*	*Range*	*Mean*	*Range*	*Mean*	*Range*	*Mean*
Indore (7)	15-21	18	42-55	52	19-21	20	0.28-0.53	0.4
Akola (11)	14-26	20	29-54	46	18-24	21	0.19-0.80	0.26
Dantiwada (15)	15-24	21	44-64	52	19-31	23	0.17-0.80	0.35
Varanasi (5)	17-24	20	36-52	47	16-26	20	0.14-0.25	0.19
Solapur (9)	12-16	14	22-58	44	11-26	20	0.12-0.65	0.32
Rajkot (3)	15-20	17	43-52	47	18-20	19	0.22-0.42	0.34
Kovilpatti (5)	18-23	20	51-58	54	20-24	22	0.19-0.33	0.25
Anantapur (9)	14-30	20	49-56	53	19-25	23	0.32-0.68	0.39
Phulbani (4)	15-20	17	53-55	54	27-34	31	0.21-0.69	0.42
Bijapur (8)	17-21	19	42-51	46	17-20	19	0.31-0.65	0.44
Hisar (3)	16-20	18	42-52	47	21-26	23	0.23-0.27	0.25
Rajahmundry (3)	17-18	18	42-46	44	25-27	26	0.26-0.36	0.31

Locations	100 seed wt. (gm)		Kernel in seed (%)		Oil in seed (%)		Azadirachtin in kernel (%)	
Hyderabad (29)	*11-23*	*15*	*43-62*	*51*	*14-29*	*22*	*0.21-0.95*	*0.55*

**Figures in parenthesis represents number of samples analysed.*

TABLE 2: *Plus trees of neem identified based on azadirachtin estimation in samples collected from selected semi-arid areas of India*

Tree No.	Azadirachtin in kernels (%)	Oil in seeds (%)	Kernel in seeds (%)	100 seed wt. (g)	Other characters
HY-33/94	*0.55*	*25*	*48*	*17*	*22 yr. old, excellent bole, high yielding, compact crown*
HY-95/C	*0.61*	*29*	*60*	*15*	*9 yr. old, high yielding, good synchrony in fruit ripening, late bearing, spreading crown*
DNT-95/2	*0.81*	*28*	*58*	*20*	*15 yr. old, high fruit yield, good bole (2.1 m), late flowering*
ATP-9/95	*0.7*	*26*	*51*	*19*	*35 yr. old, early bearing, medium fruit yield, extended fruit bearing*
CRI-8/97	*0.91*	*21*	*56*	*15*	*2 m clear bole, 130 cm GBH, compact crown, 28 kg seed yield, extended fruit bearing.*
CRI-14/97	*0.96*	*22*	*53*	*13*	*2.15 m clear bole, 106 cm GBH, spreading crown, 20 kg seed yield, early bearing*

Marked variation for other parameters was also found but usually two-folds. Per cent kernels in seed varied from 25-55, test weight from 14-30 and oil per cent oil from 17 to 34. In other words, the variation in all these parameters was nearly two times in contrast to azadirachtin which showed nearly five fold variation (0.2 to 1.0). Based on data on different parameters and the gross morphology of trees, five plus trees were identified which are listed in Table 2. Though the variation for aza content in the samples studied was high, most samples fall in the range of 0.15 to 0.45. About 85 samples (out of hundred) fall in the range of 0.15 to 0.45 and 32 within the range of 0.25 to 0.35. In other words, the randomly collected neem seed from the market is likely to have a mean azadirachtin in the range of 0.3 to 0.4 only and collection from any specific geographical location may not offer any specific advantage.

It is well known that many species including neem show year to year variation for seed yield. In order to verify whether a given high aza plus tree will remain so consistently over the years, three trees were marked and their aza content was estimated for 2 consecutive years. Data indicated that the aza per cent remained more or less similar during both the years. Surprisingly, marked variation in the total seed yield was observed in some of the trees from year to year during this period. These results show that the azadirachtin per cent is independent of quantum of seed yields and likely to remain constant in a given tree over the years. Similar results were reported in Australia from a 3-year study on 4 neem clones (Baily, 1994).

Stage of fruit ripening is reported to influence the aza content (Rangaswamy and Parmar, 1994) with green unripe fruits having the least per cent. We evaluated five stages of ripening along with a standard mix sample which represents the neem fruit traded in the market yards. Except the hard green fruits there was no significant difference in the aza content as the fruit ripened from green to yellow to brown. For example the aza content was 0.894, 0.913 and 0.906 in yellow green, yellow, and brown stages respectively in fruits of clone No. CRI-8/97 studied at Hyderabad. The mixed fruit sample collected from the ground also had similar aza per cent during 1996. Neem fruits thus can be collected at any stage of ripening once they became soft but it is important to avoid fungal infection and rancidity development of kernels.

IMPACT OF ENVIRONMENTAL AND TREE VARIABLES ON AZA CONTENT

Extensive variation in the azadirachtin content of neem ecotypes was reported by many workers (Ermel *et. al.*, 1994, Ganesh Kumar *et. al.*, 1995, Rangaswamy and Parmar, 1995; Venkateswarlu *et. al.*, 1997). No definite explanation could be offered for such high variability; however some reports indicated that extreme aridity and high rainfall are associated with high aza content (Ermel *et. al.*, 1984; Rangaswamy and Parmar, 1995). The data were not conclusive in these studies due to small sample size. In the present study the variability could not be ascribed to any of the parameters like rainfall, soil type and water availability period. For example, mean aza data of samples from Hisar, receiving only 300 mm annual rainfall and Varanasi receiving 1400 mm were similar. Samples from Phulbani, another high rainfall location however had higher mean aza than Varanasi. Even the classification of locations based on climate (arid, semi arid and subhumid) could not explain the trend in aza or oil data in seeds. In fact, high and low aza samples could be found in all climate zones viz. arid, semi-arid and sub humid with equal frequency. Soil type also did not have any impact on aza content. High and low samples were found in all three types of soils red, black and lateritic. A comparative analysis of the frequency distribution of samples in terms of concentration ranges between a red soil and black soil location of comparable rainfall did not bring out any trend.

We made an attempt to correlate the aza content with other parameters like GBH, oil per cent, 100 seed wt. and seed kernel ratio in order to predict the aza content by determining these parameters. Data collected from the 29 trees growing at Hayatnagar Research Farm at Hyderabad (loamy sand soil with 770 mm annual rainfall) was used for this analysis to eliminate rainfall and soil type variables. None of the parameters showed any relationship (positive or negative) with aza content. The previous reports on aza content did not evaluate the impact of tree age. In the present study, we divided the trees growing at same location (Hyderabad) into five GBH groups and the aza data was analysed. No definite trend was observed on aza content in relation to tree age. Since the number of trees analysed in higher girth ranges were small, further study will be needed in this regard.

From the data collected at Hyderabad where all the trees were growing under similar rainfall and soil type conditions, a 5 fold variation in aza content was found which indicates that individual trees differ significantly despite growing under similar environment. Similar data were obtained with 13 samples analysed from Dantiwada. These results point that aza content may be controlled genetically and selection of plus trees might be useful for providing high aza containing seed. Our preliminary results on the analysis of progeny seed of tissue culture raised plants from high aza plus trees also confirm this hypothesis.

Development of Micropropagation Protocol

Among a number of media and hormone combinations tried, MS with 1 mg/l of BAP was found to be the best for bud induction from primary explants. The months of June to August were the optimum time for collection of primary explants with as high as 90% response.

The 0.5-10 cm long micro shoots after two weeks transferred to the same medium showed the highest rate of multiplication achieving a ratio of 1:4 in 6 to 8 weeks. Both *in vitro* and *ex-vitro* rooting were tried with different hormone treatments. Though the success rate with *in vitro* rooting was higher, there was a high mortality during hardening. The *ex vitro* rooting tried by using soil rite (exfoliated vermiculite) medium resulted in 70% rooting in 20-25 days. There was 10-15% mortality during primary hardening in the mist chamber. Neem was found to be highly susceptible to excess humidity at all stages and regulating the humidity per cent of the green house was the most critical part in achieving good rooting and hardening. There was no mortality during secondary hardening in the shade house. Detailed data on the interaction of hormones, incubation duration, light requirements and preliminary field performance of tissue cultured plants were reported by the authors elsewhere (Venkateswarlu *et. al.*, 1998). Though there was few reports on successful micropropagation of adult neem tree (Rathore *et. al.*, 1994; Drew, 1995) this is the first report of the protocol development from plus tree with high azadirachtin.

Field Performance of Micropropagated Plants

The survival of field transferred plants was 90%. There was a steady increase in height and collar diameter of seedlings with age. Five months old plants attained a mean height of 50 cm and collar diameter of 7.0 mm. Plants were raised both under complete rainfed conditions and under protective irrigation. Irrigation had significant impact on the height and DBH. For example, tissue cultured plants under protective irrigation in summer (10 litres/week during March to June) attained a height of 4.6 m and a DBH of 8 cm after 22 months of transplanting as compared 3.9 m and 3.34 cm in unirrigated plants. Irrigated T.C. plants flowered (first flush) after 24 months while no flowering was noted in unirrigated ones. The azadirachtin content of progeny seed was similar to the mother plant (± 0.07%). These preliminary results indicate the good promise of micropropagated plants of neem to produce quality seed with high aza content. Replicated trials for field evaluation of these plants are currently being planned in order to collect detailed data on growth rate, total biomass and fruit yield, and aza content, etc. on a long term basis.

ACKNOWLEDGEMENT

Part of this work was carried out under a financial grant from AP-NL Programme on Biotechnology for Dryland Agriculture.

REFERENCES

1. Baily, I. 1994. Final Technical Report "*Preliminary studies towards the commercialisation of the neem tree in Australia*" of the Project No. DAQ-1461: 1-26
2. Drew, R. A. 1996. In: Neem and Environment (vol.2) (Eds.: R. P. Singh; M. S. Chari; A. K. Raheja and W. Krans). Oxford and IBH Publishing Co. Pvt. Ltd.: 999-1005
3. Ermel, K., Pahlich, E. and Schumutterer, H. 1984. In: Natural pesticides from tree and other tropical plants (ed. H. Schumutterer and K. R. S. Ascher) Eschborn, Germany: 83-90.
4. Ganesh Kumar, M.; Jayakumar, R.; Ragupathy A. and Rajasekharan, B. 1994. Pestology, 18: 11.
5. Govindachari, T. R.; Sandhya, G. and Ganesh Raj, S. P. 1990. Journal of Chromatography, 513: 389-391
6. Gupta, V. K.; Kumar, R. V.; Solanki, K. R. and Datta, A. 1998. Abstracts of National Symposium on Multipurpose Tree Species for Agroforestry Systems (July 11-13, 1998). Organised by National Research Centre for Agroforestry and Indian Society of Agroforestry, Jhansi: 57-58.
7. Ketkar, C. M. 1976. Final Technical Report of Directorate of non-edible oils and soap industry, Khadi and Village Industries Commission, Irla Road, Vile Parle, Bombay.
8. Kearney, M. L.; Allan, E. J.; Hooker, J. E. and Mordue, A. J. 1994. Plant Cell, Tissue and Organ Culture, 37: 67-71
9. Mohan Ram, H. Y. and Nair, M. N. B. 1993. In: Neem: Research and Development (Eds.: N. S. Randhawa and B. S. Parmar). Society of Pesticide Science, India: 6-26
10. Parmar, B. S. and Ketkar, C. M. 1993. In: Neem: Research and Development (Eds.: N. S. Randhawa and B. S. Parmar). Society of Pesticide Science, India: 270-283
11. Rangaswamy, S. and Parmar, B. S. 1994. Pesticide Research Journal, 6 (2): 193-194
12. Rangaswamy, S. and Parmar, B. S. 1995. Pesticide Research Journal, 7 (2): 140-148
13. Rathore, T. S.; Parveen, U. K.; Tomar, C. and Vyas, C. S. 1994. Neem. 1:9-11
14. Schmutterer, H. 1995a. The Neem Tree, VCH Publishers, Weiheim.

15. Schumutterer, H. 1995b. In: The Neem Tree (ed. H. Schumutterer), VCH Publishers, Weiheim:1-34.
16. Venkateswarlu, B.; Katyal, J. C.; Choudhuri, J. and Mukhopadhyay, K. 1997. Neem Newsletter. 14 (1): 7-11
17. Venkateswarlu, B.; Katyal, J. C.; Choudhuri, J. and Mukhopadhyay, K. 1998. The Indian Forester, 124 (7): 537-543

Vegetative Propagation of *Albizia amara*

A. K. HANDA, K. R. SOLANKI, S. K. SHUKLA, A. K. SHANKER AND S. P. S. CHAUHAN
National Research Centre for Agroforestry, Jhansi (UP)

ABSTRACT

Vegetative propagation is an important aspect for genetic improvement of a species and become necessary to develop quick and economic methods of producing planting material of desired characters. The main advantage of the vegetative propagation is that the selected individuals of many species can be multiplied into an infinite number of plants and plantations with a high degree of homogeneity. In utilizing these techniques, the gains from tree improvement research, whether provenance selection, progeny testing or hybridization, can be immediately capitalized and used in tree plantation establishment. The value of vegetative propagation in forestry is well understood. The transfer of non-additive characteristics through seed is difficult. However, it is routinely possible through vegetative propagation. Therefore, this can be used for achieving gains, such as growth, which has low heritability.

INTRODUCTION

Large scale plantation forestry and afforestation of degraded lands is the solution to ensure future sufficiency in biomass production. Selection of the species capable of generating maximum benefits to the rural masses is of great importance. *A. amara* is one of the important multipurpose tree species which can be used extensively in various agroforestry systems. The *A. amara* belongs to family papilionaceae and produce an excellent fuel with high calorific value

(Troup, 1921). The wood is lustrous and gives smooth finish. Although the propagation of the species through seed is easy but the germination capacity of seed collected from 15 year old coppice trees was found to be only 22 per cent and that from old trees 41 per cent (Ram Prakash *et. al.*, 1991). Keeping in mind the importance of the species, the present investigation deals with the standardization of vegetative propagation through stem cuttings. Reports are available about vegetative propagation of other *Albizia* species (Ghani *et. al.*, 1993), but literature regarding vegetative propagation of *A. amara* is very scarce.

MATERIALS AND METHODS

The experiment was conducted in the nursery of National Research Centre for Agroforestry, Jhansi during 1996. The cuttings were taken from the mature trees of the species during 1st week of March and repeated during rainy and winter season. The total duration of the experiment was for three months. The stem cuttings were upto 20 cm in length so as to include three-four nodes and 0.5 to 1.5 cm in diameter. These cuttings were treated with various concentrations of the growth regulators by giving a quick dip for 10 seconds. The details of the treatments are as under: T1 = Control; T2 – IBA 500 mg/l; T3 = IBA 1000 mg/l; T4 = IBA 1500 mg/l; T5 = IBA 2000 mg/l; T6 = NAA 500 mg/l; T7 = NAA 1000 mg/l; T8 = NAA 1500 mg/l; T9 = NAA 2000 mg/l. The stem cuttings were planted in polybags in RBD with 20 cuttings in each treatment. Each treatment was replicated three times. The data was recorded for sprouting per cent, length of shoots (cm) per cent rooting and average length of roots. The data was analysed adopting Randomised Block Design (Panse and Sukhatme, 1967).

RESULTS AND DISCUSSION

The effect of different auxin treatments on rooting in *A. amara* revealed that hormonal treatments have enhanced the rooting in the species. The highest sprouting (65.83 %) and rooting (61.00 %) in *A. amara* were recorded for the treatment IBA 1000 mg/l, which differed significantly from all other treatments. All the treatments significantly increased sprouting and rooting per cent than that of the water treated control. The minimum sprouting (9.86 %) was registered in control treated cuttings which failed to achieve any rooting success. The maximum shoot (20.2 cm) and root (15.1 cm) length were recorded in IBA 500 mg/l and IBA 1000 mg/l, respectively. The minimum average shoot length (5.8 cm) was registered in control treated cuttings.

In the present studies it was observed that the exogenous application of auxins increased rooting per cent significantly as compared to the control. It may be attributed to the fact that auxins play multifarious roles related to the division and elongation of meristem, differentiation of cambial initials into root primordia and the mobilisation of reserve food materials by enhancing the activity of hydrolysing enzymes. Role of auxins in enhancing callusing and rooting has been reported by Haissig (1971); Nanda *et. al.* (1974) and Loach (1988). According to Kralik and Sebanek (1983) auxin treatment cause considerable cell elongation and proliferation in the cortex, phloem and cambium and results in breakage of continuous sclerenchymatous rings. According to Haissig (1986), auxins induce hydrolysis and mobilisation of nutritional factors to the site of application, promoting thereby root initiation.

CONCLUSION

The present studies indicated that a large number of genetically superior plants of *A. amara* can be multiplied through stem cuttings with the application of auxins. The cuttings treated with IBA 1000 mg/l exhibited significantly higher rooting than the control.

REFERENCES

1. Ghani, A. K. M. O., Sarker, A. G. and Haque, M. A. 1993. Bangladesh Journal of Forest Science. 22 (½): 44-51.
2. Haissig, B. E. 1971. Ann. Bot. 132: 263-267.
3. Haissig, B. E. 1986. In: (ed. M. B. Jackson) New root formation in plants and cuttings. Martinus Nijhoff Publishers, Boston. 141-189.
4. Kralik, J. and Sebanek, J. 1983. Acta Univ. Agri. Brno. 31 (½): 5-11.
5. Loach, K. 1988. Acta. Hort. 226: 126-32.
6. Nanda, K. K.; Kumar, P. and Kochhar, B. K. 1974. N. Z. J. For. Sci. 4: 338- 46.
7. Panse, V. G. and Sukhatme, P. V. 1967. Statistical methods for agricultural workers. ICAR, New Delhi. 610.
8. Ram Prakash; Chaudhari, D. C. and Negi, S. S. 1991. Propagation practices of important Indian Trees. International Book Distributors, Dehradun, 98-100.

9. Troup, R. S. 1921. Silviculture of Indian Trees. Cleredon Press Oxford. 193.

Paulownia - A Potential MPTs in India

JAGDISH SINGH, A. K. BISARIA AND V. K. GARG
Center for Forestry Research Human Resource Development, Chhindwara (MP)

ABSTRACT

The genus *Paulownia* is fast growing indigenous tree species of China, valued for making furniture, musical instruments, timber, paper and pulp etc. In china it has been extensively used under agroforestry and afforestation programmes. Recently *Paulownia* has been introduced in India and preliminary results show that it has high potential to thrive and establish in India due to its very adaptable nature, easy propagation methods, suitable ecological conditions and extremely fast growing nature. Keeping in view its multifaceted values, the Research Organisations, Forests Departments, NGO's, entrepreneurs and farmers may contribute to promote the *Paulownia* as most important multipurpose tree species in India, thus reducing the heavy biotic pressure on the forests and may increase the income of rural people considerably.

INTRODUCTION

The genus *Paulownia* is represented by nine species of fast growing timber trees which are indigenous to china. It belongs to family scrophulariaceae. It is extremely fast growing and widely distributed both in south and north China. There is great potential to establish the trees of this genus in India. Recently, it has been introduced in India by ICFRE, Dehradun to evaluate the suitability under different areas and preliminary observations on its growth indicates a promising potential in certain localities (Shukla et.al., 1997).

The timber of *Paulownia* is light, yet strong, dries easily, has a beautiful grain, does not warp or crack or deform easily. The wood is easy to work with and suitable for carving and insulation. It is widely used for house construction, paper pulp, furniture veneering, farm implements, handicrafts, musical instruments. Leaves and flowers are good fodder for pigs, sheep and rabbits. Leaves are rich in nitrogen and make excellent manure. They also posses many medicinal uses. Flowers are good source for bee flora. It can also be planted for reclamation of mined areas and for landscaping urban and industrial areas.

DISTRIBUTION

Genus *Paulownia* consists of nine species, all of which are confined to china except *P. fortunei* which extends into Vietnam and Laos, and *P. tomentosa* which also grows in Korea and Japan. It grows well in plains and hills up to 2000 m altitude.

ECOLOGICAL REQUIREMENTS

Different species vary in their ecological requirement. *P. tomentosa* can withstand temperature as low as -20°C. Paulownia has very large leaf area with a high transpiration rate, and has a well developed root system. Therefore sufficient moisture is very important for its growth. However, annual precipitation in areas where it grows, varies from as low as 500 mm to a maximum of 2000-3000 mm.

Paulownia prefers deep, loose, moist, well drained soil. It is sensitive to shallow water table and salinity. Water table should be below 15 m depth. Water stagnating for 3-4 days can prove fatal. Growth is seriously impaired if the total salt content in the soil reach one per cent. *P. elongata and P. tomentosa* grow at pH 5-8.9, *P. fortunei* at pH 5-8 and other species at pH 5-6.

PROPAGATION

Paulownia can be easily propagated from seeds, root cuttings or stem cutting. Seeds are very small; one kilogram may contain 4-5 million seeds. It is grown in nursery beds and can be outplanted within one year. The height of saplings reaches 4-7 m in the nursery in one year. Outplanting is done during early spring or winter.

GROWTH RATE

Under normal conditions, a 10 year old *Paulownia* tree can attain 3–40 cm DBH and will have 0.3-0.5 m^3 timber. Under optimum conditions they can produce useful timber within 5-6 years. In Yijang country, China an 18 year old *P. fortunei* averages 21.5 m height, 100.5 cm DBH and 6.65 m^3 timber, thereby having an annual increment of 0.37 m^3 per tree. With this high growth potential the cultivation of *Paulownia* can alleviate the problem of timber shortage in the country.

In India introduction trials of different *Paulownia* species are being conducted to evaluate the suitability for different areas. *P. fortunei* has shown great potential in both growth and adaptability. 30 cm tall plants at time of out-planting attained a height of 7 m in one year in Allahabad. It has attained a height of over 7.5 m in 1.5 years in FRI Dehradun. It has great potential to produce timber at a high rate in Indo-Gangetic plains. *P. tomentosa* showed poor growth in Allahabad, but it is growing well at Dehradun.

TABLE 1: *Growth of four Paulownia species (5 year old) in Wu Gueng, China*

Species	D.B.H. (cm)	Height (m)	Timber volume (m^3)
P. elongata	18.0	11.1	0.1191
P. tomentosa	15.3	9.5	0.0716
P. fortunei	14.7	9.3	0.0696
P. glabrata	12.9	7.9	0.0433

AGROFORESTRY

Paulownia has late leaf emergence and leaf fall, and the root system is deep. The branching is sparse and so also the leaf arrangement. For these reasons, *Paulownia* is an exceptionally good tree for agroforestry. Intercropping of agricultural crops with *Paulownia* has transformed the agricultural areas in China into exemplary areas of agroforestry. Intercropping with *Paulownia* improves the micro-climate for crops and the yield increases considerably. Under intercropping condition, 10 year old *Paulownia* trees yields timber volume of 0.4-0.5 m^2/ha, with better ones upto 1.5 m^2/tree. If 50 trees are planted per hectare, 20-25 m^3/tree of timber will be produced in ten years. This is of great significance for increasing the income and improving the living standard of the rural people.

Paulownia is grown at (i) 5 □ 5-10 m spacing if primary objective is silviculture and inter-cropping is secondary, (ii) 5 □ 15-25 m when intercropping and silviculture are equally important (iii) 5 □ 30-50 m when intercropping is more important and silviculture is secondary.

AFFORESTATION

Paulownia is very adaptable except for places with too clayey, dry and poor soil, too strong wind, highly saline and alkaline soil, bad drainage, high water table and extremely low minimum temperature (below- 20° c). It is best to plant them in places with fertile, deep, loose sandy or loam soil with good drainage and a water table deeper than 15 m. *Paulownia* can be planted on home side, village side, road side and water side, thus making full use of scattered waste land thus mitigating the fuel and timber shortage. It is also used as an ornamental, in shelter belts (Dickerman *et. al.*, 1981) and in reclamation of surface mines (Carpenter, 1983).

REFERENCES

1. Carpenter, S. B. 1977. Amer. For. 83:22-23.
2. Dickerman, M. P.; Ducncan, D. P.; Gallegor, C. M. and Clark, F. B. 1981. J. For. 79:71-79.
3. Shukla, N. K.; Dhiman R. C. and Rajput, S. S. 1997. Van Vigyan 35 (1): 41-45.

Non Timber Produce of Anjan (*Hardwickia binata*)

A. K. BISARIA AND R. TIWARI
National Research Centre for Agroforestry, Jhansi (UP)

ABSTRACT

The present paper addresses the non-timber produce of anjan (*Hardwickia binata*). The non-timber produce were derived from the pruned material of six years old trees. The pruned biomass comprised of about 48-49% leaves. The leaves are palatable nutritious and rated is a very good fodder. The bark extracted from the pruned material of trees from an agrisilvicultural system. The ribbon of bark was subjected to chemical retting using sodium hydroxide at 1-10 per cent concentration. The maximum and finest fibre was obtained from the bark subjected to 6 per cent concentration of sodium hydroxide after three days. The fibre could not be obtained in control (bark dipped in distilled water) and the bark remained rough and un-retted. The experimental significance suggested that the significant quantum of fodder and fibre are obtained from the pruned material of *H. binata* which are of paramount economic importance to farmers.

INTRODUCTION

Hardwckia binata Roxb. Vern 'anjan' of family Leguminosae and sub-family Caesalpiniaceae is the tree species of acknowledged merit and of economic importance. It is a valuable indigenous tree species of Indian Union and is found throughout the drier parts of the country accending upto an elevation of 1000 m (Troup, 1921). The tree is known is `anjan' in Sanskrit, Hindi and Marathi Kamra in Kannad, Acha in Malyalayam Acha calam in Tamil and Yapi in Telgu (Anonynous, 1962). It grows best in areas with annual rainfall of 500-1000 mm but can be in areas growing with 200/500 mm annual rainfall. The plant withstand a long dry period of drought and tolerate high temperature upto 48□ C. Frost damage seedlings but saplings and trees can with stand moderate frost. Indeed the tree has been known survive in temperature as low as -5□ C (Mac Dicken, 1994). The tree grows on wide range of soil types including very shallow soils overlying fissured rocks but does best on the deep radish gravely soils. Drainage is an important factor for the successful growth and development of this species.

H. binata is a medium sized deciduous tree, leafless for a short time or nearly evergreen. It attains a height to 15-35 m depending on the site and environmental condition. It has a clear and cylindrical trunk and yield extremely hard very heavy and durable timber which is known as in the treads as anjan. The timber is resistant to rot and white ants. The wood is used for tool handles, poles, beam, tent peg, bridges and variety of agricultural tools. It is used for carving and ornamental work. It is good fuel wood and is used for producing charcoal. The leaves are palatable, nutritious and rated as a very good fodder during dry season. The leaves of the tree contains 8.9% Crude protein, 30.4% Crude fibre, 49.7% Nitrogen free extract and 0.13% Phosphorus. The bark of trees is dark grey rough and peels off in papery flakes it produces a red brown fibre after beating and jerking which is used to make rope (Deshkar *et. al.*, 1990)

Keeping in view the aforesaid importance, the prime objective of the present study was to elucidate the extraction of non timber produce of *H. binata*. In an agrisilvicultural experiments, where in trees were planted in three population densities viz. 200, 400 and 800 trees ha^{-1} during August 91 and crops were grown in interspaces. These trees were pruned ever year. Pruning operation yields huge amount of biomass in the form of branches, twigs and leaves.

MATERIALS AND METHODS

The present investigations were carried out during June, 1997. Trees of *H. binata* at the age of, six years were pruned up to 50% of their height. The biomass was collected for three densities of trees viz. 200, 400 and 800 trees ha. After weighing of the total biomass leaves were detached from the twigs and small branches. Further bark was striped from the twigs and small branch. The quantum of yield of leaves, bark and wood was determined. The bark was subjected to retting. The sample of barks were dipped on aqueous solution of sodium hydroxide (communal grade) at six concentrations viz. 1, 2, 4, 6, 8 and 10 per cent. Bark dipped in distilled water was considered as control. Retting was observed for ten days. Fibre obtained after retting was placed on filter paper to remove excess of water. The fibre yield per hectare from the pruned matter was calculated from the data of fibre/tree multiply with the population of trees ha^{-1}.

RESULTS AND DISCUSSION

The different components viz. Leaves, bark wood, fibre of the pruned materials are presented in the Table 1. The production of pruned biomass with increasing tree population both in silviculture and agroforestry system.

However, the different population density for silviculture and agroforestry system did not exhibit significantly difference. Leaf biomass which is source of fodder showed the similar trend (Ponnammal *et. al.*, 1993). Higher amount of bark was obtained from the trees of 800 trees/ha and lowest from the tree of 200/tree/ha. Some of the biomass lost due to striping of twigs and branches of *H. binata*. The data of this study removed that the density of tree had significant influence on biomass production but their mean value did not exhibit significant difference.

The process of retting accelerated with the increasing concentration of sodium hydroxide (Table 2). The maximum amount of fibre was obtained from the 6 per cent concentration of sodium hydroxide and minimum at 10 per cent concentration. It is to mention here that the loss in fibre was significantly higher at 8 and 10% concentration of sodium hydroxide. The production of fiber per hectare was calculated and the data are presented in Table 3. The production of fiber/ha was significantly higher in higher density of trees. The fiber production was found proportional to the population of trees. The fiber yield was significantly higher in agroforestry system then silvicultural system. The higher fiber production in agroforestry system was due to better growth of trees. The significance of this study suggests that pruned material of *H. binata* is a sustainable source of green fodder and fibre.

TABLE 1: *Different Constituents of the pruned material of H. binata*

System and density of trees (Trees ha^{-1})	Attribute (kg /tree)					
	Total biomass	Leaf biomass	Wood with bark	Wood	Bark Ribbons	Loss due to strapping
Silviculture 800	9.81	3.95	5.86	3.80	1.49	0.57
400	8.62	3.72	4.90	3.15	1.27	0.48
200	7.35	3.20	4.15	2.70	1.05	0.40
Mean	8.59	3.62	4.97	3.21	1.27	0.48
Agroforestry 800	9.97	4.30	5.67	3.70	1.54	0.43
400	8.75	3.90	4.85	3.00	1.46	0.39
200	7.60	3.50	4.10	2.45	1.22	0.33
Mean	8.77	3.90	4.87	3.05	1.40	0.38

TABLE 2: *Influence of sodium hydroxide (NaOH) on the retting of bark of H. binata*

NaOH Conc. (%)	Retting (days)	Fresh fibre (g/200 g bark)	Dry fibre (g/200 g bark)
0	10	-	-
1	3	42.58	8.61
2.	3	50.76	10.04
4	3	56.87	11.00
6	3	69.14	12.50
8	2	33.52	6.00
10	2	15.40	3.55

TABLE 3: *Influence of population density trees on the yield of fibre in H. binata*

System and tree density (Tree ha^{-1})	Fresh Fibre Yield (g/ tree)	Dry Fibre Yield (g/ tree)	Fresh Fibre Yield (Kg/ ha)	Dry Fibre Yield (Kg/ha)
Silviculture 800	310.64	69.89	248.00	55.80
400	180.27	40.56	144.00	32.21
200	104.35	23.47	83.10	18.76
Mean	189.42	44.64	158.46	35.59
Agroforestry 800	365.00	82.12	292.00	65.71
400	215.50	48.48	170.80	38.43
200	112.10	25.22	90.00	20.25
Mean	236.20	51.94	184.26	41.46

REFERENCES

1. Anonynous, 1962. The wealth of India (Raw Materials), C.S.I.R., New Delhi.
2. Deshkar, A. M., Bokade, S. S. and Dara, S. S.1990. Water research oxford.24:8, 1011-1016.
3. Mac Dicken 1994. Selection and management of nitrogen fixing trees FAO, Rome: United Nation Food and Agricultural Organisation. pp:149-151.
4. Ponnammal, N. R, Arjunan, M. C. and Antony, K. A.1993. Indian Forester.119:1, 59-62.
5. Troup, R. S.1921. The silvicultural of Indian Trees. Vol. II. Reprinted 1986 by International Book Distributers, Dehradun.

Management of MPTs in Agroforestry Systems for Problematic Soils of North-Western Hilly Region

PRATAP NARAIN, K. S. DADHWAL, V. P. S. TOMAR AND H. P. S. TOMAR
Central Soil and Water Conservation Research and Training Institute Dehradun (UP)

ABSTRACT

Agroforestry with multipurpose trees (MPTS) has been widely recognized as viable option for conservation of natural resources and sustaining production of food, fuel, fodder and fibre etc. Commonly grown MPTS for north western hilly region are short listed and their management practices in agroforestry systems for problematic soils of the region are described. On a class V river bed with 60% boulders, root management by planting in auger holes, long plastic tubes and trenching embedded with polyethylene increased the wheat and fodder yields up to 18-23 per cent. Orientation of tree rows in east-west direction also improved the yield by 17% over north-south. Shoot management by lopping, pollarding or even as hedges reduced competition between trees and crops.

Alley cropping through paired rows of *Leucaena* hedges and vegetative barriers of *Panicum maximum, Eulaliopsis binata* and *Veteveria* reduced runoff to 50% and soil loss up to 70% in corn grown on 4% slope. As much as 60 t/ha earth was deposited along the contour rows by redistribution of conserved sediment. This led to formation of compartmental bunds and natural bench terraces even on mild slope. Conservation efficiency, USLE cropping management factor 'C' and future strategies for agroforestry land uses are discussed in the paper.

INTRODUCTION

In the management of natural resources of the country, serious concerns are being expressed for degraded lands (107.43 m ha) occupying nearly 1/3rd of geographical area (MOA, 1994). Water erosion accounting for about 65% of land degradation is induced by deforestation, shifting cultivation with short fallow cycles and exploitation of fragile lands. Forests occupy 19.27% of total geographical area of the country, while dense forest with more than 40% canopy is hardly 11.17% (FSI, 1997). Out of 95 hill districts of the country, only 32 have 2/3rd or more forest cover as required by National Forest Policy (1988). Stabilization of steep slopes, conservation of soil and water, improvement of soil fertility and utilization of degraded torrent beds (which constitute about 1/3rd area in valley region) are major soil related problems in the hills. These problems need bio-engineering remedial measures, wherein engineering component is too expensive and beyond the reach of poor hill farmer. Hence, vegetative means, which are cost-effective and productive while being protective, seems to be logical course on private lands to protect geologically weak hill ecosystem.

Agroforestry systems with multi-purpose trees (MPTS) are potentially viable option for management of natural resources in hilly region, soil and water conservation and sustaining production of food, fuel, fodder and fibre in hilly region. Agroforestry will also reduce tremendous pressure on our limited forest resources thereby containing serious erosion and land degradation problems. Intervention of trees with crops automatically invites above and belowground competitions. Woody perennial being more aggressive in nature might affect annual crops adversely. Baring few cases, several studies support adverse effect of trees on crops when both are grown together due to competition for light, water, nutrients and allelochems (Khybri *et. al.*, 1992, and Narain *et. al.*, 1998b). The extent of damage depends on nature of tree, crop, tree intensity, direction of planting, soil working and agro-climatic situation. The adverse effects can be minimized by management practices suitable for growing MPTS with crops and agroforestry may then turn an acceptable proposition for adoption. The paper deals with some of these management practices of MPTS in agroforestry systems for problematic soils in north western hills of Himalayas.

CHOICE OF MPTS FOR AGROFORESTRY IN NORTH WESTERN HILLS

Considering characters of MPTS and commonly grown agricultural crops in north western hills, important native and some exotic MPTS, whose management practices have been studied and are given in Table 1. A Diagnosis and Design Survey of 400 farm families in Doon valley (Dadhwal *et. al.*, 1985) brought out agri-horti system to be the first preference of farmer despite a severe reduction of crop yields by fruit trees after 4-5 years. Commonly grown fruit trees are: *Mangifera indica* (mango), *Litchi chinensis* (Litchi), *Psidium guajava* (Guava), *Prunus persica* (Peach), *Pyrus communis* (Pear), *Artocarpus integrifolia* (Jack fruit), *Carica papaya* (Papaya) and *Citrus species* (Citrus).

Diversity in germplasm of MPTS demand use of modern tools like biotechnology, tissue culture and vegetative propagation for homogeneity and improvement of species. Simultaneously management and silvicultural practices suitable for MPTS and crops are required to harness full potential from the system synergistically. The management practices of some MPTS on arable (land capability class I to IV) and non arable (LCC V to VII) land are presented in following section.

SOIL WORKING TECHNIQUES

MPTS have generally been planted by conventional pit planting in 30 to 50 cm^3 pit filled with excavated soil with 1 to 2 kg FYM and 50 g BHC per pit. On slightly acidic soils application of 60 g CaO/pit equivalent to 5 tonnes/ha lime helps better establishment of pH sensitive species like Leucaena etc. The soil working should be chosen for a given annual rainfall class, soil texture and drainage requirements of the MPTS (Seth, 1960). A study on class V lands at Dehradun (Puri *et. al.*, 1995) containing 30% boulder up to 10 cm and 55-85% up to 100 cm depth revealed that the auger hole plantation technique recommended for Sodic Soil (Abrol and Sandhu, 1980) produced height, DBH and wood of *Eucalyptus hybrid* similar to pit planting. However cost of auger holes was only about 1/3rd to pits, hence former should be preferred. Degraded riverbeds developed about 147% more root biomass compared to normal deep soils in order to produce identical volume of above ground biomass (Dhyani et al, 1996). Root growth, their coiling was an essential mechanism to survive in harsh environment with impediments and deficient in water and nutrients. Root management appears to be meaningful in such environment as well as in agroforestry systems.

TABLE 1: *MPTS species for Agroforestry in north western Himalayas*

Species	Local name	Uses
Acacia catechu	*Khair*	*Fuel, fodder, industrial uses (katha, agriculture implements, gums)*
Albizzia lebbeck	*Siris*	*Fuel, fodder, furniture and building material*
Bauhinia species	*Kachnar*	*Timber, fuel, fodder, fibre, vegetables and industrial uses (agriculture implements, cherry gum, dye)*
Celtis australis	*Kharik*	*Fuel, fodder, vegetable, medicine and industrial use (agriculture implements)*
Eucalyptus species	*Safeda*	*Furnitures, building material, industrial uses (oil, paper pulp)*
Grewia optiva	*Bhimal*	*Fuel, fodder, fibre, basket making*
Leucaena leucocephala	*Soobabul*	*Fuel, fodder, green manure, industrial uses (paper pulp, gum hard board)*
Melia azadirch	*Bakain*	*Furniture, fuel, fodder, industrial uses (pulp, roofing material, agriculture implements, toys)*
Morus species	*Shahtoot*	*Fuel, fodder, fruits, furniture basket making and sericulture*
Moringa oleifera	*Senjan*	*Fuel, fodder, vegetables*
Populus species	*Poplar*	*Fuel, fodder, timber and packing cases, paper pulp and match boxes etc.*
Paulownia species		*Fuel, fodder, softwood for paper pulp and match boxes*
Sesbania aegyptica	*Daincha*	*Fuel, fodder, green manure*

At Chandigarh, *Eucalyptus tereticornis* grown in 6 cm diameter 120 deep auger holes filled with 1 kg FYM together with *Eulaliopsis binata* at 50 □ 0 cm yielded 9.5 t/ha/yr tree biomass + 4.55 t/ha airdry grass (Agnihotri and Grewal, 1993). The net value in 1993 obtained was Rs. 35228 in 8 years with B: C ratio of 2.04 (Table 2). This planting method may minimize root competition upto 60 cm depth by preferential vertical growth of tree roots. Raising seedling in 60 cm long poly begs, which can be planted in auger holes by removing the bottom may further reduce root competition of trees and crops in mixed agroforestry system. The technique may also improve availability of soil moisture from deeper layers effecting tree growth favourably.

TABLE 2: *Airdry biomass of Eucalyptus and Bhabar (t/ha) under various soil working techniques*

Soil working	Stem + branch wood	Leaves + twigs	Mean grass yield	Net value (1993) (Rs.)	B:C ratio
30 □ 30 □ 30 cm pits	*31.51*	*2.07*	*4.01*	*18,968*	*1.66*
60 cm auger holes filled with soil	*61.22*	*3.44*	*3.98*	*25,800*	*1.84*
120 cm auger holes filled with soil	*68.05*	*4.59*	*4.11*	*28,371*	*1.88*
120 cm auger holes filled with soil + 1 kg FYM	*75.27*	*4.98*	*4.55*	*35,228*	*2.04*

ROOT MANAGEMENT OF MPTS

In 8-year-old boundary plantation of *Eucalyptus hybrid* at Dehradun, 0.5 m wide and 1 m deep trench dug to eliminate root competitions improved soil moisture of adjoining maize-wheat field. Though the maize yield were not significantly influenced, however, wheat yields improved by about 30 per cent in 0-2 m segment from tree row (Dadhwal *et. al.*, 1977). Studies have been conducted (1986-1997) on class V land, torrent bed, having 60 per cent boulders and gravel, very low W.H.C. (18%) and extremely poor fertility (0.26% carbon, 0.03% N, 6.7 ppm Av. P and 21 ppm Av. K). *Eucalyptus hybrid* was planted in systematic plus design (N-S and E-W directions) having 15 plants on each arm 1.25 m apart. One third arm was trenched on both sides, 1/3 on one side and 1/3 was left untrenched. A polythene barrier was embedded to 1 m depth in the trench and covered with soil. In another study of *Grewia optiva* and *B. purpurea* were planted in star design on north-south, east-west, north, east-south west and north south-west north directions without any barrier.

Root barrier on one side of Eucalyptus row increased the wheat yield by 8% and fodder yield of jowar by 15%. This increase shot up to 18% in wheat and 23% in jowar fodder due to trenching on both the sides of tree row without any adverse effect on tree growth. The decrease in yield up to about 4 m distance from tree rows was reduced to 2 m only (Table 3) due to trenching. Khybri *et. al.* (1992) have reported 28 to 34 per cent decrease in the yield of paddy and wheat up to 5 m distance from trees due to *G. optiva, M. alba* and *Eucalyptus hybrid* on table land without any boulders. The effect of root pruning by trenching caused drastic reduction of root biomass, specific root length and root binding capacity of tree species. Interestingly roots were noticed up to 10 m in non trenched plot against 6 m in both side trenching. On normal land *B. purpurea* and *G. optiva* showed much less root length density in 1 m depth compared *L. leucocephala* and *E. hybrid* (Singh *et. al.*, 1998). Therefore former two indigenous species are likely to be less competitive with crops in agroforestry system as compared with fast growing *Leucaena* or *Eucalyptus* species. *G. optiva* is the most commonly grown species on terrace risers in hills.

TABLE 3: *Effect of root pruning of Eucalyptus hybrid on yield of wheat and Jowar fodder (Av. of 4 years q/ha)*

Distance from tree base lime (m)	Without trench		One side trench		Both side trench	
	Wheat	Jowar	Wheat	Jowar	Wheat	Jowar
0-2	6.5	70.28	7.5	79.95	8.7	93.38
2-4	8.8	83.3	9.9	99.01	11.4	102.41
4-6	10.1	76.09	11.0	102.69	11.8	108.78
6-8	11.0	98.49	11.9	109.69	12.1	115.85
8-10	10.9	105.70	11.1	119.0	13.9	121.12

DIRECTION OF PLANTING

Earlier agroforestry studies by Khybri *et. al.* (1992) and Narain *et. al.* (1998b) were with north-south orientation of tree rows. East-west planting proved superior by increasing of wheat yield by 6% and jowar fodder by 9%. Further north east - south west orientation improved grain yield marginally but the yield of fodder by about 17% over east-west orientation. As moisture is not a limitation in kharif, light seems to influence yields of crops. Orientation of MPTS rows has bearing on the success of agroforestry system.

TREE INTENSITY

Number of trees per unit area is crucial when trees are grown unpruned on crop land. Maximum reduction in wheat yield was by 41-61% due to 7-9 years old Eucalyptus at an intensity of 100 trees ha^{-1}. The average decrease in the yield of wheat and paddy due to *G. optiva, M. alba* and *Eucalyptus hybrid* (Table 4) was 28-34 per cent (Khybri *et. al.*, 1972). The effect decreased with increasing distance from tree row. In degraded lands the reduction of crop yields were observed clearly upto 4 m distance from the tree row.

TABLE 4: *Effect of trees on average crop yields (kg ha^{-1})*

Species	Wheat (1978-1990)	Rice (1978-1990)
G. optiva	1606 (29)	1642 (28)
M. alba	1592 (29)	1494 (34)
Eucalyptus	1611 (28)	1586 (30)
Control	2249 -	2275 -

Ong *et. al.*, (1992) reported 46-55% reduction in maize yield with upper storey of unpruned Leucaena. A reduction up to about 50% or more is likely to occur after sixth years of growth of Eucalyptus or Leucaena at 555 plants ha^{-1} (Narain *et. al.*, 1998b). About 50-100 plants as boarder rows or paired rows in east-west direction with partial lopping may be ideal for agroforestry system in this region.

SHOOT MANAGEMENT

Management of above ground parts of woody perennials seems to be envitable if trees and crops have to co-exist. Management of trees as hedge rows pruned at 15 cm height twice a year before sowing of maize and wheat showed negligible effect of trees on crop (Narain *et. al.*, 1998b). In a fuel-food study on degraded riverbed (Vishwanathan and Joshie, 1995), 50% lopping of *Bauhinia purpurea* produced 4 q/ha airdry fodder without adversely affecting *Chrysopogon fulvus* (10.7 q/ha) and *Eulaliopsis binata* (21.7 q/ha). The intensity of lopping may have to be increased, if trees are to be grown on crop land. Farmer resorts to 80-90 per cent lopping of Grewia in fodder scarcity winter period.

There was about 35 to 40% reduction of grain and 25 to 35% fodder yields up to 2 m distance from the tree rows as compared to bulk field. Reduction in grain yield was 35 to 40% up to 2 meters and 20% between 2-4 meters from the tree line. The decrease in case of jowar fodder was 20% and 15% respectively. The higher reduction in 0-2 m distance is primarily attributed to acute competition at the tree crop interface and partly due to poor approach of agricultural implements up to the tree stem.

SOIL AND WATER CONSERVATION

Agroforestry through MPTS has emerged a powerful tool for sustaining production through natural resource conservation. Upto 30 percent slope in humid, subhumid and subtropical climate, alley cropping with *Leucaena, Gliricidia* hedges and grass barriers on contour are effective for mulching and erosion control. Application of nitrogen rich prunings of shrubs like Leucaena and gliricidia on inter barrier space will protect land against water erosion and improve its fertility. In North-Western Outer Himalayas, Dehradun (1740 mm rainfall), contour paired rows of leucaena hedges, leucaena and Eucalyptus trees and 0.75 cm wide grass barriers at 1.0 m V.I. in corn (Narain and Grewal, 1994) brought down runoff from 40 to 30 percent of rainfall and soil loss from 21 to 8 t ha^{-1} y^{-1} (Table 5). Amongst the grass barriers, the order of efficiency was Panicum > Veteveria > Eulaliopsis. The grass barriers were more efficient than hedge rows in arresting soil loss in spite of higher runoff yield.

The barriers influenced redistribution of sediment in the catchment resulting in deposition of 30-60 t ha^{-1} yr^{-1} sediment on contours and forming bunds/terraces naturally (Table 6). The total earth deposition is much higher than the soil loss from corn as measured at the plot outlet, showing higher gross erosion as compared to measured net erosion (Narain *et. al.*, 1998a). An excellent example is utilization of Guatemala grass (*Tripsacum laxum*) on lateritic soils of Southern hill region at Ooty for formation of Puerto Rico or California type terraces through vegetative barrier (Chinnamani and Rege, 1965).

TABLE 5: *Effect of paired rows of barrier hedges, grass strips and trees on runoff and soil loss in corn at 4% slope.*

Treatment	Runoff (%)	Soil Loss (t ha^{-1} y^{-1})
Maize on contour	40	21
Maize + Leucaena hedges	21.3	12.1
Maize + Panicum (0.75 m wide)	36.7	7
Maize + Eulaliopsis (0.75 m wide)	42.7	10
Maize + Veteveria (0.75 m wide)	39.6	8.1
Maize + Leucaena trees (6-8 yrs)	20.4	8.4
Maize + Eucalyptus trees (6-8 yrs)	16.3	5.8
Average for agroforestry land uses	30	8.7

TABLE 6: *Sediment deposition along vegetative barriers at Dehradun*

Vegetative barrier at 1.0 m V.I.	No. of years in existence	Sediment deposited (t ha^{-1} yr^{-1})	Annual deposition (t ha^{-1} yr^{-1})
Leucaena hedges in turmeric	3	47.31	15.77
Leucaena hedges in maize	3	183.98	61.33
Leucaena trees in maize	9	256.54	28.5
Eucalyptus trees in maize	9	185.58	20.62
Leucaena trees in turmeric	9	90.06	10.01
Eucalyptus trees in turmeric	9	103.7	11.52

In Kerala, grass farming effectively reduced erosion to 0.4 t ha^{-1} and runoff to 96 percent rainfall from 352 t ha^{-1} y^{-1} soil loss and 1259 mm runoff in cultivated fallow plot. Casava cultivation was deleterious from soil erosion view point. Replacing 10 percent Casava population with grass strips reduced erosion by 41 percent in comparison with maximum casava treated plots (5 plants on mounds between 4 trees (Gopinath and Sreedharan 1993). Vegetally guarded conservation ditches (VGCTR) which involve reinforcing sensitive points of hedges by locally available stones, opening a trench on upstream side, converting vegetal hedges into ridges with dugout material and dumping organic trash in trench have been reported to bring down soil loss by 97-100 percent from 252 t ha^{-1} y^{-1} from pineapple fields in Kerala (Gopinath, 1994). Such VGTR are reported to have more acceptability as compared to hedge rows of Leucaena, Gliricidia or grasses.

CONSERVATION EFFICIENCY OF SOME MANAGEMENT PRACTICES

Based on the erosional losses of runoff, sediment and nutrients some of the management practices/land uses could be ranked for their conservation efficiency in the order: sole tree > grass + tree > grass > turmeric + tree/hedgerows > maize + tree/hedgerows > maize > cultivated fallow (Narain *et. al.*, 1998a). This is supported by decreasing order of USLE conservation management factor 'C' which is 0.008 for trees, 0.048 for grass and 0.425 for maize (Table 7) against 0.4 value of 'C' factor for maize in Doon valley (Narain *et. al.*, 1994).

TABLE 7: *USLE crop management ('C') factor for different land uses at Dehradun*

Sequence I	C-factor	Sequence II	C-factor	Average ('C')
Sole Leucaena	*0.01*	*Sole leucaena*	*0*	*0.009*
Sole eucalyptus	*0.01*	*Sole eucalyptus*	*0*	*0.008*
Grass + leucaena	*0.109*	*Turmeric + leucaena*	*0.242*	
Grass + eucalyptus	*0*	*Turmeric + eucalyptus*	*0.219*	
Maize wheat + leucaena	*0.257*	*Maize wheat + leucaena*	*0.304*	*0.278*
Maize wheat + eucalyptus	*0.16*	*Maize wheat + eucalyptus*	*0.225*	*0.189*
Grass	*0.112*	*Turmeric + leucaena hedge*	*0.246*	*(0.048)*[a]
Maize	*0.425*	*Maize + leucaena hedge*	*0.377*	

[a] *Excluding establishment period.*

FUTURE STRATEGIES

- ☐ Inventorise traditional agroforestry systems and evaluate for their conservation and production efficiency.
- ☐ Site specific research on sharing of above and below ground light, water and nutrients by MPTS and crops is required to adjust management practices synergistically.
- ☐ The management practices such as soil working, root and shoot management, processing of harvest etc. should be evolved for growing MPTS with crops.
- ☐ Sloping Agriculture Land Technology (SALT) developed in Philippines (Tacio, 1993), a modified version of alley cropping, needs to be tested under socio-economic conditions of hilly regions in India.
- ☐ Large-scale testing of technology on pilot project basis with peoples' participation for better adoption rate.

ACKNOWLEDGEMENTS

The authors express their deep sense of gratitude to Dr. J. S. Samra, Director, CSWCRTI, Dehradun. The painstaking efforts of Ms. Mamta Negi, Stenographer for typing and preparation of the manuscript on computer are thankfully acknowledged.

REFERENCES

1. Abrol, I. P. and Sandhu, S. S. 1980. Indian Fmg. Sept.: 10-20.
2. Agnihotri, Y. and Grewal, S. S. 1993. Indian J. Soil Cons. 21 (2): 79-84.
3. Chinnamani, S. and Rege, N. D. 1965. Agric. Res. 5: 154.
4. Dadhwal, K. S.; Dhruva Narayana, V. V. and Narain, P. 1985. Annual Report, All India Coord. Research Project on Agroforestry, 1985: 1-25.
5. Dadhwal, K. S.; Dhruva Narayana, V. V. and Narain, P. 1977. Indian Fmg. April: 43-44.
6. Dhyani, S. K.; Puri, D. N. and Narain, P. 1996. The Indian For. 122.
7. Forest Survey of India 1977. State of Forest Report Min. of Environment and Forests, Dehradun.
8. Gopinath, R. and Sreedharan, C. 1993. In: Meteorology and Agroforestry, Proc. Intl. Workshop on Application of meteorology to agroforestry systems planning and management, Nairobi, 9-13 Feb., 1987.
9. Gopinath, R. 1994. VGCTR - Farmer contributed vegetative techniques for soil and water conservation in steep hill slopes, presented at ICAR-UNDP Workshop held on 4-5th March, 94 at Dehradun, India.
10. Khybri, M. L.; Gupta, R. K.; Sewa Ram and Tomar, H. P. S. 1992. Agroforestry Systems 17: 193-204.
11. MOA 1994. Report of land degradation in India. Soil and Water Conservation Division, Deptt. of Agric. and Coop. Min. of Agric., New Delhi.
12. Narain, P.; Singh, R. K.; Sindhwal, N. S. and Joshie, P. 1998a. Agroforestry Systems. 39: 175-189.
13. Narain, P.; Singh, R. K.; Sindhwal, N. S. and Joshie, P. 1998b. Agroforestry Systems. 39: 191-203.
14. Narain, P.; Khybri, M. L.; Tomar, H. P. S. and Sindhwal N. S. 1994. Indian J. Soil Cons. 22: 1-9.
15. Narain, P. and Grewal, S. S. 1994. Agroforestry for soil and water conservation - Indian experience. 8th ISCO Conference, New Delhi: 1407-1417.
16. Prajapati, M. C.; Verma, B.; Mittal, S. P.; Nambiar, K. T. N. and Thippanwar, B. S. 1971. Annals Arid Zone 10 (2 & 3): 186-172.
17. Prasad, S. N.; Srivastava, A. K.; Narain, P. and Bhola, S. N. 1985. Indian J. Soil Cons. 13 (1): 41-46.
18. Puri, D. N.; Narain, P. and Dhyani, S. K. 1995. Eucalyptus hybrid. 121, 600-607.
19. Singh R. P.; Ong. C. K. and Saharan, N. 1989. Agroforestry Systems 9: 259-274.
20. Seth, S. K. 1960. Ind. For. 86 (5).
21. Tacio, H. P. 1993. Agroforestry Systems, 22: 145-152.
22. Vishwanathan, M. K. and Joshie, P. 1994. Annual Report, CSWCRTI, Dehradun: 31

Exploitation of Problematic Soils through MPTs under Agroforestry Systems

P. RAI
National Research Centre for Agroforestry, Jhansi (UP)

ABSTRACT

The available agroforestry technologies have shown promise in increasing the productivity of problematic areas such as saline alkaline soils, acidic soils, ravine areas, areas affected with brackish water, swampy and wet lands, mines affected area, cho and riverbed affected areas, shallow gravelly red soils and rocky areas in semi arid regions, arid desert and sand dunes as well as, cold desert areas. Inclusion of multipurpose tree species provides nutritious leaf fodder especially during lean periods, fuel wood for energy and timber for various purposes in addition to conserving and improving the soils of the area. The grasses and legumes also provide the higher quantity of green and dry forage as well as check the soil erosion and improving the nutrient contents of the soils. The major difficulties in adoption of technologies in rural farming communities include lack of proper technology transfer mechanism and availability of suitable planting material of improved species of MPTS, grasses and legumes.

INTRODUCTION

India with a total geographical area of 329 m ha, about 158 m ha area has been categorised as different types of problematic area. These areas are formed by various factors such as sheet erosion, wind erosion, stream erosion, water logging, salinity and alkanity, shifting cultivation and shifting sand dunes. These areas include hot desert and sand dunes, rocky and gravelly land, cold desert, ravine areas, saline alkaline areas, swampy and wet lands, cho and riverbed affected areas etc. Extreme of climatic, soil conditions, biotic factors and inconsistent rainfall are the main characteristics of these sites. For this reason alone these lands require specific management technology.

At present forest area is about 19.47% of the total geographical area of the country (Annon, 1996). The decline in forest cover has led to shortage of leaf fodder, fuel wood and timber. The conversion of forestland into agricultural land use by other development project has led to soil erosion, ground water depletion, frequent floods and droughts. Under the above circumstances, we have no alternative but to follow integrated agroforestry systems such as agrisilviculture, agrihorticulture, agrihortisilviculture, silvipasture, hortipasture, silvihortipasture, boundary plantation, alley cropping etc. using multipurpose tree species (MPTS) including fruit trees along with crops and pastures in the problematic areas of the country.

Incorporation of MPTS along with grasses and legumes into these problematic areas are very important to solve some of the problems including amelioration of polluted environment and also to mitigate the national target of 33% area under tree cover to get fodder, fuel, fruit and timber of our large human and livestock population. However, the author will restrict to the package of practices for different problematic areas through silvipasture only.

DIFFERENT PROBLEMATIC AREAS AND THEIR PRODUCTION POTENTIAL

Saline Alkali Soils

In India about 8 million hectare area is infested by saline alkali soils and in Uttar Pradesh alone about 1.2 m ha area is lying as 'Usar'. Hence, to increase the total forest cover and to reclaim such type of lands plantation of suitable MPTS (*Prosopis juliflora, Acacia nilotica, A. catachu, A. tortilis, Cassia siamea, Albizia, lebbeck, A. procera, Butea monosperma, Dalbergia sissoo, Casuarina equisetifolia, Tamarix articulata, Emblica officinalis, Pongamia pinnata, Salvadora spp., Eucalyptus hybrid, Syzygium cumini, Terminalia arjuna, Leucaena leucocephala, Tamarindus indica, Zizyphus mauritiana etc.)* along with grasses (*Leptochloa fusca, Brachiaria mutica, Chloris gayana, Bothriochloa intermedia, Sporobolus marginatus, Paspalam spp. etc.)*

and legumes (*Atriplex spp, Sesbania spp., Rhynchosia minima, Clitoria ternatea, Desmanthus virgatus etc.*) have been found beneficial for increasing the productivity of such lands. Misra (1990) reported that due to plantation of *L. leucocephala* in usar soils the condition of soil improved during 4th year. The organic carbon increased from 0.278 to 0.436% and pH value reduced as low as 8.38. Similarly, observation on 18 years growth of 10 MPTS in usar soils showed very good growth of lateral roots. The maximum number of lateral roots (200) recorded with *P. juliflora* followed by *A. catachu* (177) and *D. sissoo* (100). While maximum length of lateral root was recorded with *P. pinnata* (9.1 m) followed by *D. sissoo* (8.2 m), *A. procera* (7.4 m) and *T. arjuna* (6.5 m). Rai (1990) reported that planting of *B. mutica, B. intermedia, Panicum maximum, and Setaria sphacelata* in 9.5 to 10.5 pH soil after application of 12.5 t/ha gypsum with 25 kg/ha Zink sulphate at the farm of N.D. University of Agriculture and Technology, Faizabad produced 21.5 to 42.1 t/ha green forage in second year of establishment.

In a field trial on an alkali soil at Karnal, Singh (1995) reported that growth (height 7.1 m, CD 28.1 cm and DBH 23.3 cm) and biomass production (37.7 t/ha) of *P. juliflora* planted at 5 □ 3 m spacing was higher in sole stand than when interplanted with *L. fusca* grass (height 6.5 m, CD 25.3 cm, DBH 20.3 cm and biomass yield 30.9 t/ha) at six years of establishment. The green forage yield reported was 46.5 t/ha in a period of 50 months in 15 cuts. On an average pod production recorded with *P. juliflora* was 699.6 kg/ha in sole trees and 635.1 kg/ha in tree + grass. But the soil improvement in grass + tree treatment was higher than sole tree. Initial soil pH and Ec was 10.3 and 2.2 ds/m, respectively, both under sole *Prosopis* and *Prosopis* + grass but it reduced after 74 months (pH 9.3 in sole tree and 8.9 in tree + grass, Ec 0.46 ds/m in sole tree and 0.36 ds/m in tree + grass). The increase in organic carbon was from 0.18 to 0.43% in sole tree and from 0.19 to 0.58% in tree + grass. Similarly increase in NPK was higher in tree + grass than sole tree.

Acidic Soils

The huge area affected by acidic soils is also available in North-Eastern States of our country. These areas can be suitably developed with plantation of MPTS such as *Bucklandia pupulrea, Alnus nepalensis, Ficus nemoralis, Albizia chinensis, Morus cerrata, Ficus auriculata, Artocarpus lakoocha, Bauhinia variegata, Erythrina arborescens* etc.

These MPTS can be lopped after 4-8 years of establishment depending on site condition and 300-700 kg/tree leaf fodder may be obtained (Singh and Rai, 1990). In such type of soils, grasses like *Pennisetum polystachyan, P. pedicellatum, P. clandestinum, Setaria sphacelata, Thysanolaena maxima* etc. and legumes such as *Centrosema pubescens, Calopogonium muconoides, Stylosanthes guianensis* can be easily established and 5-6 t/ha dry forage yield may be obtained. Singh *et. al.* (1995) reported that a silvipastoral system consisting of *Ficus auriculata* (Nevaro) as tree + *Thysanolaena maxima* (broom grass) as grass produced maximum forage yield of 9.62 t/ha/yr (7.78 t/ha from grass + 1.84 t/ha from tree, average of 8 years) in acidic soil (pH value (4.5) as compared to other silvipastoral systems such as Nervo + Hybrid napier, Nervo + Nandi Setaria and Nervo + Palm grass in Eastern Himalaya.

Areas Affected with Brackish Water

Area affected with brackish water in our country is mostly found in some parts of Uttar Pradesh, Gujarat, etc. These areas can also be developed with plantation of some of the MPTS like *P. pinnata, Cordia dichotoma, C. siamea, Haplophragma adenophyllum, P. juliflora, A. nilotica* (Chaturvedi, 1990). Suitable grass species found in such areas (*Dactyloctanium aegypticum, D. sindicum, Chloris gayana, Panicum antidotale, Eleusine compressa, Sporobolus marginatus, L. fusca*) are established and productivity can be enhanced many times through silvipastoral systems. Recently, Singhania *et. al.* (1997) reported that out of 5 grasses, *Chloris gayana and Panicum antidotale* showed very good performance in saline water (Ec 12 ds/m) and produced on an average (1986-87 to 1989-90) green forage yield of 41.5 and 33.1 t/ha, respectively at Jobner (Rajasthan).

Ravine Areas

Ravines present the worst from erosion, which completely destroy the land. About 4 m ha area is ravinous confined largely to the state of Uttar Pradesh, Madhya Pradesh, Gujarat and Rajasthan. These lands are far below their economic utilization. The ravine lands are potential for producing, fuel, fodder and fibre and generation of additional employment. Small to medium gullies of these ravines can be put to silvipasture or hortipasture or silvihortipasture. Suitable MPTS for ravine areas are *A. nilotica, A. catechu, A. tortilis, Dichrostachys cinerea,*

D. sissoo, Azadirachta indica, Dendrocalamus strictus, Albizia species, Eucalyptus tereticornis, Grewia tenax, Zizyphus numularia, Capparis zeylanica, Aegle marmeolas, E. officinalis, Limonia acidissima (Kaitha), Z. mauritiana, Ficus species, Prosopis species. In addition of MPTS, grasses (*Cenchrus, species, Dichanthium annulatum, Eulaliopsis binata, Pennisetum pedicellatum, Panicum antidotale* etc.) and legumes (*Macroptilium atropurpureum, Rhynchosia minima, Clitoria ternatea, Alysicarpus monilifer, S. hamata* etc.) to be sown for getting higher biomass as well as controlling the soil erosion loss in these areas. A fifteen years studies in Yamuna ravine near Agra (U.P.) Revealed that a silvipastoral system consisting of *A. nilotica + C. ciliaris and A. tortilis + C. ciliaris* planted at 3 □ 3 m spacing produced the average total dry biomass yield of 5.0 (2.5 t fuel + 2.5 t forages)t/ha/year and 6.9 (4.2 t fuel + 2.7 t forage) t/ha/yr, respectively (Prajapati *et. al.*, 1993). Similarly, studies at CIRG, Farah, Mathura revealed that a silvipastoral system consisting of *A. nilotica, L. leucocephala and C. ciliaris* could increase forage production, energetics and protein potential of ravinous grazing land by 3.3, 2.6 and 8.0 folds, respectively without affecting the normal growth of trees (Bhattacharya and Sharma, 1993). In Gujarat, in Mahi ravines, Mishra *et. al.* (1988) reported that *E. binata* (Bhaber grass) has been found to be most promising and profitable fibre grass with yield potential of 9.0 t/ha. It is used a paper pulp and a material for country ropes. Employment generated is 400 man days/ha/year. In case of *D. strictus*, average annual yield of bamboo works out to be 1000 culms/ha from 7th year onwards in Mahi ravines which gives annual income of Rs. 2000-3000/ha.

Chos and Riverbeds Affected Area

The shallow hill torrents (Streams) along the foot hills have seasonal flows which are popularly known as Chos in Panjab and 'Raos' in Doon valley. These areas are largely observed in the sub mountain regions of Himalaya. Besides, there are large riverbed affected areas which are created due to shifting of major rivers of spillover river sections. These lands are known as 'Bet', 'Khadder', 'Chaur', 'Diara' etc. This type of affected area is found in eastern states of our country (U.P., Bihar etc.). In India the total area affected by this problem is 2.73 m ha (Mishra *et. al.* 1988). These area can be developed through plantation of MPTS (*A. catechu, A. nilotica, D. sissoo, Salix sp., Bahunia purpurea, L. leucocephala, Grewia optiva, A. lebbeck, Agave spp., Z. mauritiana,*

Psidium guajava, Morus sp.) along with grasses (*Eulaliopsis binata, Arundo donax, Saccharum munja, S. spontanium, P. purpureum, C. fulvus, P. pedicellatum*) and legumes (*Calopogonium muconoides, Stylosanthes guianensis, Pueraria tuberosa, P. hirsuta, P. phasedoides, P. lobata*). The productivity of such lands can be increased and soil erosion can be minimised by adopting such systems.

Studies conducted from 1980-93 at Dehra Dun in degraded river bed bouldery land with 4 trees (*A. lebbeck, G. optiva, B. purpurea and L. leucocephala*) planted at 4 □ 4 m spacing and managed with 2 lopping intensities of the crown (50 and 75%) and two grasses (*C. fulvus and E. binata*) planted at 1.0 □ 0.5 m spacing in between the tree species revealed that on an average 4.55 t/ha (0.29 t leaf fodder + 0.58 t fuel wood + 3.68 t grass forage) biomass can be obtained from silvipastoral system on such type of bouldery lands (Arora and Vishwanathan, 1995). In another study evaluation of 3 intensities of harvest of *Agave sisalana* leaves (25, 50 and 75%) with and without *E. binata* for economic utilization of bouldery waste lands from 1975 to 1986 revealed that 50% intensity of harvest of *Agave* leaves for fibre (0.17 t/ha) along with *E. binata* (3.58 t/ha forage) is the best and economically viable combination, since benefit cost ratio (BCR) was 1.67 and internal rate of return (IRR) was 32.8 (Arora and Vishwanathan, 1995). These systems also helped in restoring soil losses and conserving water through infiltration, run off reduction and better intake of water into the soil.

Mines Affected Areas

In our country lot of areas are available which has been badly affected by mining for coal, limestone, marble and other building materials. Engineering conservation measures like construction of a series of gabion structure has rehabilitated mine-spoil area and loose rock filled check dams (Mishra *et. al.*, 1988). In between the structures, quick growing MPTS like *Bauhinia ratusa, L. leucocephala, Lanea grandis, Grewia optiva, Erythrina suberosa, Acacia species, Albizia species, Salix tetrosperma, Eucalyptus hybrid, D. cinerea, Dendrocalamus strictus, Ipomoea carnea, Agave spp. Vitex negundo, A. indica, Z. mauritiana along with grasses* (*Chrysopogon fulvus, Eulaliopsis binata, Bothriochloa pertusa, Dichanthium annulatum, Pennisetum purpureum, Arundo donax, Saccharum spontaneum*) and legumes (*Macroptilium atropurpureum, Stylosanthes hamata, Pueraria, hirsuta, P. tuberosa, P.*

labata can be successfully introduced and maximum benefit can be obtained from such areas. Mishra, *et. al.* (1988) reported that application of above technology in Mussoorie hill areas, with in a year, about 17000 tonnes of debries has been retained by the structures resulting in a saving of about Rs. 64000/- to PWD and reduction in slope from 38% to 25% was observed in three years. Due to suitable soil conservation measures and revegetation with grasses, legumes and trees, the soil loss brought down from 55 to 8 t/ha and run off reduced from 57 to 37% with in a period of 6 years (Juyal *et. al.*, 1995). In lower reach of mines water shed area, Dadhwal and Katiyar (1996) reported that plantation of *Pennisetum purpureum* produced 24.2 t/ha green forage yield in two cuts in control and yield enhanced upto 50.5 t/ha with application of 50 kg N + 50 kg P_2O_5/ha.

Swampy and Wet Lands

The extent of wetlands is more than 6 m ha apart from the permanent water bodies passing different kind of problems in our country. Marshlands and swamps are usually found in southern and eastern India. Such type of affected areas can be improved by growing of suitable MPTS like *Barringtonia acutangula, Salix tetrasperma, Bischiffia javanica, Lagerstroemia flosreginae, Eucalyptus robusta, E. rudis, Dalbergia latifolia, Glyricida maculata, Populus euphratica, Syzygium cumini.* To get higher production from these areas in addition to MPTS, suitable grasses (*Brachiaria mutica, B. decumbens, Iseilema laxum, Dichanthium caricosum, Paspalum nolatum*) and legumes (*Sesbania microcarpa, Lotononis bainesii, Glycine wightii*) to be established and 20 to 40 t/ha green forage yield can be obtained (Patil and Pathak, 1981).

Shallow Gravelly Red Soils and Rocky Areas of Semiarid Region

In our country, such type of areas occupy about 35 m ha which are found mostly in the State of U.P., M.P., Maharashtra, Karnataka, Andhra Pradesh. These areas can be developed through simple protection, eradication of unwanted bushes and plantation of suitable MPTS, grasses and legumes. It has been found that the land producing hardly upto 1 t/ha/year can be improved through silvipastoral systems to produce upto 10 t/ha/year at a 10 year rotation (Pathak *et. al.*, 1996). Besides yield improvement by 8-10 times, the quality of forage has also improved by 6-7 times MPTS evaluated in red gravelly soils under rangelands condition at NRCAF, Jhansi revealed that *A. amara, A. lebbeck, A. procera, H. binata, D. sissoo, L. leucocephala, D. strictus, D. cinerea, E. officinalis, A. nilotica sub sp. cupressiformis* were found suitable (Rat *et. al.*, 1995). Suitable grasses and legumes growing in rocky gravelly areas are found to be *Chrysopogon fulvus, Sehima nervosum, Hetropogon contortus, Stylosanthes hamata, Macroptilium atropurpureum, Atylosia scarabaeoides* etc. Rai (1998) reported that on an average (1990-1996) dry forage yield of 5.06 t/ha/year can be obtained in a silvipastoral system consisting of *A. amara* + *L. leucocephala* as tree, *D. cinerea* as shrub, *C. fulvus* as grass and *S. hamata* + *S. scabra* as legumes. Besides forage, 1.66 t/ha/year fuelwood can also be obtained through pruning of trees upto 50% of their height.

Arid Desert and Sand Dunes

About 30 m ha area under arid Zone of the hot thar desert is sandy plains, sandy hummocks and sand dunes which are mostly covered the states of Rajasthan, Gujarat and Haryana. Indian desert is one of the most thickly populated deserts of the world having a population of over 19 m people with an average density of 61 person/km^2 as against 3 person/km^2 in other deserts (Paroda *et. al.* 1980). These areas can be developed with plantation of suitable MPTS (*Prosopis cineraria, Acacia tortilis, Tecomella undelata, Colophospermum, mopane Albizia lebbeck, Hardwickia binata, Acacia senegal, Zizyphus numularia, Azadirachta indica*), grasses (*Lasiurus sindicus, Cenchrus ciliaris, C. setigerus*) and legumes (*Clitoria ternatea, Lablab purpureus, Atylosia scarabaeoides*) and the productivity of these lands may be enhanced many times with improved management practices. On an average, forage production of *C. ciliaris* when grown in association with *A. tortilis* was 0.88 t/ha while yield with *C. mopane* and *H. binata* were 0.8 to 1.0 t/ha, respectively when these trees planted at the spacing of 10 □ 10 m (Deb roy, 1993). Forage production under 14-18 years old plantation of 4 desert trees (*P. Cineraria, T. undulata, A. lebbek* and *A. senegal* was recorded maximum of 1.5 t/ha with *P. cineraria* and *T. undulata* and minimum was 0.6 t/ha with *A. senegal* (Ahuja, 1980). Recently, Sharma *et. al.* (1994) reported that the productivity of Kailana arid land can be increased 2-3 times by replacing natural grass cover with *C. ciliaris* and introduction of top feed species (*Z. numularis and Grewia tenax*) under silvipastoral system.

Cold Desert

About 10 m ha area of cold desert lies in the north of Great Himalayas. Human population in Ladakh is only one person/km^2. The native grazing lands which are found over 4500 m above sea level have only short lived species which provide 1-2 months grazing during summer. Several indigenous species of grasses (*Agropyron, Agrostis, Alopecurus, Poa, Phleum, Bromus* etc.), legumes (*Lespedeza, Medicago, Melilotus, Trifolium* etc.) and few trees such as Seabuckthorn are found in Ladakh region. Recently, a potential hay species namely *Pronges pabularis* is identified which produces 1-1.5 t/ha with 10.4 per cent protein (Misri, 1997).

REFERENCES

1. Anonymous 1996. India 1996: Publication Division, Ministry of Information and Broadcasting, Govt. of India, New Delhi: 151.
2. Ahuja, L. D. 1980. In: Khejri (*Prosopis cineraria*) in the Indian desert its role in agroforestry (Eds.: H. S. Mann and S. K. Saxena), Monograph 11, CAZRI, Jodhpur: 28-30.
3. Arora, Y. K. and Vishwanathan, M. K. 1995. In: Torrent Menace: Challanges and opportunities (Eds.: G. Sastry; V. N. Sharda; G. P. Juyal and J. S. Samra) CSWCR&TI, Dehradun: 156-163.
4. Bhattacharya, N. K. and Sharma, K. 1993. In: Agroforestry in 2000 AD for the semiarid and arid tropics. (Eds.: A. S. Gill; R. Deb Roy and A. K. Bisaria), NRCAF, Jhansi: 1-12.
5. Chaturvedi, A. N. 1990. In: Silvipastoral Systems in India. (Eds.: P. S. Pathak and P. Singh), RMSI, IGFRI, Jhansi: 23-28.
6. Dadhwal, K. S. and Katiyar, V. K. 1996. Indian Forester 122 (11): 1078-1079
7. Deb Roy, R., 1990. In: Multipurpose tree species for Agroforestry systems. (Eds.: P. S. Pathak; R. Deb Roy and P. Singh), RMSI, NRCAF, Jhansi: 9-14.
8. Deb Roy, R. 1993. In: Agroforestry in 2000 AD for semiarid and arid tropics (Eds.: A. S. Gill; R. Deb Roy and A. K. Bisaria), NRCAF, Jhansi: 10-30.
9. Juyal, G. P.; Katiyar, V. S.; Dadhwal, K. S.; Joshie, P. and Arya, R. K. 1995. In: Torrent menace: Challanges and opportunities (Eds.: G. Sastry; V. N. Sharda; G. P. Juyal and J. S. Samra), CSWCR&TI, Dehradun: 119-145.

10. Misra, C. M. 1990. In: Multipurpose tree species for Agroforestry system (Eds.: P. S. Pathak; R. Deb Roy and P. Singh), RMSI, NRCAF, Jhansi,: 133-135.
11. Mishra, A. S.; Bhatt, P. N. and Venkataramanan, C. 1988. Soil Conservation Prospects and Challanges. CSWCR&TI, Dehradun: 12.
12. Misri, B. 1997. *Pronges pabularis Lindt.* A potential hay species in Ladhak (Unpublished personal communication).
13. Paroda, R. S.; Mann, H. S. and Verma, C. M. 1980. Management of Indian Arid Rangelands. Tech. Bull. No. 4, CAZRI, Jodhpur.
14. Pathak, P. S.; Gupta, S. K. and Singh, P.1996. IGFRI approaches: Rehabilitation of degraded lands, IGFRI, Jhansi: 1-23.
15. Patil, B. D. and Pathak, P. S. 1981. International Journal of Ecology and Environmental Sciences. 7: 151-156.
16. Prajapati, M. C.; Nambiar, K. T. N.; Puri, D. N.; Singh, J. P. and Malhotra, B. M. 1993. In: Agroforestry in 2000 AD for the semiarid and arid tropics. (Eds.: A. S. Gill; R. Deb Roy and A. K. Bisaria), NRCAF, Jhansi: 33-34.
17. Rai, P. 1990. Wasteland News 6 (7): 44-46.
18. Rai, P.; Deb Roy, R. and Rao, G. R. 1995. Range Management and Agroforestry 16 (2) 103-113.
19. Rai, P. 1998. Wasteland News (in press).
20. Sharma, S. K.; Singh, R. S.; Tiwari, J. C. and Uday Burman. 1974. In: Agroforestry systems for degraded lands (Eds.: P. Singh; P. S. Pathak and M. M. Roy), RMSI, IGFRI, Jhansi, Vol. 2: 749-754.
21. Singh, G. 1995. Agroforestry Systems 29: 61-75.
22. Singh, K. A. and Rai, R. N. 1990. In: Multipurpose tree species for agroforestry systems (Eds.: P. S. Pathak; R. Deb Roy and P. Singh), RMSI, NRCAF, Jhansi,: 29-35.
23. Singh, K. A.; Balaraman, N.; Bhusan, B. and Pradhan, P. 1995. Range Mgmt. and Agroforestry 16 (2): 115-121.
24. Singhania, R. A.; Bhati, D. S.; Singh, G. D.; Sharma, H. S.; Yadav, B. L.; Kameria, N.; Vyas, K. K. and Somani, L. L. 1997. Range Mgmt. and Agroforestry 18 (1): 31-34.
25. Young, A. 1989. Agroforestry for soil conservation. Wallingford, U.K., CAB International Science and Practice of Agroforestry No. 4: 276.

Characterisation and Management of Limestone Minespoils of Outer Himalaya through MPTs and other Plant Species

K. S. DADHWAL[1], G. P. JUYAL, P. JOSHIE AND V. S. KATIYAR
Central Soil and Water Conservation Research and Training Institute, Dehradun (UP)

ABSTRACT

Studies were conducted from 1983 till date on characterisation and management of limestone minespoils of outer north-west Himalayas of U.P. Mine spoils were found mainly composed of shales, schists, chert, loose rocks and limestone mixed with soil and neutral to alkaline in reaction (pH 6.9 to 9.0), poor in nutritional status, sandy loam in texture, high in $CaCO_3$ and bulk density, low in water holding capacity and had much higher infiltration rates (almost double) as compared to adjacent normal soils. To stabilise them, measures adopted were a combination of mechanical and biological (MPTS, shrubs, grasses, climbers etc.) techniques to arrest soil, water and nutrient loss on the one hand and increasing productivity and combating environmental problems on the other. Measures comprised construction of drop structures upto 40% slope, silt detention basins upto 20% slope, construction of log wood, brushwood and loose rock filled check dams, crib structures especially on slide faces on steeply sloping areas, toe and retaining walls, terracing, contour trenching, contour wattling, mulching, normal soil placement and mixing in trenches and pits, fertility management through leaf manures, farm yard manure and fertilisers, direct seeding, pitting and planting of suitable MPTS and other plant

[1] *Present address: National Research Centre for Agroforestry, Jhansi-284 003 (UP), India*

species etc. Apart from this, measures like use of geo-jute, netlon, geogrids, excelsior matting, mulching with native plant material etc. proved effective for steep slopes (30- 70%) stabilisation. The MPTS and other plant species identified and performed best based on experimental findings are *Leucaena leucocephala, Salix tetrasperma, Eucalyptus hybrid, Ficus spp., Erythrina suberosa, Bauhinia retusa, Acacia catechu, Grewia optiva, Sapium insignii, Bauhinia veriegate, Toona ciliata, Lannea grandis, Ipomoea carnea, Vitex negundo, Agave americana, Eulaliopsis binata, Pennisetum purpureum, Saccharum spontaneum, Arundo donax, Pueraria hirsuta, Apluda mutica, Eriophorum cosmosum, Thysanolaena maxima, Adhatoda species, Wood fordia fruiticosa,* etc.

With continuous biotic protection together with rehabilitation measures and protection against grazing and cutting of grasses and MPTS the retrogression and erosion has come to stand still. The vegetal cover has increased from 10% to > 75% over a period of 12 years. MPTS and other plant species over time have helped in improvement of spoil characteristics and water quality. Treated mine sites have now started generating additional resources of water, firewood, fodder, fibre, etc. on sustainable basis and have improved the economy of the area. Environment has improved and there is a perennial flow of water in the treated area.

INTRODUCTION

Since 1960's there has been rapid increase in the mining activity in the Himalayan region of India. Mining has been carried out indiscriminately on steep and fragile slopes without consideration for ecological and environmental protection. It is the unscientific mining which has done immense damage to the fragile ecosystem in the form of heavy soil erosion, land degradation, drying of water resources, disruption of communications, floods and consequent decline in flood and milk production in the area (Juyal et.al., 1995). Integrated land use on watershed basis is the only solution to rehabilitate the degraded abandoned mine areas of the region. Establishment of permanent vegetative cover of suitable MPTS, shrubs and grasses is an option and proposition of great promise for the rehabilitation and management of such lands in view of combating soil, water and nutrients losses on one hand and meeting the growing demand of food, firewood, fodder, fibre, environmental restoration on the other.

Keeping this in view the research efforts were attempted from 1983 till date at the Central Soil and Water Conservation Research and Training Institute, Dehradun, which have resulted in the development of appropriate technologies for the improvement of such degraded lands and biomass production. A brief account of the technologies generated and measures adopted alongwith their impact is reported in this paper.

The present paper reports on (I) Characteristics of degraded mine lands, (ii) Land use, (iii) Techniques/measures adopted for raising MPTS and other plant species for stabilisation, (iv) Impact of measures adopted, (v) Cost and (vi) Socio-economic impact of rehabilitation technologies on rural development.

CHARACTERISATION OF MINESPOILS

To define limitations for vegetation establishment, characterisation of minewastes for their geo and physico-chemical characteristics is very important. Improvement of fertility and other characteristics greatly help in quick establishment of plant species in the degraded minesites (Dadhwal *et.al.* 1988).

Geo-characteristics

Geo-characteristics of the minespoils depend upon the composition and nature of the minerals present in the material. Dadhwal et.al. (1991 c), studied some geo-characteristics of the limestone minespoils from outer Himalayas of Uttar Pradesh, found that the consistency of the debris/spoil was highly variable from loose to hard and compact at different places. Older deposits were hard and compact whereas fresh deposits were loose. Spoil material was found mainly composed of shales, chert, loose rocks, schists and of limestone mixed with soil (Table 1).

TABLE 1: *Geo-properties of limestone minespoil from Mussoorie hills*

Block	No. of Sites	Slope (%)	Compactness	Geology
Bandal	*7*	*5-25*	*Hard, mildly hard, mildly loose.*	*Shales, chert, limestone mixed with soil.*
Chamasari	*35*	*5-25*	*Overburden dump, loose, old compact.*	*Loose rock, shales, schists, limestone mixed with soil.*

Block	No. of Sites	Slope (%)	Compactness	Geology
Rajpur	*4*	*2-10*	*Compact and loose, old compact.*	*Shales and limestone mixed with soil.*
Rikholi	*6*	*5-15*	*Mildly loose, mildly compact, recent loose.*	*Marble, limestone mixed with soil.*

Earlier Negi (1982) also described the geology of limestone minespoil from outer Himalayas of U.P. derived mainly from the units comprising limestone sequences (Krol formation) and shales (Tal formation).

Physico-chemical Characteristics

Accurate determination of particle size distribution is essential in planning vegetation programme because texture is used extensively as a guide to evaluate soil-water storage, water availability, surface erodibility and land stability. Therefore, to know the composition of different particle sizes, minespoil contained in a 1 m^3 pit at Sahastradhera (Dehradun) was passed through different mesh sieves. It was observed that only 4% of the material passed through 2 mm sieve and 60% of the material is larger than 16 mm sieve (Dadhwal, 1984).

The analysis of 52 limestone minespoil samples from 20 minesites of outer Himalayas of Uttar Pradesh for important physico-chemical characteristics is given in Table 2 (Dadhwal et.al., 1992). The analysis reveals that minespoil material (< 2 mm) is sandy loam in texture, alkaline in reaction, poor in fertility parameters, highly calcareous, high in bulk density and low in water holding capacity. Due to the presence of high gravel content and low fine material, the infiltration rate of the spoil is high (almost double as compared to adjacent normal soil) and water-holding capacity low (Dadhwal et.al., 1989).

LAND USE

Before Mining

There was luxuriant vegetation of tropical dry mixed deciduous forest type on account of high rainfall and favourable edaphic factors before mining (Champion and Seth, 1968). Dry mixed deciduous forests beared the following species with thin grass cover forming a three tier system (Joshie et.al., 1995):

- ☐ **Upper canopy of tree species (3-8 m height):** *Acacia catechu, Bauhinia retusa, Bauhinia purpurea, Cassia fistula, Erythrina suberosa, Lannea coromandelica, Mallotus philipinensis, Emblica officinalis, Sapium insignii, Toona ciliata, Bombax ceiba, Ficus roxburghii, Dalbergia ericea, Wendlandia exserta, Phoenix acaulis*, etc.
- ☐ **Middle canopy shrubs:** *Coculus laurifolius, Vitex negundo, Murraya koengii, Woodfordia fruiticosa, Mimosa himalayana, Adhatoda zeylanica, Rumex nepalensis, Euphorbia royaleana, Lantana camara, Solanum hispidum*, etc. The climbers are: *Pueraria tubersosa, Bauhinia vahlii, Acacia pinnata*, etc.
- ☐ **Ground flora of grasses and herbaceous weeds:** Grasses belong to Chrysopogon - Eulaliopsis association of Themeda - Arundinella grassland cover type with species of *Apluda mutica, Chrysopogon fulvus, Cymbopogon nardus, Heteropogon contortus, Sachharum spontaneum, Eriophorum cosmosum, Themeda anathera*, etc. Herbaceus weeds are *Ageratum conyzoides, Cassia tora, C. occidentalis, Fumaria parviflora, Eupatorium odoratum*, etc.

After Mining

With biotic interference and accelerated soil erosion four degraded stages has been recognised.

- ☐ **5B/C_2-DS_1:** It is a low broken soil cover of shruby growth, 3-6 m high, malformed due to biotic interference. The grass cover occurs throughout. Dry deciduous shrub species are: *Woodfordia fruiticosa, Carissa opaca, Mallotus philipinensis, Acacia catechu, Mimosa himalayana, Euphorbia royaleana, Lannea coromandelica, Toona ciliata, Bauhinia retusa, Bauhinia vahlii, Erythrina suberosa, Bombax ceiba and Rumex nepalensis.*
- ☐ **5B/C_2-DS_2 - Dry savannah:** Trees stand far apart singly or in groups or patches with more or less heavy grass. *Phoenix acaulis, Bombax ceiba* and *Toona ciliata* are characteristic species of this type.
- ☐ **5B/C_2-DS_3 - *Euphoribia* scrub**: This shows extreme degraded stages of erosion with exposure of rocks. Important species are

Euphorbia royaleana, Woodfordia fruiticosa, Eriophorum cosmosum and *Adhatoda zeylinica. Lanatana camara* invades secondarily.

- **5B/C_2-DS_4 - Grassland:** *Aristida cynantha, Saccharum spontaneum, Heteropogon contortus, Cymbopogon nardus and Eriophorum cosmosum.*

Under rocky phase conditions *Saccharum spontaneum, Carissa opaca and Woodfordia fruiticosa* reflect degraded condition of DS_3 and DS_4 reveals that the job of revegetation of minespoils with MPTS, Shrubs and grasses is very difficult and require suitable techniques and dedicated efforts.

MEASURES

Mechanical

In highly degraded mined areas, where establishment of vegetation is very difficult due to moisture stress, absence of fertile soil, movement of minespoil and rocky material on steep slopes, mechanical measures are essential for controlling erosion, conserving soil, water and nutrients and helping the vegetation to establish (Dhruvanarayana et.al., 1987; Dadhwal and Katiyar, 1990). Studies conducted by CSWCR&TI, Dehradun reveal that structures like cross barriers/check dams, log wood barriers, toe walls, gabion structures, loose check dams, etc., arrest the movement of large amount of loose rocks and spoil down the unstable and uneven steep slopes of the area and also through the drainage channels. Based on the research and experience the prominent mechanical measures to be adopted consist of planned placement of over burden in hilly terrain, provision of road drains and culverts, proper designing of water courses, construction of trenches on denuded long slopes, terracing of minespoil deposited in heaps, construction of gabion drop structures up to 40% slope, construction of toe and retaining walls, chute spill way with cement concrete where the vertical fall is more than 3 m etc. These stated measures have been found very useful in grade stabilisation, retention of spoil particles, reduction in slope, retention of run off and establishment of vegetation (Dhruvanarayana et.al., 1987; Katiyar et.al., 1988; Dadhwal and Katiyar, 1990 and Dadhwal, 1994).

Biological

It is necessary to put the land to productive use (agriculture, agroforestry, forestry or recreation) and to maintain aesthetic beauty and visual impact. Therefore, biological measures with proper scientific techniques must be undertaken alone or with mechanical measures. Though biological measures are cheaper than to mechanical but in certain situations they have to be adopted simultaneously depending upon the site and physiographic conditions. Techniques adopted must have positive effect on spoil properties and plant growth. The selection of MPTS and other plant species should be proper keeping in view the economical and ecological considerations (Dadhwal et.al., 1988). Selected plant species should provide quick and immediate carpet and for this grasses are the best proposition initially. The selected MPTS, shrubs and grasses should preferably be native of the area, lime loving and leguminous with better root soil binding characteristics. The measures emerged from research and experience area as under:

TABLE 2: *Characteristics of limestone minespoil*

Characteristic	**Mean of 52 samples from 20 minesites**
pH (1:2.5 soil:water susp.)	8.0
EC (mmhos cm^{-1})	0.13
Organic Carbon (%)	0.23
$CaCO_3$ (%)	61.60
Total N (%)	0.02
Available P_2O_5 (Kg ha^{-1})	4.04
Available K_2O (Kg ha^{-1})	56.40
Mechanical Composition (%)	
Sand	67.4
Silt	18.3
Clay	14.3
Undispersed silt + clay (%)	8.2
Bulk density (Mg m^{-3})	1.69
Water holding capacity (%)	16.6
Dispersion ratio	27.9

1. Since minespoils are very poor in fertility status (Table 2), their fertility management require due attention for quick establishment of vegetation. Studies reveal that 50 kg N ha^{-1} and 20 Mg FYM ha^{-1}

are the economic and ideal dose for *Eulaliopsis binata* in minespoil (Dadhwal and Katiyar, 1989; Dadhwal et.al., 1991 a). Similarly, for the establishment of MPTS, application of dimmonium phosphate @ 10 g tree^{-1} has given encouraging results.

2. Leaf litter/mulches are very useful in moisture retention and fertility build up. Application of leaf litter chopped @ 20 Mg ha^{-1} of the nearby existing plant species like *Sapium insignii, Wild kudzu, Erythrina suberosa*, grasses, etc. has given better growth of *Eulaliopsis binata* in inaccessible degraded steep sites (Dadhwal et.al., 1993a).

3. For establishment and management of MPTS and other vegetation, placing normal soil of the unaffected adjacent sites up to a depth of 30 cm in the first year was found effective for *Leucaena leucocephala, Bauhinia retusa, Eulaliopsis binata* and *Chrysopogon fulvus* whereas in the second year mixing up to the same depth was found better over placing (Dadhwal et.al., 1991b; Dadhwal et.al., 1993b). In another study mixing normal soil and minespoil (1:1) proved better for the establishment of *Ipomoea carnea and Agave americana* (Dadhwal and Singh, 1989). Similarly, for *Eucalyptus hybrid, Grewia optiva, Bauhinia retusa, Bombax ceiba* and *Leucaena leucocephala* better heights, basal diameters and biomass yields were recorded in 1:1 mixture (Dadhwal and Singh, 1993b). Therefore, these studies suggest that normal soil available nearby can be used in pits or holes for better establishment and management of MPTS and other plant species.

4. Select preferably leguminous and MPT Species for such affected sites which may add N. For example, *Leucaena leucocephala and Pueraria hirsuta* when planted in minespoil at Sahastradhera gave enough foliage (having 2-3% N). This foliage is now being used as an organic manure/mulch for other MPTS and vegetation planted in minespoil area. Apart from organic manure *Leucaena leucocephala* also provides fuel and fodder.

5. Techniques to be adopted for minespoil degraded sites depend mainly upon site and physiographic conditions in the hilly terrain. Direct seeding, contour wattling, mulching, planting slips, stumps, cuttings and bulbils, pit planting, contour trenching and planting, sowing and planting in holes etc. are the measures recommended for the minesites of Mussorie hills (Dhyani et.al., 1988). In highly

steep lands spraying over the dumps with seeds mixed with cowdung, hay, and soil made into slurry is suggested (Dadhwal and Katiyar, 1990).

6. Different geotextiles *viz.* Geojute, Netlon, Geo-cell, Geo-grid, Excelsior matting, Wire mesh (in 2 □ 4 m strips) proved effective to arrest soil and to support vegetation on steep slopes (30-70%). Plant species like *Thysanolaena maxima, Saccharum spontaneum, Pennisetum purpureum, Eulaliopsis binata, Ipomoea carnea* and *Vitex negundo* performed well (Juyal and Dadhwal, 1996).

7. The length and thickness of plant roots in binding the soil particles is very important. Studies conducted in the minespoil of Mussoorie hills recorded highest soil binding with *Leucaena leucocephala* (502), *Agave americana* (664) and *Pennisetum purpureum* (1736) among MPTS, shrub and grass species, respectively. Total root length was also recorded highest in the said species (Dadhwal and Singh, 1993a).

8. Different MPTS, shrub and grass species evaluated for their suitability in the minespoils of the outer Himalayas are: *Leucaena leucocephala, Grewia optiva, Bauhinia retusa, Bombax ceiba, Acacia catechu, Erythrina suberosa, Bauhinia variegata, Salix tetrasperma, Sapium insignii, Ficus rumphii, Melia azadirach, Grevillea robusta, Dalbergia sissoo, Syzygium cuminii, Toona ciliata, Cassia fistula, Mallotus phillipinensis, Dodonea viscosa, Populus deltoides, Dendrocalamus strictus, Callistemon viminalis, Lannea grandis, Albizia procera, Terminalia belerica, Psidium gujava, Citrus* species and *Mangifera indica*. The best performed MPTS based on experimental findings are: *Leucaena leucocephala, Salix tetrasperma, Eucalyptus hybrid, Erythrina suberosa, Bauhinia* spp. etc. and among shrub and grass species are *Ipomea carnea, Vitex negundo, Agava americana, Eulaliopsis binata, Chrysopogon fulvus, Pennisetum purpureum, Saccharum spontaneum, Arundo donax* etc. Other species which are coming nicely in degraded sites are *Pueraria hursuta, Apluda mutica, Eriophorum cosmosum, Thysanolaena maxima, Woodfordia fruiticosa* and *Adhatoda* spp.

IMPACT OF MEASURES

Impact of MPTS, Shrubs and Grasses

With continued biotic protection together with rehabilitation measures and protection against grazing and cutting of grasses and MPTS the retrogression and erosion has come to stand still. The vegetal cover increased from 10% to > 75% due to reforestation over a period of 12 years (Dadhwal et.al., 1997). The oven dry grass yield in geojute area with bouldery fragmental soil after mulching was 394 kg ha^{-1} in the upper reach, 640 kg ha^{-1} in the middle and 1960 kg ha^{-1} in the lower middle reach (Joshie et.al., 1995). Measures which performed best for the establishment and management of MPTS and other plant species are given in Table 3 (Dadhwal et.al., 1998).

TABLE 3: *Measures for rehabilitation of minespoils*

Reaches in hilly terrain	Measures adopted	Planting technique	MPTS, Shrubs and Grasses which performed best
Upper	** Holes (30 cm deep) and small trenches,* *** Contour trenching (0.7 m wide □ 0.7 m deep).*	*Sowing seeds, planting with slips, stumps and cuttings, pit planting with addition of good soil, application of manure/leaf litter/fertiliser.*	** Ipomoea carnea, Arundo donax,* *** Vitex negundo, Ipomoea carnea, Eualaliopsis binata, Pennisetum purpureum.*
Middle	** Direct seeding on moderate slopes, direct planting of grasses, seeds shurry.* *** Holes in debris (30-60 cm), pits, trenches and spreading with ball earth on moderate slopes, mixing and placing of normal soil with debris in pits.*	*Seeds slurry prepared with cowdung and hay in precipitous slopes, direct seeding and planting of grass slips in slide face, stump in trenches, slips cutting, etc., application of mulch, manure/fertiliser, leaf litter, etc.*	** Arundo donax, Pennisetum, Ipomoea cernea, Leucaena spp., Grewia optiva, Eucalyptus spp.* *** Pennisetum, Arundo donax, Ipomoea carnea, Grewia optiva, Bauhinias, Eulaliopsis binata, Chrysopogon fulvus, Agave spp., etc.*
Lower	** Holes, pits, trenches.* *** Holes, pits,*	*Planting of root slips and cutting at places, sowing of seeds, stump planting,*	** Ipomoea carnea, L. leucocephala, P. purpureum, E. binata.*

Reaches in hilly terrain	Measures adopted	Planting technique	MPTS, Shrubs and Grasses which performed best
	trenches, normal soil placement and mixing, checkdams retaining walls, etc.	*vegetative check dams, planting seedlings on the ridge and stream bank, application of leaf litter, manure/fertiliser, etc.*	*** Bauhinia spp., Dendrocalamus, Ficus spp., Salix tetrasperma, Arundo donax, E. binata, Erythrina suberosa, Saccherum spp., Agave spp., etc.*

Impact on Spoil Characteristics

Impact of rehabilitation measures on some important characteristics over time was seen in treated watershed. On an average, after 12 years, pH of the spoil came down by 0.6 units (8.1 to 7.5), organic carbon increased from 0.13 to 0.42%, whereas $CaCO_3$ (%) decreased from 54.6 to 31.0 and bulk density (Mgm^{-3}) from 1.63 to 1.47 which shows that planting of MPTS, shrubs and grasses over time have helped in improvement of minespoil characteristics.

Impact on Water Quality

Runoff samples collected from treated and untreated mined areas were analysed for calcium, magnesium, sulphates and bi-carbonates (Table 4). Runoff water from treated minesite was found within the permissible water quality limit (Juyal et.al. 1993).

TABLE 4: *Water quality parameters (ppm)*

Site	Ca	Mg	SO_4	HCO_3
Treated oldmine	74	34	138	131
Untreated oldmine	188	39	240	98
Untreated fresh mine	389	120	756	77

COST AND SOCIO-ECONOMIC IMPACT

Mined area rehabilitation cost Rs.15,000 ha^{-1} which also includes expenditure on research (hydrological monitoring, surveys, etc.). But the average cost of plantation of MPTS, shrubs and grasses varied from Rs.5000 to Rs.10,000 ha^{-1} depending upon the location. The treated minespoil sites has now started generating additional resources of water for irrigation, fuel, fodder, fibre, timber, etc. (for the use of farmers) on sustainable basis and have improved the socio-economic status of the area.

ACKNOWLEDGEMENTS

Authors are grateful to the Director of CSWCR&TI, Dehradun for guidance and encouragement during the course of study. Technical help rendered by Mr. R. K. Arya and Mr. Devendra Singh is also acknowledged. Help rendered by Sh. Deepak Vij in typing the manuscript is also acknowledged.

REFERENCES

1. Champion, H. G. and Seth, S. K. 1968. Revised survey of forest types of India. Govt. of India Publication: 200-209.
2. Dadhwal, K. S. 1984. Soil Conserv. Newsletter, 3 (2): 4.
3. Dadhwal, K. S. 1994. Agroforestry systems for degraded lands (Eds.: Panjab Singh, P. S. Pathak and M. M. Roy), Vol. I, Oxford & IBH Pub. Co. Pvt. Ltd., New Delhi, 63-69.
4. Dadhwal, K. S. and Katiyar, V. S. 1989. J. Indian Soc. Soil Sci., 37 (2):412-415.
5. Dadhwal, K. S. and Katiyar, V. S. 1990. Himalaya (Man and Nature), 13 (9): 14-18.
6. Dadhwal, K. S. and Singh, B. 1989. Indian J. Soil Cons., 17 (3): 45-50.
7. Dadhwal, K. S. and Singh, B. 1993a. Indian Forester. 119 (1): 71-74.
8. Dadhwal, K. S. and Singh, B. 1993b. J. Indian Soc. Soil Sci., 41 (4): 738-744.
9. Dadhwal, K. S.; Joshie, P. and Juyal, G. P. 1997. Indian J. Soil Cons., 25 (1): 33-40.
10. Dadhwal, K. S.; Juyal, G. P.; Katiyar, V. S. and Joshie, P. 1998. Soil and water conservation - challenges and opportunities (Eds.: L. S. Bhushan, I. P. Abrol and M. S. R. Rao), Vol.2; 8th ISCO Conf., New Delhi, pp.1345-1354.
11. Dadhwal, K. S.; Singh, B.; Katiyar, V. S. and Dwivedi, V. K. 1991c. Indian J. Soil Cons., 19 (3): 79-82.
12. Dadhwal, K. S.; Narain, P.; Singh, B. and Singh, R. 1989. J. Indian Soc. Soil Sci., 37 (2): 223-228.
13. Dadhwal, K. S.; Singh, B. and Narain, P. 1991a. J. Indian Soc. Soil Sci., 39 (4): 786-788.
14. Dadhwal, K. S.; Singh, B. and Narain, P. 1991b. J. Indian Soc. Soil Sci., 39 (4): 751-756.

15. Dadhwal, K. S.; Singh, B. and Narain, P. 1992. Indian Forester, 118 (9): 650-658.
16. Dadhwal, K. S.; Singh, B and Narain, P. 1993a. Indian J. Soil Cons., 21 (2): 52-57.
17. Dadhwal, K. S.; Singh, B. and Narain, P. 1993b. Indian Forester, 119 (12): 1011-1019.
18. Dadhwal, K. S.; Singh, B.; Narain, P. and Katiyar, V. S. 1988. Indian J. Soil Cons., 16 (2): 51-62.
19. Dhruvanarayana, V. V.; Katiyar, V. S. and Dadhwal, K. S. 1987. Indian J. Soil Cons., 15 (2): 82-90.
20. Dhyani, S. K.; Dadhwal, K. S. and Katiyar, V. S. 1988. Indian Farming, 38 (7): 15-18.
21. Joshie, P.; Dadhwal, K. S.; Juyal, G. P. and Katiyar, V. S. 1995. Adv. For. Res. India (Ed.) Ram Prakash, IBD, Dehradun, Vol. XII: 65-95.
22. Juyal, G. P. and Dadhwal, K. S. 1996. Indian J. Soil Cons., 24 (3): 179-186.
23. Juyal, G. P.; Katiyar, V. S.; Dadhwal, K. S. and Joshie, P. 1993. Wastelands News, 9 (1): 57-61.
24. Juyal, G. P.; Katiyar, V. S.; Dadhwal, K. S.; Joshie, P. and Arya, R. K. 1995. Sustainable Reconstruction of Highland and Headwater Regions. (Eds.) R. B. Singh and Martin J. Haigh, Proc. 3rd International Symp. On Headwater Control, Oxford & IBH Pub. Co. Pvt. Ltd., New Delhi: 441-450.
25. Katiyar, V. S.; Dadhwal, K. S. and Dhyani, S. K. 1988. Wastelands development and their utilisation (Ed.) K. A. Shankar Narayan, Scientific Publishers, Jodhpur, India: 226-236.
26. Negi, S. S. 1982. Environmental Problems in the Himalaya. Bishan Singh, Mahendra Pal
27. Singh, Dehradun, India, 188 p.

Management of MPTs for Salt Affected Soils in Agroforestry System

GURBACHAN SINGH
Central Soil Salinity Research Institute, Karnal (Haryana)

ABSTRACT

Large tracts of salt-affected soils in India are confined to villages as community lands, as public properties near cities, along the railway tracks and roads, and government land reserved for specific purposes. Establishment of permanent vegetative cover of suitable trees and herbaceous crops is an option of great promise for the use of such lands in view of the growing demand for fuelwood and fodder as well as environmental considerations. Research efforts of more than two decades at the Central Soil Salinity Research Institute, Karnal have resulted in the development of appropriate technologies for biomass production and improvement of such wasted lands for village development. A brief account of these technologies alongwith their socio-economic impact on rural inhabitants are reported in this paper. The present paper reports on (i) biomass production potential of promising forest, fruit and herbaceous species in salt lands (ii) agro-techniques and cultural practices for raising trees (iii) a silvo-pastoral model for development of Panchayat lands (iv) agroforestry models for moderately alkali or reclaimed lands (v) amelioration of alkali soils by tree plantations for growing crops (silvi-agri model) (vi) economics of afforestation and (vii) socio-economic impact of biomass production technologies on rural development.

INTRODUCTION

Recent surveys have indicated that good forests with more than 30 percent cover are hardly 11 percent of the geographical area of the country. Continued deforestation at an alarming rate in the recent past has created a wide gap between the supply and demand of fuelwood, industrial wood, timber and other forest products. The gap in fuelwood supply and demand is being filled with alternative energy sources, such as animal dung and agricultural wastes. It has been reported that about 73 million tonnes of animal dung and 41 million tonnes of agricultural wastes are burnt as substitutes for firewood in India. The projected fuelwood demand is 225 million m^3 by 2000 AD. This target can be achieved by increasing either the area under fuelwood plantations or the productivity per unit area. There seems to be little scope for extending area under fuelwood plantations because of preference for productive lands for agricultural crops as well as encroachment with the ever increasing urbanization. Thus, the only possibility to increase area under fuelwood plantations is to extend their cultivation on wasted lands constrained by the stresses of aridity, salinity, extremes of pH etc. Wastelands are constituted to cover nearly 93.6 million ha. Out of this, nearly 8.5 million ha are salt affected.

RESULTS AND DISCUSSION

Biomass Production Potential of Forest, Fruit and Herbaceous Species in Salt Lands

The most promising forest species identified for planting in salt-affected soils are: Valayati Babul (*Prosopis juliflora*), Babul (*Acacia nilotica*), Faras (*Tamarix articulata*), Casuarina (*Casuarina equisetifolia*), Arjun (*Terminalia arjuna*) and Kanji (*Pongamia pinnata*). Relative growth and biomass yields of selected tree species 10 years after planting are presented in Table 1. The fruit species which can be grown in alkali soils are: Ber (*Zizyphus mauritiana*), Guava (*Psidium guajava*), Jamun (*Syzygium cuminii*), Imli (*Tamarindus indica*), pomegranate (*Punica granatum*), Karaunda (*Carrissa carandus*) and Aonla (*Emblica officinalis*).

Suitable grasses which can be grown in alkali soils without amendments in order of salt tolerance and biomass production are: *Leptochloa fusca, Chloris gayana, Bricharia mutica, Cynodon dactylon* and *Panicum laevifolium*. Certain *Atriplex* species such as *Atriplex nummularia* are reported to perform better at high pH than in normal soils. Babuna or German Camomil *(Matricaria cammomila)* and Isabgol *(Plantiago ovata)*, aromatic and medicinal high value crops can also be cultivated in high pH soils. More recently, fruit (1271, 1280 and 1287), fodder (1270) and vegetable (1308) clones of Cactus (*Opuntia ficus* Indica) have been introduced, screened for sodicity tolerance and multiplied for exploiting salt-lands of arid and semi-arid regions. These edible clones can be cultivated for biomass production with little or no inputs of water and nutrients in arid environments. The fruits of cactus commonly known as *tuna* are edible and can be eaten raw. It can also be used in the preparation of juices, jams, pulps, *salad* etc. It is considered excellent for production of biomass and ethanol.

TABLE 1: *Growth performance and biomass production by 10 years old salt tolerant trees in a soil of pH 10.4*

Growth Parameters	Tree plantations			
	P. juliflora	*A. nilotica*	*C. equisetifolia*	*E. tereticornis*
Height (m)	12.9	11.6	14.5	14.9
DSH (cm)	15.9	15.4	15.0	13.6
DBH (cm)	12.5	13.6	12.0	11.0
Bole weight (kg/tree)	112.6	85.4	84.2	65.6
Branches + Leaves weight (kg/tree)	43.2	43.8	28.4	

Scope of genus *Prosopis*: As far as biomass production potential in waste lands is concerned no other tree can be compared with *Prosopis juliflora*. In many developed countries including USA, its wood is commercially used for making furniture, flooring, doors, sports goods and craft articles etc. In Brazil its pods are being used commercially for making biscuits, ice-cream powder etc. for human consumption. Seeing its potential for biomass production, we introduced large germplasm from Texas, USA and HDDRA, UK including thornless species and established the seed orchards. Efforts are in progress to improve its plant type for better bole, thornless and compact canopy architecture through selection and asexual propagation. There is an urgent need to create awareness among the growers, wood workers and users about the potential uses of this otherwise neglected woody genus in India.

Agro-techniques/Cultural Practices for Raising Trees

For planting forest trees in alkali soils, augerbores of 20 to 25 cm diameter and 120 to 140 cm deep are made using tractor operated diggers. These holes are filled with a mixture of original alkali soil plus 3 to 4 kg gypsum and 8 kg FYM before planting. For planting fruit trees, a pit of 45^3 dimension is dug manually and in the centre of this pit augerbores of 160 to 180 cm deep are made to facilitate rooting both in horizontal and vertical direction. These pits cum augerbores are filled back with a mixture of original soil + 7.5 kg gypsum + 20 kg FYM + 20 kg river sand. Limited irrigation during first two years, results in two fold increase in growth and biomass production of selected trees. For firewood production, 2 □ 2 m spacing between rows and plants was found optimum; whereas for agroforestry 4 □ 4 m spacing proved better. Pruning of side branches of trees during first 4 years helps the trees to grow faster and improves yield of inter-planted grasses/inter-crops.

A Silvo-pastoral Model for Development of Community Lands

A silvo-pastoral model comprising of *Prosopis juliflora* and *Leptochloa fusca* has been found most promising in terms of firewood and forage production and also for amelioration of alkali soils. *Leptochloa* in association with *Prosopis* produced 46.5 tonnes green fodder/ha in 15 cuttings over a 50 months period without any fertilizer or any other soil amendments. Similarly, *Prosopis* produced about 40 tonnes air dried firewood/ha in 6 years. This silvo-pastoral system improves the soil to such an extent that less salt tolerant and more palatable fodder species such as *Trifolium resupinatum*, *Trifolium alexandrium* and *Melilotus parviflora* can be grown after ploughing under of grass after 4 years. These crops give almost equal yield to their yield levels in alkali soils reclaimed by gypsum.

Agroforestry Models for Moderately Alkali or Reclaimed Lands

Most promising trees for agroforestry on reclaimed alkali lands identified are: poplar *(Populus deltoides)*, Safeda *(Eucalyptus tereticornis)*, Babul *(Acacia nilotica)* and Teak (*Tectona grandis*). The suitable inter-crops during establishment stage are rice-berseem and rice-wheat. However, from third year onwards, as tree canopies develop, shade loving species such as turmeric and colocasia perform better than the first group. These crops help the trees grow faster.

Populus based agroforestry is more remunerative because of its faster growth and high wood price in the market compared to *Eucalyptus* and *Acacia* based systems. However, poplar being sensitive to salt stress, its cultivation should be limited to the soils having profile pH less than 9.0. In addition to better economic returns, agroforestry practices result in better fertility and organic matter build up in the soil than agricultural practices.

Amelioration of Alkali Soils by Tree Plantations for Growing Crops (Silvi-agri. Model)

In addition to biomass production trees growth on alkali soils helps in their amelioration by improving physical, chemical and biological properties of the soils. Nitrogen fixing trees ameliorate alkali soils at a much faster rate than non-N fixing trees. The inclusion of salt tolerant grasses further hastens the reclamation process. Ameliorating effects of 20 years old tree plantations of *Prosopis juliflora, Acacia nilotica, Eucalyptus tereticornis, Albizzia lebbeck* and *Terminalia arjuna* are reported in Table 2.

TABLE 2: *Ameliorating effects of tree plantations on an alkali soil*

Species	Original		After 20 years	
	pH	Organic carbon (%)	pH	Organic carbon (%)
A. nilotica	10.3	0.12	9.03	0.55
A. lebbeck	10.3	0.12	8.67	0.47
T. arjuna	10.3	0.12	8.15	0.58
P. juliflora	10.3	0.12	8.03	0.58

The ameliorating effect on the soil appeared to be in the order: *Prosopis* > *Acacia* > *Terminalia* > *Albizzia* > *Eucalyptus*. To know whether agricultural crops like wheat and oats can be grown in alkali soils reclaimed through trees, an experiment was conducted in pots after taking soil under the canopies of above trees. Soil was also taken from the adjacent crop land field where alkali soil was reclaimed through amendments. Ten seeds each of wheat and oats were sown in a randomized way in each pot and no fertilizers were used. Mean grain and straw yields of both crops were significantly higher in the soil reclaimed through trees than through the application of gypsum (Table 3). Further, the yields of both the crops were highest in the soil taken under the *Prosopis* canopies followed closely in the soil under

Acacia canopies. The yields of both crops were in the other: *Prosopis* > *Acacia* > *Albizzia* > *Terminalia* > Cropland.

TABLE 3: *Straw and grain yields of wheat and oats in an alkali soil reclaimed through tree plantations and though gypsum application*

Tree plantations	Wheat		Oats	
	Grain Yield	Straw Yield	Grain Yield	Straw Yield
P. juliflora	61.7	87.5	87.9	111.1
T. arjuna	44.0	38.5	45.8	62.8
E. tereticornis	32.3	25.3	42.7	58.5
A. lebbeck	45.3	43.5	52.8	66.9
A. nilotica	55.7	68.8	61.6	67.5
Crop land	13.3	15.4	24.3	26.7
LSD (0.05)	2.8	2.0	7.0	9.4

Under the World Bank funded Social Forestry Project in Haryana, the State Forest Department has raised *Prosopis juliflora* plantation in a number of villages in abandoned high pH soils between 1982-1990. To study ameliorating role of these plantations on properties of alkali soils, samples were collected under 8 years old trees from nearly 100 ha area located in four villages namely Shera, Sutana, Nain, and Bhalsi of Panipat district. The data related to original pH and organic carbon status and changes occurred in 8 years is reported in Table 4.

TABLE 4: *Reclaiming effects of Prosopis juliflora plantations recorded on village community lands*

Name ofVillages	Soil Properties				
	Depth (cm)	pH		Organic Carbon (%)	
		0*	8*	0*	8*
Shera	0-15	10.0	8.8	0.13	0.48
	15-30	10.1	9.6	0.06	0.21
Sutana	0-15	10.2	8.5	0.20	0.40
	15-30	10.6	9.4	0.07	0.09
Nain	0-15	9.7	8.6	0.08	0.45
	15-30	10.2	9.2	0.10	0.12
Bhalsi	0-15	10.0	8.8	0.22	0.65
	15-30	10.5	10.1	0.08	0.12

*0 and 8 refer to original and 8 years after planting

Economics of Afforestation

Rough estimates were made about the likely benefits of raising *Prosopis* plantation in salt lands. The major expenditure for establishing mesquite plantation on salt-affected soils is on site preparation. Expenditure on site development is more in alkali than in saline soils. In alkali soil, site preparation involves land leveling, making strong bunds around the periphery for in-situ rain water conservation, digging of augerholes, cost of amendments and providing for large number of irrigations. In saline soils, the preparation involves digging of channels only for planting mesquite and no expenditure is involved in making augerbores and amendment application. Planting one hectare of an alkali land through the augerhole technique, at a planting density of 1250 plants/ha, will approximately cost Rs.12600 (based on 1990 prices). The net income/ha/annum, based upon 8 years rotation works out to be Rs. 8175 for alkali soils and Rs.3507 for saline soils (Table 5).

TABLE 5: *Economics of raising Prosopis plantations in salty soils*

Expenditure/Income	Alkali soils	Saline soils
Total expenditure in 8 years/ha	*Rs. 32000*	*Rs. 22400*
Gross income after 8 years/ha	*Rs. 98000*	*Rs. 51000*
Net income/ha/year	*Rs. 8175*	*Rs. 3587*

SOCIO-ECONOMIC IMPACTS OF BIOMASS PRODUCTION TECHNOLOGY ON RURAL DEVELOPMENT

Meeting Fuelwood and Fodder Demand

A large chunk of salt lands in the country, mostly the village community lands have been revegetated in the recent past. Earlier, these community lands were lying barren and were not contributing anyway to the village economy. Most of these lands are now supporting good plantations of *Prosopis* and *Prosopis* plus grasses. Amongst the village communities, most beneficiaries of this land development programme are the small and marginal farmers and landless labourers. Owing to acute firewood and forage shortages a village woman has to travel miles together to collect fuelwood and forage to meet daily needs of her family. After the establishment of tree plantations on the village abandoned community lands affected by salts two basic needs of low income groups in the villages have been fulfilled.

Employment Generation

Raising tree plantations generates round the year employment opportunities in villages. Roughly 216 man days/ha are needed for initial establishment of trees on alkali soil. For raising trees and grasses together for 8 to 10 years nearly 900 to 1050 man-days are generated. Owing to generation of employment opportunities in the village itself, there is tremendous decrease in labour flow from the villages to the cities.

Replacement of Cow Dung and Saving of Nutrients

As a substitute for firewood, a large quantity of cow dung is burnt as fuel in rural India. It has been estimated at CSSRI that 90 to 120 t cow dung can be saved by raising one ha forest plantations. The nitrogen saved in the form of animal dung can meet the demand for this nutrient for about 20 to 26 t/ha. By using saved cow dung in farm fields, nearly 8 tonnes of food grains can be augmented by raising one ha *Prosopis* or *Acacia* plantations.

Income Generation for Village Development

In addition to meeting fuelwood and forage needs of less privileged sections in the villages, these plantations generate sufficient income after 8 to 10 years for the Panchayat, to provide for overall development of the village. It was worked out that one ha. *Prosopis* plantation (1250 trees/ha) can generate revenue of Rs. 64000/- after 8 years. Further, sensitivity analysis made at CSSRI also showed that afforestation is financially feasible even when the cost rises by 10 percent and returns remain normal.

Increase in Land Value

Trees growth for a certain period of time reclaims the alkali soils to such an extent that these can be used for growing all kinds of agricultural crops after removing the trees. Therefore, after reclamation, land value increases significantly.

REFERENCES

1. Singh, Gurbachan; Singh, H. and Bhojvaid, P. P. 1997. Land Degradation and Development. (In press).
2. Singh, Gurbachan,; Singh, H.; Dagar, J. C.; Singh, N. T. and Sharma, V. P. 1997. Agroforestry Systems 37 (3):279-295.
3. Singh, Gurbachan.; Dagar, J. C. and Singh, N. T. 1997. Land Degradation and Rehabilitation 8:257-268.
4. Singh, Gurbachan. 1996. The role of *Prosopis* in reclaiming high pH soils and in meeting firewood and forage needs of small farmers. Keynote paper presented in the International *Prosopis* Workshop in Washington. DC (USA), March 13-15, 1-27.
5. Singh, Gurbachan. 1995. Agroforestry Systems, 27: 61-75
6. Singh, Gurbachan. 1995. Forest Ecology and Management 80: 267-278.
7. Singh, Gurbachan.; Singh, N. T. and Abrol, I. P. 1994. Land Degradation and Rehabilitation; 5: 223-242.
8. Singh, Gurbachan.; Gill, H. S.; Abrol, I. P. and Cheema, S. S. 1991. Field Crops Research 26: 45-55.
9. Singh, Gurbachan.; Abrol, I. P. and Cheema, S. S. 1989. Exptl. Agric. 25: 401-408.
10. Singh, Gurbachan, Abrol, I. P. and Cheema, S. S. 1989. Forest Ecology and Management 29: 1-14.
11. Singh, Gurbachan.; Abrol, I. P. and Cheema, S. S. 1989. International Tree Crops Journal, 6:81-99.
12. Singh, Gurbachan.; Singh, N. T. and Tomar, O. S. 1993. Technical bulletin No. 17, CSSRI, karnal, India, 65.
13. Singh, Gurbachan, Singh, N. T. and Dagar, J. C. 1993. Technical Bullettin No. 18, CSSRI, Karnal, 24.

Role of MPTs in Management of Wastelands with Special Reference to Sodic Soils

MUNISH KUMAR AND A. S. WARSI
C.S. Azad University of Agriculture and Technology, Kanpur (UP)

ABSTRACT

The studies were conducted at Regional Research Station, Daleep Nagar, Kanpur Dehat under Wasteland Development Project on sodic soils and found that the area is characterised by high concentration of sodium salt at upper surface of soil having pH value from 8.9-10.5. About 25 acre land area was covered with different multipurpose trees, *Prosopis juliflora, Leucaena leucocephala, Eucalyptus Spp., Dalbergia sissoo, Albizzia procera, Acacia nilotica, Terminalia arjuna, A. lebbeck and Azadirachta indica* found best suited to the sodic land conditions in respect of different growth parameters. The digging of pits was very costly by manual labour specially, in salt affected soils of Daleep Nagar due to presence of hard (Kankar) pan. The tractor drawn post hole auger proved useful in breaking this hard pan. This auger whole method of pit digging was found suitable and more remunerative for the area as well as farmers of the area for plantation of MPTS under agroforestry system on sodic soils.

INTRODUCTION

The substantial improvement in physico-chemical properties of sodic soil is also noticed under agroforestry system which enhances the biological activities in the soil so that the soil conditions have improved and facilitated the movement of water and reducing hardness of the soil.

Continued deforestation and land degradation poses a serious threat to the sustainability of our land resources. At present only 11.73 per cent of the total land area is under dense forest cover and there is a deficit of almost all types of woods, of which firewood alone accounts for 195 million cu. m. Though other sources of energy for cooking food have increased but still firewood remains the major source for rural households and its deficit forces use of crop residue and animal dung-cake for cooking food. This deprives land of its rich source of organic matter (Pathak, 1997). Presently, the per capita availability of the forest area in the country is only 671 square meters as against the estimated requirement of 1,605 square meters. As such, it has to be increased by 2.39 times at least. The national Commission on Agricultural has estimated that 40 m hectares will have to be brought under wood production by 2000 AD for various purposes including fuel, and 10 m hectares each for production of crops and fodder (Joshi, 1998).

Salt affected soils have considerable potential to augment food gain production to feed the fast increasing population of the country. Keeping in view the above facts, a Project on "*Rehabilitation of wastelands of various kinds-salt affected gully and ravenous dry lands for various sites*" was formulated during 1990-91 and executed for a period of four years. The sites were selected for implementation of the project are Regional Research Station, Daleep Nagar Kanpur Dehat) for sodic soil, NARP Centre, Hazaratpur (Firozabad) for gully and ravines, Regional Research Centre, Bharari (Jhansi) for dry lands and Regional Research Station, Madhurikund (Mathura) for brackish saline conditions.

MATERIALS AND METHODS

About 25 acres area was covered with different MPTS trees at RRS Daleepnagar under sodic soil conditions alongwith different fodder grasses in selected blocks of plantation.

Planting Techniques

The pits size was 60 □ 60 □ 90 cm and filling material was used mixed with 6 kg. gypsum + 6 kg FYM alongwith 100 g BHC powder for control of termite infestation per pit. The planting spacing was kept 4 □ 3 m apart for the tree species, while the different fodder grasses were planted in the block of *D. sissoo, L. leucocephala, A. indica and A. nilotica* in inter-spaces at 50 □ 50 cm, apart.

The digging of pits was very costly by manual labour especially in salt affected soils of Daleep Nagar due to the presence of hard pan (Kanpur pan) between 70 to 125 cm depth. It is very difficult to break this hard pan manually. The use of tractor drawn -Earth auger proved useful in breaking this hard pan but it has disadvantage of its low diameter (10"). Due to this low diameter the most of the soil got filled again in the pits before planting and thus limited scope for filling material i.e. gypsum, FYM and sand. To overcome this difficulty of low diameter of earth auger a new type of earth hole auger was designed and got fabricated at M/S Shakti Industries, Khatauli, Meerut, U.P. The total length of the auger was 140 cm. In which upper half having the diameter of 45 cm, while the lower half 18 cm. This Earth auger has both the advantages of breaking of Kankar pan and big size of upper portion of pits facilitating the filling of more material and its packing. The working of this earth hole auger was observed by different government agencies and farmers of the area.

Selection of Planting Material

The choice of species under various situations is based on the site characteristics such as geology, topography, climatic, soil characteristics existing indigenous plants. Further, the choice of the species for various locations and specially, for sodic conditions were decided on the following characteristics: tolerance to drought and sodicity; faster growth; nitrogen fixing plants; to fulfill the multiple needs of local areas for fuel, fodder and timber; establishment and adaptability to edaphic and climatic conditions; root characteristics and soil conservation aspects.

RESULTS AND DISCUSSION

The digging of pits was very costly by manual labour especially in salt affected soils of Daleep Nagar due to the presence of hard pan (Kanpur pan) between 70 to 125 cm depths. It is very difficult to break this hard pan manually. The use of tractor drawn Earth Auger proved useful in breaking this hard pan.

The cost of digging the pit by tractor mounted hole auger is considerably reduced compare to the number of labour engaged in annual digging. The mechanical digging helps penetrating beyond one meter depth. It breaks hard pan and facilitate the proliferation of roots to the deeper layer. The innovative technology so backed is being increasingly adopted to other places facing the similar problems. Even on the other problematic soils the pits are dug mechanically and widened manually for the purpose of plantation of horticultural crops.

Improvement of Soil Conditions

The substantial improvement in physico-chemical properties of sodic soil at Daleep Nagar is revealed from the data presented in Table 1.

TABLE 1: *Changes in physico-chemical properties of sodic soil*

Parameters	Initial	Final
pH	*10.5*	*8.6*
ESP	*65.2*	*36.8*
EC (dsm^{-1})	*2.93*	*1.92*
Organic carbon (%)	*0.48*	*0.52*

Initial values of pH, ESP and EC, the major parameters indicating the degree of sodicity have gradually reduced considerably over a period of 4 years due to extraction of salt by the plants, improvement in the physical structure of the soil through the addition to organic matter makers it easier for the roots to penetrate and improves drainage by releasing the water slowly and enhancing biological activities in the soil.

Survival and Growth Characteristics of Trees on Salt Affected Soils

The studies revealed that the maximum survival of plants in *P. juliflora* followed by *A. nilotica, D. sissoo* and least by *T. arjuna and A. indica.* The rate of growth was measured in term of plant height and collar girth, which differed, with the nature of species. *Prosopis, D. sissoo, A. procera* and *L. leucocephala* registered maximum height and the least by *T. arjuna*. Similar observations were also reported by Prasad (1988).

Forage Yield of Grasses and Salt Affected Soils

The forage yields of different grasses are presented in Table 2. The data revealed that the maximum green and dry fodder was obtained in

S. sphacelate and least in *C. gayana*. Similar results were also reported by Lal (1995).

TABLE 2: *Forage yield of different grasses (q/ha)*

Grass species	Mean of two year (1994 and 1995)	
	Green fodder	Dry fodder
Brachiria mutica	280	104
Chloris gayana	158	52
Leptochola fusca	195	68
Setaria sphacelate	340	110

Note: Uniform does of gypsum @ 4 t/ha, was applied at the time of planting grass species.

Biomass Yield

The most of the biomass obtained from salt affected soil was neither acceptable nor palatable to livestock. With the minimum earth work (mechanical changes) i.e. pit digging result in 10-16 t/ha production of green matter in comparison to 3-4 t/ha biomass from salt affected natural soils. The biomass of natural grass appreciated indicating the increase in microbial activities in soil. The water holding capacity of soil improved due to the breaking of hard pan during digging and facilitating the penetration of the tree roots in to the deeper layers of soils.

Regeneration of Natural Grasses

With the treatment of the soil, the natural grasses regenerated and covered larger areas which indicate the improvement in the soil conditions, microbial activities, enrichment of soil fertility and water holding capacity etc.

REFERENCES

1. Joshi, N. C. 1988. Employment News, New Delhi, vol. XXIII, May 1998.
2. Lal, B.1995. India J. Soil Cons. 23 (1):44-46.
3. Pathak, P. S. 1997. ICAR, News Letter, April-June: 1-3
4. Prasad, R. 1998. Technology of wastelands development. S. K. Gupta Associated Publishing Co., New Delhi: 157-1178.

Performance of MPTs in Problematic Soils of Central Plain Zone of Uttar Pradesh

S. A. KHAN AND I. N. SHUKLA
C.S. Azad University of Agriculture and Technology, Kanpur (UP)

ABSTRACT

Degradation of land through accumulation of excessive salt in the root zone is major problem which decreased the production gradually. By and large area under alkalinity and sodicity is increasing in the semi-arid areas where drainage is impeded. Keeping in view the seriousness of alkalinity and sodicity an experiment entitled "*performance of multipurpose trees in problematic soils under agroforestry system of Central plain zone of Uttar Pradesh*" was undertaken. The planting of MPTS species have improved the soil condition and reduced the soil pH and EC upto 8.10 and 0.31 mmhos, respectively.

INTRODUCTION

Land being a limited entity in a country like ours generally hinders development of programmes that aim at horizontal expansion. Considerable area not presently under cultivation but potentially arable, have been lying barren due to one or another soil constraints and the lack of economical and proven technology for their reclamation. Amongst these, degradation through accumulation of excessive salt in the root zone is major problem which decreased the production gradually. By and large, area under alkalinity and sodicity is increasing in the semi-arid areas where drainage is impeded. Out of 9.38 m ha. of salt affected land in India, 1.2 m ha, is in the state of

Uttar Pradesh, out of which 0.43 m ha falls in Central Plain Zone of Uttar Pradesh comprising Allahabad, Lucknow, Kanpur and Agra Divisions. Amongst the tree species tested the highest plant height was attained by the *Eucalyptus hybrid* followed by *Leucaena leucocephala, Acacia nilotica, Albizia procera* and *Azadirachta indica* whereas the maximum plant girth was obtained by *A. indica* followed by *Acacia nilotica and Albizia procera.*

MATERIALS AND METHODS

The field experiment was initiated in the year 1989 and 1990 under All India coordinated Research Project on Agroforestry at Regional Research Station on Sodic soil, Daleep Nagar, C.S. Azad University of Agriculture and Technology, Kanpur. The soil of the experimental field was alluvial in nature and Sodic in reaction. The initial pH and EC of the soil were 10.5 and 0.73 mmhos during the start of the project, respectively. The soil possesses poor drainage and a hard canker pan at 60- 90 cm depth, impeded infiltration and percolation of salts. The soil is low in organic carbon, available nitrogen and potash and medium in phosphorus. The experimental area falls under sub-tropical, semi-arid climate with average rainfall of 800 mm, out of which about 90% rain is received from July to October and rest 10% as winter showers. The experiment comprised of ten tree species namely *Azadirachta indica, Dalbergia sissoo, Albizia procera, Terminalia arjuna, Eucalyptus hybrid, Leucaena leucocephala, Acacia nilotica, Acacia catechu, Morus alba* and *Cassia siamea* planted in 60 □ 60 □ 90 cubic cm pits filled with 3 kg pyrite+5 kg compost+10 kg sand and soil of the same field and replicated four times in randomized block design. The pits were prepared prior to the on set of monsoon. One year old seedlings were used in plantation. The plants were irrigated as and when required and excess water was drained during high rainfall period. The plot size of each species was 144 m^2.

RESULTS AND DISCUSSION

Change in Soil pH and EC

The results indicated that there was an improvement in the soil pH and EC after the plantation of different tree species. The initial average pH and EC of the soil (in 1990) was 10.5 and 0.73 mmhos, respectively. Gradually, by the tree crop association, these values decreased annually and reached upto a pH of 8.10 and EC 0.310, respectively. These results corroborate with those of Mishra (1990).

Growth Behaviour

Plant Height

In the early stage of experiment (1990-91) the highest plant height was obtained by the *L. leucocephala* followed by *Eucalyptus hybrid, A. indica, A. nilotica, D. sissoo, M. alba, A. procera, T. arjuna* and being lowest with *C. siamea* (Table 1).

During second year of the experimentation, the highest plant height was obtained by *L. leucocephala* followed by *Eucalyptus hybrid, M. alba, A. indica, T. arjuna* and being lowest with *A. catechu*, but during the year 1992-93 to 1997-98, the *Eucalyptus hybrid* attained the highest plant height followed by *L. leucocephala, A. nilotica and A. procera, A. indica* and lowest was with *M. alba*. This increase in height of *E. hybrid* may be attributed to fast growing nature of the species.

Plant Girth

Data in Table-2 revealed that during the early growth stage (1990-91), the maximum plants girth was exhibited by *L. leucocephala* followed by *A. indica, A. procera, Eucalyptus hybrid, A. nilotica and D. sissoo* and was lowest with *T. arjuna* followed by *A. catechu, M. alba and C. siamea*. During 1991-92 to 1997-98, *A. indica* showed the highest plant girth, followed by *A. nilotica and A. procera*.

TABLE 1: *Plant height of tree species*

Name of the species	Plant height (cm)							
	90-91	91-92	92-93	93-94	94-95	95-96	96-97	97-98
A. indica	95.2	128.4	370.0	465.0	613.0	667.0	685.0	698.0
D. sissoo	83.3	139.0	294.0	459.0	562.0	582.0	590.0	610.0
A. procera	45.3	88.0	330.0	471.0	601.0	690.0	705.0	718.0
T. arjuna	40.3	117.2	240.0	405.0	560.0	602.0	618.0	624.0
E. hybrid	109.3	206.2	422.0	627.0	860.0	902.0	995.0	1020.0
L. leucocephala	126.7	244.5	419.0	552.0	711.0	775.0	800.0	815.0
A. catechu	35.5	82.1	225.0	458.0	570.0	617.0	628.0	639.0
A. nilotica	94.7	175.3	303.0	547.0	680.0	762.0	780.0	786.0

Name of the species	Plant height (cm)							
	90-91	91-92	92-93	93-94	94-95	95-96	96-97	97-98
M. alba	21.3	130.5	231.0	267.0	310.0	330.0	345.0	352.0
C. siamea	35.7	96.9	304.0	458.0	558.0	590.0	608.0	620.0

TABLE 2: *Girth of the plants*

Name of the species	Plant girth (cm)							
	90-91	91-92	92-93	93-94	94-95	95-96	96-97	97-98
A. indica	4.6	9.3	22.9	41.4	58.1	62.0	64.5	66.4
D. sissoo	3.9	6.1	15.6	28.0	40.1	41.5	43.0	45.5
A. procera	4.2	7.3	19.2	37.8	50.3	51.3	53.5	56.6
T. arjuna	2.9	7.3	17.2	31.9	48.4	49.7	51.5	53.2
E. hybrid	4.2	7.6	15.9	31.5	41.6	43.4	46.5	47.2
L. leucocephala	5.3	8.2	16.3	33.5	46.7	48.0	49.5	50.8
A. catechu	2.9	6.1	11.3	37.0	43.0	43.4	45.0	45.2
A. nilotica	4.1	7.9	15.4	36.2	48.4	52.5	54.0	56.8
M. alba	3.1	5.6	11.3	15.0	29.0	31.5	32.5	33.4
C. siamea	3.2	6.1	13.7	37.0	41.5	42.5	44.0	45.0

CONCLUSIONS

It can be concluded that under Sodic soils condition out of ten species evaluated, *A. nilotica* followed by *A. indica, Eucalyptus hybrid and L. leucocephala* proved to be the most suitable trees in respect of growth and soil rehabilitation for Central Plain Zone of Uttar Pradesh.

REFERENCES

1. Khan, S. A.; Shukla, I. N. and Yadav, V. K. 1995. In: Workshop on Technology Extension Present Status and Future Strategy held at India International, New Delhi from June 6-7: 1-5.
2. Mishra, C. M. 1990. Proceedings of the National Workshop on Multipurpose Trees Species for Agroforestry System held at IGFRI, Jhansi from February 1-3: 133-135.
3. Singh R. C. and Pathak, P. S. 1990. Proceeding of the National Workshop on Multipurpose Tree Species for Agroforestry System held at IGFRI, Jhansi from February 1-3: 127-132.

Sustenance of Soil Fertility through Adoption of MPTs

R. S. YADAV, MUNNA RAM, K. S. DADHAWAL AND K. R. SOLANKI
National Research Centre for Agroforestry, Jhansi (UP)

ABSTRACT

Multiple purpose tree species play a crucial role in sustenance of soil productivity through litter production, regular supply of available nitrogen over a longer period by nitrogen fixation, fine roots turn over and nutrient cycling and improvement in soil structure. The major role of agroforestry, to maintain soil fertility, is based primarily on observations of higher crop yields in association with MPTS.

INTRODUCTION

The amount of nutrients provided by prunings and/or litter fall are determined by the production rate and nutrient concentration, both depending upon climate, soil type, tree species, plant part, density and pruning regime. A large number of screening and alley cropping trials in different climate-soil environments indicate that pruning of several tree species contain sufficient amount of nutrients to meet crop demand, with the notable exception of phosphorus. This review examines the role of MPTS in the build up and maintenance of soil fertility.

ROLE OF MPTS IN SOIL FERTILITY BUILD UP

The studies have reported that there is an increase in nutrient content of soil (Table 1) under *Prosopis* species in arid environments of India

(Singh and Lal, 1969). A significant improvement of nitrogen status, modest in phosphorus and decline in potash status was noticed in red soils (Alfisols) at Jhansi, (Hazra, 1993). However, agrisilviculture system may sustain nitrogen and phosphorus levels but there is need to add potash fertilizer (Table 2).

In peninsular India, alley cropping proved beneficial in terms of soil and water conservation (Table 3) with less run off and soil loss with 3 m alleys then with 5.4 m alleys (Rao *et. al.*, 1991). The canopy ground cover, mulching and residue management have greater role to play in soil and water conservation (Young, 1989). Soil analysis after seven years of experimentation, revealed that tree species (up to 6 m distance) lowered soil pH by 0.3 units, improved organic carbon content by 0-15 per cent and decreased bulk density by 0.1 Mgm^{-3} (Tomar *et. al.*, 1994).

TABLE 1: *Nutrient content of soil under two Prosopis species in arid region of India*

Tree	Site	Org. C (%)	Total N (%)	Total P	Total K
P. juliflora	*Under Tree*	0.73	0.075	37	296
	Open Area	0.25	0.027	31	294
P. cineraria	*0-30 cm*				
	Under Tree	0.37	0.045	3.82	12.2
	Open Tree	0.25	0.038	1.52	7.52
P. cineraria	*31-60 cm*				
	Under Tree	0.11	0.02	1.95	9.31
	Open Tree	0.04	0.01	1.23	6.36

Source: Singh and Lal, (1969).

TABLE 2: *Effect of hedgerows on soil fertility status on Alfisol*

Hedge/rows		Av. Nutrients (Kg./ha)	
	N	P	K
Leucaena+crops	168	13.6	176
Sesbania+crops	159	14.0	185
Leucaena hedge	183	13.6	182
Sesbania hedge	178	14.3	193
Forage crops	145	14.5	198

Source: Hazra, (1993).

TABLE 3: *Effect of sole crop, Leuceana and alley cropping systems on run off and soil loss on shallow Alfisols*

Treatments	Run off (mm) Unmulched Mulched	Soil loss (t/ha) Unmulched Mulched
Sole annual crop	*44 6*	*0.45 0.22*
Alley cropping (5.4 m)	*30 3*	*0.33 0.10*
Alley cropping (3 m)	*20 1*	*0.22 0.06*
Sole Leuceana	*9 0*	*0.04 0.02*

Source: Rao et. al., (1991)

TABLE 4: *Nutrient required by a crop of maize compared to the nutrients in 4 t of organic inputs*

A. Nutrient required by a maize crop

		Nutrient (kg/ha			
	N	P	K	Ca	Mg
Grain (2 t)	*50*	*12*	*30*	*6*	*4*
Stover (3 t)	*30*	*6*	*36*	*9*	*6*
Total	*80*	*18*	*66*	*15*	*10*

B. Nutrient added in 4 t of leaves of various organic inputs

Species		Nutrients (kg/ha		
	N	P	K	Ca Mg
L. leucocephela	*154*	*8*	*84*	*52 13*
Erythrina poeppigina	*132*	*7*	*46*	*61 ----*
Inga edulsis (fertile)	*142*	*11*	*40*	*45 6*
I. edulsis (infert)	*127*	*9*	*50*	*30 7*
Senna siamea	*105*	*6*	*44*	*110 7*
Dactyladenia barteri	*60*	*4*	*31*	*40 8*
Grevilea robusta	*52*	*2*	*24*	*60 7*
Maize stover	*40*	*8*	*48*	*13 8*

Source: Palm, (1995)

Puri *et. al.*, (1994), revealed that soil contents of N, P, K, soil moisture, and organic carbon were higher under tree canopies of *Prosopis cineraria* than away from the influence of trees. According to them, *P. cineraria* benefits understorey crop growth and yield due to improvement of soil fertility and conservation of moisture. In another study, *Leucaena leucocephela* provided 2901 kg oven dry foliage in 4 year for mulching, and its incorporation added 91.9, 4.2 and 44.3 kg/ha of N, P and K, respectively, to the soil (Grewal *et. al.*, 1995).

Agroforestry, however, does not supply most of the phosphorus input required by crops. Deep capture of P is negligible due to very low concentration of available phosphorus in the sub-soil (Table 4). Leguminous mulches applied at a realistic rate of 4 t/ha provides 8 to 12 kg P/ha, about half the requirement of a maize.

SOIL AMELIORATIVE EFFECTS

Studies on ameliorating effect of 20 years old plantations of *Prosopis juliflora, Acacia nilotica, Eucalyptus tereticornis, Albizzia lebbeck* and *Terminalia arjuna* (Singh and Gill, 1992) raised on a high soil pH (10.2-10.5) indicated that the average pH of 0-120 cm profile dropped from 10.2 to 8.0 in case of *Prosopis* (Table 5.) Organic carbon of the profile increased several folds of the original soil, the highest being under *Prosopis* and lowest under the *Eucalyptus* tree. The extent of soil improvement with various trees was in the order of *Prosopis*> *Acacia* > *Terminalia* > *Albizzia* > *Eucalyptus.* This study further suggested that tree plantations can reclaim alkali soils to such an extent that arable farming can be practised on such lands after the trees are felled upon reaching the maturity. Similarly, the soil pH and EC in the 0-15 cm. layer decreased to 8.9 and 0.36 dS/m from 10.3 and 2.2 dS/m respectively, within 74 months in *P. juliflora* and *Leptochloa* grass treatment (Singh, 1995).

TABLE 5: *Effect of 20 years tree growth on properties of alkali soil*

Tree Species	Soil depth (cm)	pH	ECdS/m	O.C. (%)	Avail. P (kg/ha)	Avail. K (kg./ha)
Original soil	0-15	10.3	1.5	0.24	32	308
A. nilotica	0-15	8.4	0.25	0.85	59	458
	0-120	9	0.53	0.55	33	298
Eucalyptus	0-15	8.5	0.44	0.66	33	358
	0-120	9.2	0.6	0.35	34	260
P. juliflora	0-15	7.3	0.51	0.93	110	701
	0-120	8	0.41	0.58	50	330
T. arjuna	0-15	79	0.32	0.86	67	409
	0-120	8.2	0.25	0.58	41	235
A. lebbeck	0-15	79	0.32	0.62	42	387
	0-120	8.7	0.51	0.47	44	225

Source: *Singh and Gill (1992).*

FUTURE THRUST AREAS

The foregoing discussion indicates that introduction of trees on farmlands improves / sustain the soil fertility which eventually may result in higher sustained crop yields. These benefits may be exploited for organic farming to minimise the use of chemical fertilizers and laying the future research thrust on:

- ☐ Standardisation of number of trees/ha (spacings) which may maintain soil fertility without much reducing the yield of companion crops i.e. not more than 20 per cent of sole cropping.
- ☐ Evolving need based nutrient package for different tree based agroforestry systems depending upon the nutrient replenishment capacity of trees.
- ☐ Synchronisation of nutrient release from tree litter and meeting crop demand.

REFERENCES

1. Grewal, S. S.; Singh, K.; Juneja, M. L. and Singh, K.1995. Indian J. Agril. Sci., 65: 3, 191-195.
2. Hazra, C. R. 1993. Indian J. Soil Conserv.21: 1-7.
3. Palm, C. A. 1995. Agroforestry systems.30: 105-124.
4. Puri, S.; Kumar, A. and Singh, S. 1994. J. Arid Environment.27: (1)85-98.
5. Rao, M. R.; Ong, C. K.; Pathak, P. and Sharma, M. M. 1991. Agroforestry Systems, 15: 51-63.
6. Singh, G. 1995. Agroforestry systems, 29: 61-75.
7. Singh, G. and Gill, H. S. 1992. Indian J. Agri. Sci., 62: 144-146.
8. Singh, K. S. and Lal, P. 1969. Ann. Arid Zone, 8: 33-36.
9. Tomar, V. P. S.; Dadhwal, K. S. and Singh, R. K. 1994. Proc. of 8th ISCO conference 'Soil and water Conservation: Challenges and opportunities' (Eds.: L. S. Bhusan; I. P. Abrol, and M. S. Rama Mohan Rao) Vol. II: 1484.
10. Young, A.1989. Agroforestry for soil conservation, C.A.B. International, Wallingford, Oxon (U.K.)

Soil Improvement through *Hardwickia binata* based Agroforestry System

MUNNA RAM, R. S. YADAV, K. S. DADHWAL, A. K. BISARIA AND K. R. SOLANKI
National Research Centre for Agroforestry, Jhansi-284 003 (UP)

ABSTRACT

Anjan tree (*Hardwickia binata*), a lesser known, unexploited, potential MPTS has been found to occur naturally on a wide range of soils and climatic conditions. It is capable of enriching the soil, which in turn favourably affect the associated vegetation. The per cent return of Ca, Mg, P and K through canopy flux is higher as compared to N which is least. The tree has characteristic to retain low amount of nutrients and return major portion to the soil. The N, Ca and P concentration in leaf litter of the tree is 1.45, 2.28 and 0.13%, respectively. A substantial increase in organic carbon, available N and P, per cent pore space and field capacity was noticed whereas pH, EC and bulk density decreased under *H. binata* when pruned biomass was returned to soil system whereas, no perceptible changes in soil variables were noticed when pruned biomass was used for fodder purpose.

INTRODUCTION

The role of vegetation in soil formation and its effect on soil properties was realized even in early stages of soil development. The productivity level of a soil at any time is the resultant of the interaction of degradative processes and conservation practices that are occurring simultaneously in any system. For any system to be viable, the benefits from the conservation processes should be equal to or greater than the negative effects of the degradative processes. Further, the most compatible system is one that optimises the use of limited resources without degrading the soil quality. The trees and annuals/perennials are the essential components of the agro-forestry systems that help in maintaining and regenerating the soil fertility. Addition of nutrients can surpass the removals in properly managed systems.

Integration of tree with annuals/perennials helps in improving the soil productivity. The tree based land use systems have unique fertility restoring capacity through organic recycling mechanism. The bulk of organic matter and nutrients added to the soil are located in the top soils. The trees are capable of taking up of nutrients and water from the layers that are usually not utilized by herbaceous crops. The nutrients stored in various plant parts are recycled with in soil-plant system. The relative magnitude of improvement in soil properties depend on tree species, duration of tree have been grown, antecedent soil properties and prevalent climatic conditions.

SOIL AND NATURAL HABITAT

H. binata, a lesser known, unexploited, potential MPTS has been found to occur naturally on a wide range of soil and climatic conditions (Table 1).

TABLE 1: *Distribution of Hardwickia binata in different Agro-ecoregions of India*

Locations	Soil Types	Physiography	Climate	Soil related constraints
Anantpur (AP) Bellary, Karnataka (3)	*Mixed Red and Black SoilsShallow and medium, slightly acidic and noncalcareous loamy red soils and deep, clayey slightly alkaline and calcareous black soils with medium to high AWC*	*Deccan Plateau*	*Hot arid*	*High runoff and erosion hazard, prolonged dry spell during cropping period and sub-soil sodicity affecting soil structure and drainage in sub dominant black soils.*
Gwalior (MP), Banda and	*Alluvium Derived Soils Deep,*	*Northern plain and central high lands*	*Hot semi-arid*	*Coarser soil texture and low AWC, low*

Locations	Soil Types	Physiography	Climate	Soil related constraints
Jhansi (UP) (4)*	*loamy, alluvium derived soils with medium AWC and shallow to medium, sandy to loamy, gray brown soils with low to medium AWC.*	*including parts of Gujarat plains*		*water table and imperfect drainage at some places leading to surface and sub-surface salinity/sodicity.*
Indore, Khandawa, East and West Nemar (MP) (5)	*Medium and Deep black soils Medium and deep, fine loamy and clayey, slightly alkaline, calcareous, black soils having typical swell and shrink properties with medium and high AWC*	*Central (Malawa) Highlands and Kathiawar peninsula*	*Hot semi-arid*	*Droughtiness, imperfect drainage limiting optimum root ranification, salinity and alkalinity hazard under irrigated condition.*
Belgaun and Dharwad (Karnataka) Buldhana and Nasik (Maharashtra) (6)	*Shallow and Medium (inclusion of Deep) Black soilsShallow and medium, loamy to clayey, highly calcareous, Black soils with low and medium AWC, inclusion of deep, clayey, calcareous and moderately alkaline, black soils with swell and shrink properties and high AWC.*	*Deccan plateau*	*Hot semi-arid*	*High runoff and heavy soil loss, prolonged dry spell and nutrient imbalance (deficiency of NP and Zn).*
Kurnool, Cuddapah and Nellore (AP) (7)	*Red and Black soils Medium, non-calcareous and neutral red*	*Deccan Plateau and Eastern Ghats*	*Hot semi-arid*	*High runoff and severe soil loss, frequent droughtiness,*

Locations	Soil Types	Physiography	Climate	Soil related constraints
	loamy soils with low and medium AWC and deep, clayey, calcareous and strongly alkaline black soils with marked swell and shrink properties and high AWC			*nutrient imbalance and sub soil salinity and sodicity in black soil area due to high ground water table.*
Mysore (Karnataka) North Arcot and Salem (Tamil Nadu) (8)	*Red Loam soils Shallow to deep, loamy, non-calcareous, slightly acidic to calcareous, moderately alkaline, red soils low to CEC with low to medium AWC.*	*Eastern Ghats (TN uplands) and Deccan Plateau*	*Hot semi-arid*	*High runoff causing severe erosion, droughtiness, coarse texture and nutrient imbalance (deficiency of NP and Zn).*
Hoshangabad (MP) (10)	*Medium and Deep Black soils Medium and deep, clayey, calcareous and slightly alkaline black soils with distinct shrink-swell properties and medium and high AWC*	*Central Highlands (Malawa and Bundelkhand)*	*Hot sub-humid*	*Droughtiness, narrow workable moisture and nutrient imbalance (deficiency of NP and Zn).*
Nagpur (Maharashtra)*	*Red and Black SoilsMedium, loamy, neutral, red soils and slightly alkaline, black soils and medium and high AWC.*	*Deccan plateau and central highlands (Bundelkhand)*	*Hot sub-humid*	*Soil erosion, narrow workable moisture and imbalance of N, P and Zn).*
Palamu (toward some river and Kymore hills)	*Red and Yellow Soils Medium and deep, loamy,*	*Eastern Plateau (Chhatisgarh region)*	*Hot sub-humid*	*Erosion hazards, sub soil graveliness and coarse texture*

Locations	Soil Types	Physiography	Climate	Soil related constraints
*Bihar, Mirzapur and Sonebhadra** (UP) (12)*	*non-calcareous, neutral to slightly acidic red and yellow soils with medium AWC.*			*and nutrient imbalance (N, P, Zn and B).*
Parts of East Godawari (AP) (13)	*Red Loamy Soils Shallow and medium, loamy red soils and deep, clayey, lateritic, non-calcareous, slightly to moderately acidic soils with low CEC and low to medium AWC.*	*Eastern (Chhota Nagpur) Plateau and Eastern Ghats*	*Hot Sub-humid*	*Soil erosion, droughtiness, sub soils graveliness and coarse texture, nutrient imbalance (N, P, Zn and B) and moderate P fixation in some red and lateritic soils.*
Parts of East Godavari, West Godavari, Krishna and Nellore (AP) (19)	*Alluvium-Derived Soils Deep, loamy, coastal and delatic alluvial soils with medium and high AWC, soil reaction, slightly to strongly sodic.*	*Eastern coastal plain*	*Hot Sub-humid*	*Imperfect to poor drainage and soil salinity.*

** Artificially introduced, ** Added on farmers perception, AWC-Available water capacity, Number in parenthesis indicate agro-ecological regions.*

The landscapes on which tree occurs naturally are red loamy soils, red and yellow soils, red and black soils, mixed red and black soils, shallow and medium black soils, medium and deep black soils and alluvium derived soils developed in arid to semi-arid and hot sub-humid climatic conditions (Sehgal *et. al.*, 1990). In its natural habitat the best growth is attained on geological formations such as sand stones, conglomerate, quartzite, granite, schist, shales, gneiss and metamorphic rocks which disintegrates into gravelly sand, sandy loam to loam than stiff soils (Troup, 1921 and Singh, 1982). The tree can grow on dry shallow soils and rocky grounds where most of the other tree species can not grow. It avoids clayey soils unless covered with some depth of sand debris.

TABLE 2: *Physico-chemical properties of soil after two years of berseem cropping with and without tree association*

Soil Characteristics	Without Tree	With Tree	% increase/decrease
Soil pH	7.8	7.6	-2.6
EC (dS/cm)	0.35	0.28	-20.0
% organic carbon	0.38	0.44	15.8
Available N kg ha-1	190	210	10.5
Available P kg ha-1	9.1	10.7	17.6
% pore space	40	42	5.0
Bulk Density (g/cc)	1.53	1.48	-3.4
Field Capacity	13.0	13.2	1.5

Source: Hazra and Tripathi (1996), Age of tree 4-5 years and height 4-6 m.

TABLE 3: *Influence of H. binata on soil properties under silvipastoral system (grasses)*

Soil Characteristics	Initial	After 7 years		% increase/decrease	
		Open Pasture	*H. binata*	Open Pasture	H. binata
Soil pH	8.3	7.8	7.6	-4.8	-8.4
E.C. (dS/cm)	0.49	0.39	0.32	-20.4	-34.7
%organic carbon	0.21	0.40	0.49	90.4	133.3
Available N kg ha-1	89	202	229	126.9	157.3
Available P kg ha-1	8.6	12.7	15.6	47.7	81.4
% pore space	31	41	42	32.2	35.5
Bulk density (g/cc)	1.78	1.51	1.47	-15.1	-17.4
Field capacity (% w/w)	11.5	13.3	13.7	15.6	19.1
Mean Dry forage yield t ha-1	-	6.58	5.96	-	-9.4

Source: *Hazra, (1988)*

The tree is fairly drought resistant and thrives in hostile environmental conditions such as low and erratic rainfall and intense solar radiation where evapo-transpiration exceeds the precipitation during most parts of the years.

Comparative study (Mohan *et. al.*, 1990) on fertilizer response of selected forest species under varying soil texture and its impact on biomass yield at seedling stage indicated that *H. binata* produced higher above and below ground biomass in clayey soil but showed decrease in biomass production due to fertilizer application (Table 4A and B). The study suggest that anjan tree can grow on a well drained heavy textured soils whereas Gupta *et. al.*, (1988), reported that tree prefers medium textured soils.

TABLE 4A: *Physico-chemical properties of experimental soil*

Soil Properties	Sand	Loam	Clay
PH	8.6	7.3	8.45
EC (dS/cm)	4.54	2.95	14.29
% organic carbon	0.12	0.94	0.47
Free $CaCO_3$ (%)	0.6	0.2	10.2
CEC (cmol/kg)	8.4	15.5	47.5
Moisture retention at 15 bore (%)	3.03	10.4	24.7
Exchangeable cations (me%) Ca	1.24	2.40	9.0
Mg	1.34	1.84	2.9
Na	0.79	0.41	2.41
K	0.23	1.10	0.12
Mechanical composition (%)	76.3	38.8	22.3
Silt	8.4	32.4	20.4
Clay	14.8	28.8	56.8

Source: *Mohan et. al., (1990)*

TABLE 4B: *Influence of soil texture on biomass production by H. binata with and without fertilizer application*

Treatment	Sand	Loam	Clay	Mean
Leaf biomass (g/pot)				
Control	4.32	4.67	7.53	5.51
Fertilizer	1.86	2.42	5.36	3.21
Mean	3.09	3.54	6.54	-
Shoot biomass (g/pot)				
Control	7	5.88	5.5	6.13
Fertilizer	1.0	1.65	3.50	2.05
Mean	4.0	3.77	4.50	-
Above ground biomass (g/pot)				
Control	11.32	10.55	13.03	11.64
Fertilizer	2.86	4.07	4.86	5.26
Mean	7.09	7.31	10.95	-
Below ground biomass (g/pot)				

Treatment	Sand	Loam	Clay	Mean
Control	8.38	8.50	7.13	8.0
Fertilizer	7.13	9.05	15.25	10.48
Mean	7.76	8.78	11.19	-

Source: Mohan et. al., (1990).

NUTRIENT FLUX

Anjan tree is capable of enriching the soil which became available to associated vegetation. The percentage return of Ca, Mg, P and K through canopy flux is also higher as compared to N which is least. The tree has characteristic to retain low amount of nutrients and return major part to the ground. The nutrient concentration in leaf litter is found as N-1.45, Ca-2.28% and P-0.13% (Prasad *et. al.*, 1974).

Agri-silvi-culture System

In this system, tree components are introduced on farm land. An analysis of soil after 2 years of berseem cropping with and without tree indicated an improvement in soil productivity in the form of lowering the bulk density, increased porosity, higher water retention capacity, lowering the pH and EC and increasing organic carbon, available N and P status of soils (Hazra *et. al.*, 1996). Further, study suggests that amongst the tree tried, *H. binata* is most suitable tree species in allowing higher radiation (80% and above) to the crop underneath and help in higher biomass production of understorey crops. To some extent the increasing level of P application reduces the ill-effect of shade on understorey crop (Table 2).

In another set of experiment with winter maize, Hazra *et. al.*, (1996) observed that after three years of cropping, organic carbon and field capacity increased where as bulk density and available N, P, K and S decreased over the initial values of tree and non-tree but there was no change in soil pH. Tripathi *et. al.*, (1997) also reported an improvement in organic carbon and field capacity and decrease in bulk density, pH and available N, P, K, and S content of soil in both the system being lowest under tree cover. Yadav and Bisaria (1998) reported that no perceptible changes in pH and organic carbon were observed after sixth year of experimentation in Anjan based agroforestry system with a population density of 200, 400 and 800 trees/ha in semi- arid region (Table 5). The pH and organic carbon varied between 7.23 to 8.03 and 0.41 per cent to 0.52 per cent, respectively. The little variation between tree spacings may be attributed to non-incorporation of pruned biomass into the soil. The flow of organic material towards control plot and water logged condition during rainy season might have resulted into higher content of organic carbon i.e. 0.52 per cent.

TABLE 5: *Influence of anjan Tree and its spacings on organic carbon and pH*

Tree Spacing (m)		O.C. (%)		pH	
		A	B	A	B
5□2.5	Tree	0.45	0.48	7.70	7.63
	Tree+Crop	0.48	0.44	7.45	7.23
5□5	Tree	0.47	0.42	7.72	8.03
	Tree+Crop	0.52	0.47	7.55	7.72
5□10	Tree	0.47	0.41	7.77	7.76
	Tree+Crop	0.48	0.50	7.60	7.47
Control		0.52		7.4	

Source: *Yadav and Bisaria (1998)*

A = Soil samples taken between the tree row. B = Soil samples taken within the tree row.

Silvipastoral System

Hazra (1988) investigated the fertility status of soil under anjan tree after seven years in silvipasture system (Table 3). It was concluded that comparatively high organic matter available N and P, per cent pore space and field capacity and lowering the pH, EC and bulk density under *H. binata* promoted the growth and yield of crops grown in its vicinity indicating thereby an appreciable improvement in soil fertility and crop productivity. The relative efficacy of different tree species in augmenting the soil productivity was in the descending order: *L. leucocephala, A. nilotica, H. binata* and open pasture. Singh (1997) reported a substantial increase in fertility indices i.e. organic carbon, available K and soil moisture and decrease in soil pH and EC under pure plantation and silvipasture system of arid conditions in Rajasthan. A mature stand of anjan tree (20 years or older) maintained in association of pasture grasses on a degraded land in semi-arid Bundelkhand region enhance the fertility status of soil in term of organic carbon by 4.7-22.7 %, nitrogen by 4.3-81.3 %, potassium by 16.9-36.1 % and calcium by 0.9-6.5 % over the initial values but its magnitude varied depending on the tree density (Roy, 1995 and 1996).

FUTURE THRUST

- The brief review of work done on Anjan based agroforestry system suggests that future line of research should lay emphasis on-
- Standardisation of tree density which may sustain the soil productivity to be worked out.
- Nutrient budgeting (addition/removal) per unit area and per unit time to be worked out for the system.
- Synchronisation of nutrient release from decomposing Anjan biomass should be carried out to meet the crop demand.
- Systematic studies should be undertaken to monitor the changes in soil quality indices.

REFERENCES

1. Gupta, G. N.; Prasad, K. G. and Mohan, S. 1998. Van Vigyan, 86: 35-42.
2. Hazra, C. R. 1988. In: Pasture and forage crop research (ed. P. Singh). Range Mgmt. Society of India, IGFRI, Jhansi: 230-238.
3. Hazra, C. R. and Tripathi, S. B. 1996. Forage Research, 22 (2 & 3): 71-75.
4. Mohan, S.; Prasad, K. G. and Gupta, G. N. 1990. Indian Forester 116 (1): 49-57.
5. Prasad, D. A.; Reddy, K. S. and Subba Reddy, C. S. 1997. Indian J. Animal Sci. 44 (3):150-152
6. Roy, M. M.1995. Ph.D Thesis. Bundelkhand University, Jhansi, India.
7. Roy, M. M.1996. Agroforestry Today, 1996, (3): 12-13.
8. Singh, H. P. 1997. (Personal Communication), CAZRI, Jodhpur.
9. Singh, R. V.1982 Fodder trees in India, Oxford, IBH Publisher, New Delhi.
10. Sehgal, G. L.; Mandal, D. K.; Mandal, C. and Vedivelu, S. 1990. Tech. Bulletin, NBSS Pub. 24, 73
11. Tripathi, S. B. and Hazra, C. R. 1997. Journal of Indian Society of Soil Science (Accepted).
12. Troup, R. S. 1921. Silviculture of Indian tree. Oxford University Press Vol. II: 341-360.
13. Yadav, R. S. and Bisaria, A. K. 1998. Annual Report, NRCAF, Jhansi,: 13-14.

Agroforestry Systems in Agricultural Dominated Ravenous Watershed

R. A. SINGH
C.S. Azad University of Agriculture and Technology, Kanpur (UP)

ABSTRACT

The different systems of agroforestry have been introduced in the agriculture dominated ravenous watershed on farmers' fields. The watershed site is located in village Rendhar, district Jalaun of U.P. The village Rendhar typically represents the edaphic, climatic and socio-economic conditions of Bundelkhand region. The watershed area is composed of highly eroded rakar, parwa, kawar and mar soils. The entire degraded area of 747.83 ha of watershed was treated with different soil conservation measures.

INTRODUCTION

The grain crops like cereals, pulses and oilseeds grown by farmers under fragile conditions are just enough to meet their family requirements but the products of fruit trees give cash to them. Some fruit trees leaves contain good percentage of crude protein which proves beneficial as forage. Apart from leaves the fruit trees also yield wood from the annual pruning, which can be used as fuelwood. Likewise, the forest trees maximise the per unit production of fodder, fuel, wood and other forest products. In addition to this, the forest trees also optimise the productivity of biological and physical resources viz. land, labour, livestock, soil moisture and solar radiation. The forest trees also preclude the soil erosion, conserve the soil moisture, increase the soil fertility, help in eco-friendly farming and maintain the ecological balance.

It is only recently that alternate land use systems viz. agro-forestry, conservation forestry, forage forestry, farm forestry etc., are receiving attention in dryland agriculture. This is because these systems give stability to the resource poor, small and marginal dryland farmers. With the view to accord the benefits from the fragile forestry and horticulture in agricultural dominated watershed Rendhar, the systems of agroforestry have been deployed.

MATERIALS AND METHODS

The operational area of 747.83 ha of watershed is situated in Rendhar village, Konch tehsil of Jalaun district of Bundelkhand (U.P.). The site is located 149 m above sea level. The site of watershed typically represents edaphic, climatic and socio-economic conditions of Bundelkhand region. The mean annual rainfall of watershed was 880 mm. The watershed soils belong to class II, III and IV of land capability class and were suitable for cultivation. The predominant hazards recognized in the watershed area were erosion due to runoff of excess water, root zone and climatic limitations. The watershed area was composed of highly eroded rakar, parwa, kawar and mar soils. The whole watershed area was treated with different soil and water conservation measures. The rain water was harvested in between twenty five check dams and five big farm ponds.

The different systems of agroforestry were laidout in the operational area of watershed. *Acacia nilotica* and *Dendrocalamus strictus* have been planted on slopes and gully bottoms under conservation forestry. *Zizyphus mauritiana and Psidium guajava* have been planted under conservation horticulture. *A. nilotica* and *D. strictus* have been planted under farm forestry system. Pearlmillet in association of clusterbeans and cowpea was planted as green fodder between rows of the *Mangifera indica* under forage horticulture system. Grasses under hortipastoral system have been planted in between rows of *M. indica* and *M. latifolia*. Similarly *Z. mauritiana, A. nilotica* and *Prosopis juliflora* have been planted on gully banks and road sides under browse forestry system. Under extension for agroforestry system *A. nilotica, Cassia siamea, Euycalyptus* spp., *Azadirachta indica* have been planted on community assets. Under silvi-pastoral system *A. nilotica* was grown as a forest tree for wood production and grasses raised in the vacant space left in between trees for rearing of domesticated farm animals.

RESULTS AND DISCUSSIONS

Conservation Forestry System

Under this programme the *A. nilotica* and *D. strictus* have been planted slopes and gully bottoms of ravinous land for preventing soil erosion and runoff and productive utilization of land. These perennial forest plants have intercepted rainwater, reduced the impact of falling rain drops on the soil and alimented the ground water table. For afforesting in ravenous area of watershed, *D. strictus* have been found most promising.

Conservation Horticulture System

On class IV land *Citrus* spp. and in ravines, *Z. mauritiana, P. guajava*, etc. have been planted under conservation horticulture programme. These fruit trees are thriving well and protecting the degraded land of watershed. These trees have started giving fruits. The farmers of watershed area are also using the wood as a fuel from the annual pruning of these trees.

Farm Forestry System

Under this system, the *A. nilotica* and *D. strictus* have been planted around the fields on the mends at recommended spacings. These forest trees are thriving well and no adverse effect was seen on crops up to 5-6 years of plantation. The pods of *A. nilotica* raised under this system are being used as nutritive feed to the browsing animals.

Forage Horticulture System

Pearl millet in association of clusterbeans and cowpea was planted as a green fodder between the *M. indica* rows and this system of agroforestry gave about 3000 kg/ha mango fruits to the farm families and 360-370 kg/ha green fodder to the animals.

Horti Pastoral System

The grasses have been grown between the rows of *M. indica* and *M. latifolia* during rainy season, this system has also given 3000 kg/ha fruits of *M. latifolia* or 3500 kg/ha fruits of *M. indica* and 175-225 q/ha green grasses.

Browse Forestry System

Under this system, the horticultural trees like *Z. mauritiana* and forest trees like *A. nilotica* and *P. juliflora* have been planted on nala banks and roadsides. The green leaves and fruits of these trees are being used as a green fodder for browses. The *Ficus bengalensis, F. glomerata, F. lacor*, etc., have been managed with watershed technology and now its green leaves and fruits are also being used as green fodder for browses during summer season or drought period.

Extension Forestry System

The selected area of dryland watershed was treated with different soil and water conservation measures. After the treatment of operational area, the forest trees like *A. nilotica, C. siamea, Eucalyptus* spp., *A. indica*, etc. have been planted on community assets viz. nala bunds, water storage structures and nala banks for meeting the general requirements of watershed families.

Silvi Pastoral System

After the implementation of watershed technology in the operational area, the rearing of milch animals has been started by the watershed families. For meeting the needs of fodder, this system has been introduced. In this system the *A. nilotica* was grown as a forest tree on the limited land for wood production and local grasses raised in the vacant space left in between trees for rearing of domesticated farm animals.

This system has been found dependable for utilization of degraded lands of watershed because trees tolerated extreme soil and climatic conditions whereas grasses provided good land cover.

REFERENCE

Singh, R. A. 1997. Intensive Agriculture, XXXV: (9-10): 3-6.

Control of Storage Fungi Associated with Seed of Some Important MPTs

ANIL KUMAR, G. R. RAO AND A. K. HANDA
National Research Centre for Agroforestry, Jhansi (UP)

ABSTRACT

National Research Centre for Agroforestry, Jhansi has been maintaining a seed bank of important MPTS since its inception in 1988. Adverse effect on seed germination and viability was observed due to high incidence of storage fungi and insect pests in some collections. A study on identification and control of the storage fungi associated with seeds of *Acacia nilotica, Albizia lebbeck, Dalbergia sissoo*, *Leucaena leucocephala* and *Sesbania sesban* was carried out.

INTRODUCTION

Seed storage is an integral part of any tree improvement and breeding programme. Generally, the seed storage for a longer duration results in the loss of viability and attack by various diseases and pests. There is an urgent need to standardized the storage containers and duration to meet the demand of MPTS seeds during off years as most of MPTS does not produce large quantity of seeds every year. In the present investigation an attempt has been made to identify the fungi associated with seeds of important MPTS during storage, their control methods and suitable containers for their storage.

MATERIALS AND METHODS

Treatments included five types of containers (steel-box, plastic-box, polythene-bag, paper-bag and cloth-bag) and seven fungicidal treatments viz., Bavistin 50 W (methyl 2-benzimidazole carbamate), Bayleton 25 WP (triadimefon), Blitox 50 W (copper oxychloride), Dithane M-45 75% (zinc ion and manganese ethylene bisdithio-carbamate), Kavach 75 WP (chlorothalonil), PMA 1% Hg (phenyl mercury acetate) and Thiram 75% (tetramethylthiuram disulphide), each in combination with 10% BHC. There were two controls, BHC alone and no seed treatment. Seed was manually extracted from mature pods, dried in the sun after removing diseased and insect infested seeds, treated with the various chemicals @ 2 g per 100 g of seed, and stored in the various containers under room temperature conditions. Triplicate seed samples from each treatment were drawn after different periods. Per cent seed germination was determined by using towel paper method (Kumar and Rao, 1995). Agar plate method was employed to determine presence of microflora (Agarwal and Sinclair, 1987). Seeds were examined with a stereo-microscope after incubation period, and the fungi isolated were cultured on PDA slants for further identification.

RESULTS AND DISCUSSION

1. **Storage fungi:** *Aspergillus, Alternaria* and *Helminthosporium* were most commonly isolated fungi from seeds of *S. sesban, A. lebbek, A. nilotica* and *L. leucocephala. Rhizophus, Penicillium* and *Trichoderma* were also isolated in addition to some non sporulating fungi (Table 1). *Alternaria* in *S. sesban, Fusarium* in *D. sissoo* and *A. niger* in *A. lebbek* caused seedling mortality. Activity of bruchids was observed in *A. nilotica, A. lebbeck* and *S. sesban* during storage (Table 2).

2. **Control:** Problems related to storage fungi could easily be controlled by taking following precautions:

    i. Manual extraction of seed and avoidance of micro-injuries at harvest stage.

    ii. Seed sanitation i.e. removal of unsound, undersized, distorted, diseased and insect infested seeds.

    iii. Solarization after seed sanitation.

iv. Seed treatment with a suitable fungicide and insecticide. Seed treatment with Dithane, M-45, Kavach, PMA and Thiram @2% kept storage fungi completely under control. BHC kept the insect activity under check. None of tested fungicides/insecticides showed any sign of phytotoxicity. Sanitation, solarization and fungicides significantly reduced the amount of seed microflora.

v. Storage in a suitable container at dry place.

3. **Storage Period:** *S. sesban* and *L. leucocephala* seed treated with any one of above mentioned fungicides plus BHC (@ 2% each) could be stored in air-tight plastic-box for six years. *A. lebbek* seed treated with PMA plus BHC could be stored in paper or plastic boxes for five years. Best germination in *A. nilotica* was observed in cloth-bag after two years of storage.

4. **Storage Container:** Amount of microflora associated with the seed was less in non-perforated containers i.e. plastic-box, steel-box and poly-bag as compared to perforated containers i.e. cloth-bag and paper-bag.

5. **Longevity of storage fungi:** Amount of seed microflora significantly reduced with increase in storage period.

TABLE 1: *Microflora associated with seeds of S. sesban, L. leucocephala, A. lebbeck and D. sissoo*

Name of Fungus	*S. sesban*	*L. leucocephala*	*A. lebbeck*	*D. sissoo*
Aspergillus spp	+	+	+	+
Alternaria spp.	+	+	+	+
Helminthosporium spp.	+	+	+	+
Fusarium spp.	-	-	-	+
Penicillium spp.	+	+	+	+
Rhizophus spp.	+	+	+	+
Trichoderma spp.	-	+	-	-
Non-sporulating fungi	+	+	+	+

TABLE 2: *Seed born diseases and pests of some MPTS observed in seed storage and nursery*

MPTS	Fungus	Damage (%)	Control measure
D. sissoo	*Fusarium spp.*	*> 90*	*Soil drench with Dithane M-45*
S. sesban	*Alternaria spp.*	*70*	-
A. lebbeck	*Aspergillus spp.*	*> 80*	*Avoidance of excessive moisture*
A. lebbeck and A. nilotica	*Pulse beetle*	*100*	*Seed treatment with BHC*

CONCLUSION

The information emerged out of present investigation can be used to control fungi and insect pests associated with MPTS seed during long term storage under ambient conditions without using costly and cumbersome storage techniques, which are presently in vogue.

REFERENCES

1. Agarwal, V. K., and Sinclair, J. B. 1987. "*Principles of seed pathology. Vol. 2.*" CRC Press, Florida (USA).
2. Kumar, Anil, and Rao, G. R.1995. NFTRR. 13: 92-95.

Loss Assessment in *Albizia* Species due to Pod Boring Insects

CHITRA SHANKER AND K. R. SOLANKI
National Research Centre for Agroforestry, Jhansi (UP)

ABSTRACT

Albizia species is a commonly recommended agroforestry tree species for planting on degraded lands. This requires mass multiplication of seedlings. But the seeds of *Albizia* are extensively damaged by seed and pod boring insects. A study was undertaken in order to assess the extent of damage due to seed pests on three species of *Albizia* viz., *A. lebbeck, A. amara, A. procera*. The damage by seed bruchid *Bruchidius uberatus* and the seed lepidopteran was 38.63 percent in *A. lebbeck,* 21.06 per cent in *A. procera* and 17.2 per cent in *A. amara*.

INTRODUCTION

Albizia lebbeck, A. procera and A. amara are three tree species commonly recommended for agroforestry systems in arid and semi arid regions. *A. lebbeck* is also extensively planted under social forestry and other afforestation programmes for which large number of seedlings are required. But the seeds are infested by eight species of insects viz., *Bruchus bilineatopygus* Pic., *B. sparsemaculatus* Pic., *Bruchidius saundersi* Jakel, *B. schroderi, B. pisorum* Linn., *B. uberatus* Fab., *Caryedon gonagra* Fab., *Corydylocera livida* Hope in different parts of India (Singh and Bhandari, 1986; Joshi *et. al.*, 1990 and Harsh and Joshi, 1993). Therefore a study was undertaken to quantify the loss of seeds due to insects in three species of *Albizia* namely *A.*

lebbeck, A. procera and *Albizia amara*. Quantification of losses will therefore help in selection of a suitable species.

MATERIALS AND METHODS

The three tree species viz., *A. lebbeck, A. procera, and A. amara* were selected from three locations in and around the campus of NRCAF, Jhansi. Fifty mature pods were taken as sample size. Seeds were extracted and sorted based on visible symptoms such as bore holes and insect feeding and classified as damaged healthy seeds and damaged immature seeds and the percentage damage was statistically analysed.

RESULTS AND DISCUSSION

The percentage of immature seeds in *A. lebbeck, A. procera* and *A. amara* were 3.0, 35.0 and 38.0 respctively. The mean per cent damage by pod borers was observed to be maximum in *A. lebbeck* (38.63) followed by *A. procera* (21.06) and *A. amara* (17.2). Two types of borers inflicted damage viz., *Bruchidius uberatus* and a lepidopteran *Stathmopoda basiplectra*. The seed feeding lepidopteran larvae accounted for 60.68 per cent in *A. lebbeck*, 56.76 per cent in *A. procera* and 3.94 per cent in *A. amara*. The larvae preferred larger seeds which was the case in *A. lebbeck* and *A. procera* while *A. amara* had not only small seeds but the inter seed space is difficult to penetrate. The bruchid beetle accounted for 10% of the total infested seeds in *A. lebbeck* 33.78% in *A. procera* and 50.39% in *A. amara*. Size of the seed and inter seedal space influenced the extent of damage. There was no significant difference between the three locations. Up to 38.67% damage by insects in the field was reported from Jabalpur (Harsh and Joshi, 1993) and this was found to increase in storage as they were found to complete their third generation in storage (Joshi and Namdeo, 1990). Therefore it is of utmost necessity to take control measures right in the field or immediately after seed collection in order to restrict insect activity in storage.

TABLE 1: Visual classification of seeds of the three species of *Albizia*

Tree Species	Location 1			Location 2			Location 3		
	Per cent of seeds			Per cent of seeds			Per cent of seeds		
	Healthy	Immature	Damaged	Mature	Immature	Damaged	Mature	Immature	Damaged
A. lebbeck	39.3	11.9	48.8	53.3	7.96	38.75	67.6	4	28.4
A. procera	57.64	21.39	20.97	25.6	48.96	25.43	67.7	29.31	22.97
A. Amara	67.22	22.01	10.77	49.8	27.21	22.97	57.8	24.36	17.85

CD at P = 0.05% Tree Sp. Means = 3.1%; Location Means = 5.4%

ACKNOWLEDGEMENT

Authors are thankful to the Director NRCAF for the facilities provided and we also acknowledge the technical assistance provided by Rajendra Singh, T-4, NRCAF.

REFERENCES

1. Harsh, N. S. K. and Joshi, K. C. 1993. Indian Forester: 932.
2. Joshi, K. C. and Namdeo, R. K. 1990. Tech. Bulletin No. 3 Institute of Deciduous forests, Jabalpur,: 41
3. Joshi, K. C.; Meshram, R. B.; Usha Kiran and Ram Bhajan 1990. J. Trop. For.: 261- 263.
4. Singh, P. and Bhandari, R. S. 1986. Proc. National. Seminar on Forest tree seed, Indo- Danish Project on Seed procurement and Tree Improvement and Andhra Pradesh Forest Department, Hyderabad. Pp:155- 169.

Association of Pathogenic Diseases with *Dalbergia sissoo* Roxb. in North Bihar

M. S. ALI, V. K. SINGH, M. M. JHA, O. P. CHATURVEDI
S. MISHRA, AND R. C. JHA
Rajendra Agricultural University, Pusa (Bihar)

ABSTRACT

Studies on disease surveys of *Dalbergia sissoo* roxb. (Shisham) were carried out in certain parts of gangetic alluvial tract of north Bihar during 1994-1998. Damage and mortality in shisham seedlings were caused mainly by *Rhizoctonia* spp. in the nursery stocks, whereas sapling, mature and over mature shisham tree were mostly infected and damaged by *Fusarium solani* and *Ganoderma lucidum*. *F. solani* caused economic loss in term of biomass and productivity in the shisham tree ranged between 8-33%, whereas *G. lucidum* caused losses 3-5% of overall shisham population in north Bihar. Highest range per cent wilt disease incidence 22.0 to 24.0 was recorded in the districts of Gopalganj and Siwan during 1997-98. In certain older trees gummosis incidence was visible during the monsoon season and caused rotting of main trunk of the tree. All these diseases under varied stress conditions have resulted in economic loss of shisham timber in the area and affected the future plantation of the species by the farmers.

INTRODUCTION

Dalbergia sissoo Roxb. is the most common and valuable tree species found in north Bihar. It is adapted to various ecological conditions and is widely recommended for afforestation and reforestation programmes. Trees are planted with/without crops extensively under different eco-edaphic conditions viz., irrigated and riverine areas, along canals, around fields, degraded sites, wastelands, fallow uncultivated lands, low lying areas and on denuded hill sides.

During recent years, different growth stages viz., seedlings saplings and mature trees of *D. sissoo* have been victim of complex stresses. Trees have been exposed to the risk of injuries by biotic and abiotic agencies throughout their life due to poor site condition, mismatch and negligence. Among biotic stresses, pathogens adversely affect the growth patterns and development of trees causing great economic loss in term of biomass and productivity. The present study was under taken to assess the damage and mortalities by various pathogens to *D. sissoo* growing under plantation and natural conditions in north Bihar.

MATERIALS AND METHODS

The study was conducted in the north Bihar plains, broadly delineated into Zone-I (North West alluvial plain) and Zone-II (North East alluvial plain). Zone-I, comprises the districts of West and East Champaran, Gopalganj, Siwan, Saran, Sitamari, Vaishali, Madhubani, Darbhanga, Muzaffarpur and Samastipur.

Detailed study was made at Tirhut College of Agriculture, Dholi Kothi Farm located on the right bank of river Bhuri Gandak in the district of Muzaffarpur. It is situated at 26.41°N latitude and 34.6°E longitude with 52.8 m above mean sea level. The climate is semi-arid and sub-tropical, characterised with moderate precipitation, hot and dry summer and cold winter. About 90% of total rainfall of the year is received from south-west monsoon. The soil of the experimental farm is sandy loam in texture, with low organic matter.

Periodic surveys of different stages and stands of *D. sissoo* trees in the gangetic plains of the north Bihar were conducted during 1994-98 to ascertain the occurrence of different causal factors of diseases, mode of infection, extent of damage and mortality and predisposing factors associated with trees. Observation on abiotic and biotic stresses, association of diseases causing pathogens, soil types and textures were recorded. Besides, symptoms, incidence and intensity of diseases were also recorded. Diseased parts of seedlings, and available *Ganoderma spp.* were collected in the polythene bags and brought to the laboratory for isolation of the causal organisms. Different mycoflora were isolated on potato dextrose agar medium (PDA) and were purified by frequent transfer into fresh medium. Identification of casual organism was made on the basis of their microscopic and macroscopic appearance.

RESULTS AND DISCUSSION

The seedling stocks of *D. sissoo* were infected and damaged in the range 6-15% and 4-10%, respectively by *Rhizoctonia* spp. between June to November in nurseries. Different stages of *D. sissoo* were severely infected by soil born wilt pathogen, *Fusarium solani* from March to November between 1994 to 1998 which varied in the range of 8 to 36% and resulted into heavy mortalities in the range of 10 to 23%. Under monoculture plantations and natural stands of *D. sissoo* the incidence of wilt disease spreads rapidly in patches, resulting in large gaps due to mortality of wilted trees. Infected trees exhibited yellowing leaves in acropetal order from lower side to apical region of the tree. Young plants were killed soon after infection, while mature trees were wilted and killed more slowly. The leaves lost their turgidity within a few days. The stressed foliage withered off the branches and finally killed the whole stands. The vascular tissues in the outer layer of sapwood turned pink. This pathogen persists for many years and make the soil unfit for commercial crop production.

The root and basal trunk of old trees were damaged by *Ganoderma lucidum* in the uneven natural stands growing on degraded lands. Trunk gummosis was observed in mature trees. In the nursery conditions, black leaf spot, caused by *Rhyllochlora dalbergia* were observed in the foliage of seedlings and saplings in the range of 5 to 30% and caused 8 to 15% damage in pure stands.

The sequence of wilt diseases depend upon different field situations. Moderate to heavy wilt incidence were recorded from sapling to mature stage in different situation. Incidence of wilt diseases in plantations and natural stand conditions in north Bihar (during 1994 to 1998) revealed that highest wilt disease incidence (24%) in plantations and natural stands was recorded in the different localities of the district Gopalganj, followed by Siwan district (22%).

Effect of age group of *D. sissoo* on mortality due to wilt diseases revealed that 26 -30 years age group trees were most susceptible, whereas lowest mortality was observed in the age group 51-80 years (Table 1).

TABLE 1: *Effect of Fusarium solani on different age-class of D. sissoo Dholi Kothi Farm during 1994-98*

Tree age-class	No. of green tree	No. of wilted trees	Mortality
1-5	162	52	32
6-10	183	36	19.7
11-15	237	33	13.9
16-20	122	17	13.9
21-25	105	10	9.5
26-30	25	20	80
31-35	15	7	46.7
36-40	8	1	12.5
41-45	14	1	7.2
46-50	3	1	33.3
51-80	42	1	2.4
	916	179	19.5

Factors Involved in the Occurrence of Diseases

Generally shisham trees growing on marginal, degraded, wasteland and uncultivated lands have reduced vigour and immune functional system, which favours infestation by insects and infection of various pathogenic diseases causing severe damage to the species. The stress factors which influences and enhances the disease possibilities in shisham have been presented in Table 2. Most of the mortalities are related to seasonal flooding, water-logging, poor drainage, high moisture regime, close spacing, high density, soil compaction, debarking and monoculturing.

These varied disease complexes have been created due to various stresses. Bakshi and Singh (1954) reported that light soils with poor drainage of water, high pH, high temperature (30-35□) and poor nutritional status are known to favour high disease severity. In the present study, soils with high clay content and excessive soil moisture due to over watering were the main contributing factors to high diseases incidence and seedling damage and mortality. The stressed trees become susceptible to wilt disease. In this situation, trees can withstand some salinity but may fail entirely on saline soil with a shallow pan. Chaturvedi *et. al.*, (1984), reported that *D. sissoo* used 75% more water than *Eucalyptus tereticornis* but it does not tolerate water-logging which leads to stress and attack by wilt and root rot pathogens. Bakshi *et. al.*, (1957), reported heavy mortality of adult sissoo tree due to wilt and root rot. They also indicated that higher seedling mortality in nursery beds than polythene seedlings is due to the root contact of healthy and diseased seedlings.

TABLE 2: *Factors responsible for causing wilt diseases in north Bihar*

S. No.	Predisposing factors	Influencing factors	Contributing factors
1	*Climatic change*	*High rainfall*	*Fungi*
2	*Water logging*	*Poor soil structure*	*Wood and bark borer*
3	*Poor drainage*	*Problem soil*	-
4	*Nutrient deficiency*	*Nutrition upset*	*Nematodes*
5	*Poor soil drainage*	*Bark rotting*	*Virus*
6	*High moisture regime*	*High water saturation*	-
7	*Eroded root system*	*Sickened root system*	*Bacteria*
8	*Old age stands*	*Degraded land uncultivated land unprotective land*	*Domestic animals*
9	*Seasonal flooding*	*Erosion, siltation*	*Other biotic stresses*
10	*Poor site condition*	*Water stresses*	
11	*Soil compaction*	*Poor fertility*	
12	*Stagnated polluted water*	*Problem soils*	
13	*Heavy texture soil*	*Weakened sites*	
14	*Pure stand*	*High host preferences*	

ACKNOWLEDGEMENT

Thanks are due to the ICAR, New Delhi for providing funds to support this study through AICRP on Agroforestry.

REFERENCES

1. Bakshi, B. K. and Singh, S. 1954. Indian Forester 81: 276-281.
2. Bakshi, B. K.; Arora, K. K. and Singh, S. 1957. Indian Forester, 83: 555-558.
3. Chaturvedi, A. N.; Sharma, C. S. and Srivastava, R. 1984. Forestry Review. 63 (3).

Evaluation of Sustainable Productivity in MPTs based Farming System under Rainfed Condition

S. D. UPADHYAY, M. K. AWASTHI AND K. K. JAIN

Jawaharlal Nehru Krishi Vishwa Vidayalaya Jabalpur (MP)

ABSTRACT

MPTS based farming systems ensure economically viable and sustainable production in rainfed conditions. An evaluation of the role of MPTS in sustainable production has been made for *Albizia lebbeck, Leucaena leucocephala, Dalbergia sissoo and Acacia nilotica* through computing land equivalent ratio (LER). Under 3□3 m planting geometry of MPTS, the agricultural crops wheat, linseed, mustard and safflower were grown and the crop yield was estimated. The results revealed synergistic effect on tree as well as crop component in terms of total biomass production, *L. leucocephala* and mustard based cropping system found to be more sustainable having higher LER as compared to other combinations Allometric equations have also been developed for tree component for biomass estimations. Among various growth attributes (height, CD and DBH) and diameter^2 □ height (d^2h), d^2h was found to be the most appropriate predictor variable explaining 0.065 r^2 value for *L. leucocephala.*

INTRODUCTION

The most viable alternative for maintaining productivity is the deliberate incorporation of trees in farming system especially in rainfed conditions, which promote higher land equivalent ratio (LER) and increased per unit area biomass production. Keeping this in view, four prominent multipurpose trees (MPTS) of the area were selected and their growth, volume, LER and biomass values were quantified under rainfed conditions. Also, the productivity of MPTS based agroforestry system has been identified.

MATERIALS AND METHODS

The crops were grown under 3□3 m planting geometry of the MPTS for three consequent years under vertisols having pH 7.2, EC 0.11 mmhos/cm with moderate NPK. Mean annual rainfall of the area is 1354 mm. Crops were taken under rainfed conditions. Periodic observations of woody component and crop yield were recorded. Biomass estimations of woody component were developed by dimensions-weight allometric relationships. Land equivalent ratio (LER) was calculated as suggested by Rao and Coe, (1992).

RESULTS AND DISCUSSION

Among various growth attributes (height, CD and DBH) and diameter^2 □ height (d^2h), d^2h was found to be the most appropriate predictor variable explaining 0.65 r^2 value for *L. leucocephala* being higher MAI (1.12 m yr^{-1}). *L. leucocephala* also produced higher volume (216 m^3 ha^{-1}) and biomass (107.87 t ha^{-1}) followed by *A. nilotica, D. sissoo and A. lebbeck* (Table 1).

TABLE 1: *Growth of MPTS under study at the age of 10 years*

Species	d^2h (r^2 values)	Volume (m^3 ha^{-1})	Sp. gr. (gm. cc^{-1})	MAI (m yr^{-1})	Biomass (t ha^{-1})
A. nilotica.	0.45	116.00	0.68	0.83	78.88
A. lebbeck	0.52	57.00	0.71	0.60	40.47
D. sissoo	0.34	83.00	0.83	1.06	68.89
L. leucocephala	0.65	216.00	0.50	1.12	107.87

It is interesting to note that *A. lebbeck* produced wood volume only about 25% to that of *L. leucocephala,* but weight wise it produced actually about 42% owing to its higher specific gravity (0.71 gm cc^{-1}).

L. leucocephala + mustard based cropping system found to be more sustainable having higher LER (1.69) which is very well comparable to *A. nilotica* + linseed combination, *Acacia* and *Albizia* with safflower and *Dalbergia* with all four crops combinations (Table 2) were having almost equal LER (1.65).

TABLE 2: *Land equivalent ratio (LER) as sum of relative yields of the component species under woody perennials*

Woody component (MPTS)	LER crop component			
	Wheat	Linseed	Mustard	Safflower
A. nilotica	1.60	1.67	1.61	1.65
A. lebbeck	1.63	1.62	1.58	1.65
D. sissoo	1.65	1.65	1.64	1.65
L. leucocephala	1.58	1.40	1.69	1.55

Here LER is used as quantifiable productivity measurement which indicates the relative land requirements for intercrops versus monocultures (Mead and Willey, 1980). Advantageous and most relevant productivity has been observed under four different MPTS based agroforestry systems as shown by maximized LER values alongwith growth parameters of woody perennials.

REFERENCES

1. Mead, R. and Willey, R. W. 1980. Experimental Agriculture, 16: 217-228.
2. Rao, M. R. and Coe, R. 1982. Agroforestry Today 4 (1): 4-8.

Kadam - Its Prospects as Multipurpose Tree under Agroforestry Systems

M. S. HAQUE AND S. KANNAPIRAN
NABARD, Mumbai (Maharashtra)

ABSTRACT

Agroforestry systems have been practised by Indian farmers for a long time although to a limited extent, as in "*Taungya systems*". The research activities are continuing in various research institutes on agroforestry, unfortunately we have not yet been able to select a good number of tree species suitable for different agroclimatic regions of the country. As is well known, various clones of *Populus deltoides* (poplar) are one of the best tree species for agroforestry and millions of trees have been raised under this system. However it has the limitations in that it can not be grown everywhere or elsewhere. In this context *Anthocephalus cadamba* (kadam) has been identified as a promising agroforestry species. India's almost 80% of plywood industries are located in north eastern and eastern region and many are facing closure because of non-availability of raw materials. Kadam which is not only an indigenous tree but also fast growing, yielding high economic returns within a 8 year rotation, can solve partly the raw material requirements of the industry, if raised scientifically under agroforestry systems. Kadam being a commercially viable tree, the farmers get very high returns after harvesting. It has been estimated that for raising one hectare kadam plantation, the IRR works out as high as 31.41.

INTRODUCTION

Agroforestry is an age old concept that has now been renewed and invigorated by the ingress of technology. Farmers, especially those in the tropics, have a long tradition of raising food crops, trees and animals together as well as exploiting a multiple range of production from natural woodlots. But foresters and agriculturists, who have traditionally operated within their rigid disciplinary boundaries concentrating on monocultural production of their preferred commodities of crops, animals and trees, used to ignore such combined integrated production system. Agroforestry, thus, is a sustainable land use system that maintains or increase total yield by combining food crops (annuals) with tree crops (perennials) and/or livestock on the same unit of land, either alternately or at the same time, using management practices that suit the social and cultural characteristic of the local people and the economic and ecological conditions of the area.

AGROFORESTRY TREE SPECIES

There are only a few trees that have been tested and found suitable for agroforestry in India. The most widely known agroforestry species in our country is the *P. deltoides*. But it has certain limitations like, it can profitably be grown only in northern state viz. U.P., Punjab, Haryana, H.P., J&K etc. Therefore, search was going on for identifying suitable agroforestry species for the rest of the country. It has now been observed that kadam is a very good agroforestry species for almost entire country on suitable sites.

Kadam has been chosen for agroforestry because of the following reasons:

- It is indigenous fast growing species that gives early economic returns, within 8-10 years.
- The tree is straight, conical with small primary branches and there is hardly any secondary branch on the tree. Hence the land area required by the tree is much less.
- It can tolerate pruning and it also sheds off old branches on its own.
- It is an industrial tree since it produces one of the best raw materials for plywood industry. Besides, kadam wood can be used

as match splinters, packing cases, panel boards, pencils, sports goods and various other products. Thus market for kadam is assured.

- It is normally planted with wider espacement, thus enough of interspace is available for growing intercrops.
- Kadam can be grown in agricultural fields in association with many agricultural crops. Farmers can continue to grow crops during the entire project periods with suitable change in cropping pattern and thus can earn substantial income form sale of such crops besides the income from kadam tree.
- Nursery and plantation technology for kadam have been well standardized.

SITE REQUIREMENTS

Kadam grows naturally in West Bengal, Assam and many other parts of India and is therefore well suited for the country. However, the quality of the site and the type of care given to the trees will have a bearing on the growth of the trees. The site should be elevated, well drained and should not get affected by flood, loamy and well-drained soil is preferred. It can be very profitably cultivated on rainfed agricultural lands, marginal lands and also on certain kinds of wastelands viz. uplands, etc.

Inter Cropping

Dry paddy can be cultivated as an intercrop up to 3 years without much difficulty. Once the trees become big it is desirable to change the cropping pattern i.e. ginger, turmeric, arvi, etc. besides vegetable, pineapples, arhar and pulses can be taken as intercrops in kadam plantations.

UNIT AREA REQUIRED

Kadam trees are planted during the monsoon. The trees are generally planted at an espacement of 5□5 m accommodating 400 trees per hectare, thus leaving ample space for growing other crops. Since it is a well known fact that, only a few farmers will be able to spare one ha land, available unit may be one that will accommodate at least 100 trees. Where the farmers are not interested in intercropping, a closer espacement of 4□4 m may be used. However, this is suggested when wasteland are used.

COST OF CULTIVATION

The total cost over a period of 10 years for raising 400 trees in 1 ha. of land has been estimated at Rs. 33,930.00. For a minimum unit of 100 trees the cost will be Rs. 7,990.00 (assuming a wage rate of Rs. 35.00 per manday). Cost of intercropping is provided for Ist year only. From the income of intercropping, it is presumed that farmers will meet the cost of intercropping in the subsequent years.

MARGIN MONEY/BANK LOAN ETC.

National bank stipulates beneficiary's contribution to the project cost in order to ensure his stake in the involvement. Such margin money varies from 5 to 25% defending upon the type of investment and the class of borrowers. It is only 5% in case of small and marginal farmers, thus the bank loan may be 95% of the project cost.

YIELD AND ROTATION

Tree Crop (Kadam)

Kadam is a very fast growing tree especially under farm forestry conditions. It has been observed that the tree reaches height of 3-4 m in first two years and attains a girth of 30 cm. during the period. Growth statistics of kadam, collected from India and abroad, establish the fact that kadam when planted at an espacement of 5□5 m under good growth conditions on a rotation period of 10 years would grow to an average size of 30 cm diameter (over bark) at breast height and yield an average of 0.4 cubic meter peelable timber per tree. Commercial volume table of kadam under forest conditions in North Bengal (girth class 91-120 cm) indicated a volume of 0.75 cubic meter wood. On a very conservative estimate, yield of 0.4 cubic meter wood can be easily obtained from a kadam tree under rotation of 10 years by which time each tree will have a girth varying from 90 cm to 1 m. Regarding rotation, we are apprehensive that people may harvest the trees much before the stipulated 10 year rotation because of very fast growth as has happened in case of poplar plantation that have been harvested in 6th year instead of the recommended 8th year. Thus, assuming a conservative yield of 0.4 cubic meter wood per tree and at the rate of Rs. 2100 per cubic meter a return of Rs. 840 per tree is almost assured. However, we have assumed a support price of Rs. 500 per plant for working out the economics.

Inter Crops

Agricultural crops like rice, ginger, turmeric, chilly etc. can be cultivated alongwith the tree crops i.e., kadam upto 3 years till the canopy gets closed. Since agricultural crops will be cultivated alongwith tree crops we may consider only 50% of average crop yield per hectare. However, as the model is viable with the income generated from kadam trees alone, we may not consider the yield and income of agricultural crops for working out the economics.

IRR, REPAYMENT SCHEDULE ETC.

Kadam plantation under agroforestry is both technically feasible and financially viable. The IRR of such plantation has been worked out to 31.41%.

MARKETING

Kadam is considered as one of the best raw material for plywood industries. Since most of the plywood industries are facing acute shortage of raw material, plywood manufacturer s will be glad to purchase the trees at remunerative prices. In fact, in case of some sanctioned kadam projects, plywood manufacturers association have offered buyback guarantee at a price of Rs. 840 per tree based on a yield of 0.4 cubic meter wood at the rate of Rs. 2100 per cubic meter. Thus marketing of kadam trees is almost assured and no problem in marketing is envisaged.

PROJECT SANCTIONED/PLANTATION RAISED

Kadam is an indigenous tree and has been raised for many years by the forest department of different states. However, it is only recently that its potential has been fully realised and is being raised scientifically under agroforestry system. Different wood based industries have come forward to promote this tree and nodal among them are WIMCO, KITPLY industries, Woodcrafts Products, etc. Since none of these companies have their on land, the kadam plantation is being raised on individual farmers's lands with technical and extension supports by the industries. Finance required for such plantations will come as bank loan from financial institutions. Already a project worth Rs. 100 crorers (approx.) has been sanctioned to WIMCO and KITPLY for raising kadam trees in Assam, Arunachal Pradesh and N.E., states. A few projects are in the pile line for raising kadam in North Bengal and Andman and Nicobar Islands and in other N.E. States. It is presumed that, once such plantations become successful the raw material needs of the plywood industries will be fully met indigenously and the industries will not be required to import wood, thus conserving valuable foreign exchange of the country.

Performance of Multipurpose Tree Species under Various Agroforestry Systems under Semiarid Conditions

A. S. GILL
Indian Grassland and Fodder Research Institute, Jhansi (UP)

ABSTRACT

Agriculture production is low and risky in semiarid conditions primarily due to erratic and low rainfall, poor fertility status of the undulating soils and large area confined to rainfed conditions. Introduction of the tree component in agriculture have numerous service functions, which make agroforestry a profitable enterprise. In the present paper an attempt is made to highlight the results of tree-crop interface based on on-farm and on-stations trials at NRCAF, Jhansi on agrisilviculture, agrihorticulture, boundary plantations and other agroforestry systems.

INTRODUCTION

Agroforestry systems have made its inroad in Indian agriculture. For the last two decades major contribution has been on the tree-crop interface pertaining to its productive roles. There are 4 possibilities in a given tree-crop interface (Huxley, 1983):

1. Tree grows better - agricultural crop grow worse.
2. Tree grows worse - agricultural crop relatively affected.
3. Both tree and agricultural crop do better than expected.

4. Both tree and agricultural crop adversely affected to some degree or another.

Basic factors for the success of the studies will rest on:

- □ Choice of species - especially for screening of trees and agricultural crops.
- □ The design of agroforestry system.
- □ The management of agroforestry systems.

AGRISILVICULTURE

Trees are of vital importance for semiarid conditions under Indian conditions. The climate is hostile and the undulating soils are poor in fertility status. Normally the trees are found growing on field/farm boundaries, scattered here and there. Beside providing basic material for the farmers, its extends numerous other roles like providing shade to livestock and human, fodder and fuel including medicines. The farmers are reluctant to raise the trees on their fields because of the fear that they will shelter birds and their roots will effect the crop yields. Therefore, systematic agri-silvicultural studies (trees + crops) were initiated at thc National Research Centre for Agroforestry, Jhansi (Gill and Deb Roy, 1997).

Studies were carried out with twelve important multipurpose tree species. The design was split plot with treatments imposed during 1989-90 duly replicated. In main plots there were 3 MPTS spacing (2 □ 4 m, 2 □ 6 m and 2 □ 10 m), 12 MPTS in sub plots and 4 intensive crop rotations in sub-sub plots.

Grain Yield from the Interspaces of MPTS

Wheat: Significant variations in grain yield of wheat were noticed for 4 years, cxcept during 1990-91. In initial stages, lowest productivity was recorded from the interspaces of *Eucalyptus* and in the later stages from the interspaces of *Leucaena*. Due to poor growth of *Madhuca*, there was minimum competition between tree and the crop, therefore, maintained fairly good yields during all the 5 years. Productivity was also fairly good from the interspaces of *Acacia's, Casuarina* and *H. binata* (all NFT's) as well as from *Melia* (this tree tends to shed its leaves during the crop season thus allowing sufficient light penetration). Due to enlargement of plant canopy noticed in *Leucaena* and *Albizia*, wheat productivity was affected.

Chickpea: Gradual reduction in productivity was noticed with the advancement in the age of MPTS in case of Chickpea, a leguminous crop. Grain yields were significantly affected during all the 5 years, except during 1992-93. During 1992-93, 50 per cent of the MPTS were removed from all the treatments thus, increasing the plant spacing treatments from 2□4 m, 2□6 m and 2□10 m to 4□4 m, 4 □ 6 m and 4 □ 10 m respectively.

Chickpea yields were fairly high from the interspaces of *Madhuca* as expected followed by *Melia*, as earlier stated that *Melia* has the tendency to shed its leaves during the crop season (winter). Chickpea also gave better performance from the interspaces of *H. binata*, a MPTS having tiny leaves and upright growth characters (an ideal species for agroforestry systems). *Leucaena* and *Albizia*, due to enlargement in plant canopy came in the way of lowering the yields.

During the course of the trial, the MPTS were regularly pruned 50 per cent of their height to reduce the competition between trees and crops. Also during the summer season root pruning was a continuous process.

Therefore, success of the agri-silvicultural studies lies in proper selection of MPTS, its timely pruning of foliage and roots and the proper choice of crop to be raised in the interspaces. Due to hostile climatic conditions and poor fertility status prevailing in the semiarid conditions, crops do compete for nutrients, moisture and light especially during the winter season. Association of trees is an advantageous proposition which play numerous roles, in a agri-silvicultural system.

AGRIHORTICULTURE

India is one of the leading countries in fruit production only next to Brazil and China. In the semiarid conditions fruit tree species play a very important role in the farmer's economic front. In the initial years of fruit trees establishment, the interspaces can be successfully exploited for raising food grains, etc. Fruit tree species in the semiarid conditions after 4-6 years of establishment provide assured production. There are several hardy fruit tree species growing successfully in the semi arid conditions, preferably Ber (*Zizyphus*), Guava, Anar, Aonla, *Citrus* speceis, Karonda, etc.

On-Station Trial (Irrigated)

Studies at NRCAF, Jhansi were initiated to raise food grains in the interspaces of fruit trees during the establishment years (Gill and Deb Roy, 1998). Treatments comprised of three fruit tree spacings (5 □ 5 m, 5 □ 10 m and 10 □ 10 m) with four fruit trees (Guava, Ber, Anar and Kinnow planted during 1988-89) and 4 intensive crop rotations raised in their interspaces (groundnut-wheat, sorghum-wheat, groundnut-chickpea and sorghum-chickpea) in replicated split plot design. Additional control plots of sole cropping and sole fruit trees were also kept. During 1989-90, a fast growing nitrogen fixing tree species (*L. leucocephala*) was also introduced with 400, 200 and 300 plants per hectare in 5□5 m, 5□10 m and 10□10 m fruit tree spacings, respectively. *Leucaena* was coppiced each year before the sowing of rabi crops for foliage (fodder) and fuel production.

Wheat: On an average (1989-94), from the interspaces of fruit trees a grain yield of 23.5 q/ha was recorded with a reduction in yield in the order of 17.2% only i.e. a relative yield of 83.3%. Among the 4 fruit trees, maximum grain yield (24.8 q/ha) was registered from the interspaces of Kinnow with a relative yield of 88.1% i.e. only 11.9% reduction in grain yield. Similar production was recorded from the interspaces of Anar and Guava. In case of Ber, wheat grain yield was 20.9 q/ha (74.3% relative grain yield). However, this is not the true picture as during harvesting of ber fruits wheat plants are damaged.

Chickpea: On an average (1989-94), chickpea grain yield was 6.7 q/ha with a relative grain yield of 50.4% i.e. nearly 50% reduction in the production. Among the fruit tree species, the trend was similar as recorded in wheat, but the reduction in production by growing chickpea in the interspaces of fruit trees were very much high as compared to wheat. In Kinnow, Anar, Guava and Ber reduction in production was in the order of 47.2%, 47.8%, 50.9% and 52.2% respectively.

The reduction in production of wheat was compensated to a great extent with the production of fodder and fuel due to coppicing of *L. leucocephala* each year. In case of Ber, loss in production of wheat was further compensated with the production of foliage and fuel due to pruning of Ber.

Based on the five year results it is recommended to raise wheat rather chickpea in the interspaces of fruit trees. It is also desirable to grow fast growing nitrogen fixing tree species in between the fruit trees for additional benefits and profits. These trees also extend shelter to the fruit trees during the hot summer months as well as enrich the soil. Its foliage, as being NFT is enriching in protein. Among the fruit trees, it is recommended to raise Ber as it is an ideal plant for the semi arid conditions, beside fruit production it also provides foliage and fuel production. It is, therefore, much profitable to exploit its interspaces successfully and profitably.

On-farm Trials (Irrigated) at NRCAF, Jhansi

Village Karari: Agrihorticultural studies were initiated during 1989-90 with planting three fruit tree species saplings in 6 m □ 6 m and crops were raised in its interspaces in replicated randomised block design. During *rabi* wheat was raised in the interspaces of the fruit trees as per the treatments. Interestingly, on the basis of six year data highest yield of wheat was recorded from the interspaces of citrus, on an average (Gill and Rai, 1995).

Village Bhattagaon: Agrihorticultural studies were initiated during 1990-91 with three citrus species planted in 6 m □ 6 m spacing with two crop rotations in a replicated randomised block design. Mean wheat yield data of 4 years showed that maximum yield of 3.56 t/ha was obtained from the interspaces of Mousambhi followed by nibbu (3.49 t/ha) and minimum in kninnow (3.38 t/ha). In case of chickpea, 3 year data revealed that grain production was lowest in association with mousambhi in all the years. Pooled yield of 3 year showed that chickpea gave at par production from the interspaces of nibbu and kinnow and lowest production (0.76 t/ha) was recorded with mousambhi (Gill and Rai, 1995).

Village Simardha: Agrihorticultural studies were initiated during 1992 with four fruit tree species and crops in their interspaces in a replicated randomised block design. Pooled grain yield of wheat in 3 years exhibited slightly higher yield from the interspaces of citrus species (3.36 t/ha). However, significant variation was not observed for grain yield of wheat during 1992 to 1994 from the interspaces of fruit tree species (Gill and Rai, 1995).

Under Rainfed Conditions

Introducing hardy fruit tree component into the agro-horticultural system under rainfed conditions appears promising. In this direction on-station trials were carried out at NRCAF, Jhansi in 1989-90.

Aonla (*Emblica officinalis*): Four important cultivars of Aonla were tried during 1989-90 comprising of Chakaiya, Kanchan, Krishna and NA-7, alongwith and without Leucaena and kharif crops in their interspaces. Fruit bearing in aonla had started after 4 years of its planting (1992-93). The fruit yield after 8th year was highest (75-120 kg/plant) in Kanchan, followed by NA-7 (55-108 kg/plant). The lowest fruit yield was reported in Chakaiya (20-60 kg/plant). Besides fruit yield of aonla, 3-4 quintals of grain yield of Black gram was from the aonla interspaces.

Ber (*Zizyphus mauritiana*): The experiment was conducted with four varieties of ber (Seo, Gola, Banarasi Karaka and Deshi) raised under 2 spacings (6 □ 6 m) and (6 □ 12 m) and kharif crops raised in their interspaces. Fruit tree saplings were planted during the monsoon season of 1989-90. Fruit bearing in ber started during 1992-93 (4th year). Highest fruit yield was found in Banarasi karaka (10.93 q/ha) followed by Seo (10.57 q/ha). Lowest ber fruit yield was achieved in Deshi (4.73 q/ha). On the basis of average productivity of 6 years (1992-97) the system provided 8.94 to 10.27 q/ha fruit from ber and 0.68 to 3.74 q/ha grain from different crops. Besides fruit and grain production from the system, annual pruning of ber provided 3.85 to 4.75 q/ha fuelwood and 0.63 to 0.72 q/ha fodder (leaves). Among different crops black gram was found suitable intercrop with ber with a average yield of 3.74 q/ha every year.

REFERENCES

1. Gill, A. S. and Rai, P. 1995. In: Achievements - IDRC sponsored agroforestry project. NRCAF, Jhansi. pp. 5-10.
2. Gill, A. S. and Deb Roy, R. 1997. Indian Farming 47 (6): 21-22.
3. Gill, A. S.; Deb Roy, R. and Bajpai, C. K. 1997. Indian Farming 47 (6): 25-30.
4. Gill, A. S. and Deb Roy, R. 1998. Indian Farming 47 (12): 25-27.
5. Gill, A. S. 1998. Indian Farming 48 (2): 27-29.
6. Huxley, P. A. 1983. Working paper No. 13, ICRAF, Nairobi, Kenya.

Influence of *Faidherbia albida* on Growth and Yield of Short Duration Pulses

G. PRATIBHA AND G. R. KORWAR
Central Research Institute for Dryland Agriculture, Santoshnagar, Hyderabad (AP)

ABSTRACT

A field experiment on tree-crop intercropping was conducted for two years (1995-1996) on a red loamy sand soil representing Alfisol and related soil order at Hyderabad to evaluate the influence of *Faidherbia albida* on growth and yield of three short duration pulses. The trees were at two densities, viz., 625 (high) and 156 (low) per hectare. Crop yields, and growth parameters, viz., Total phytomass, Leaf Area Index, Crop Growth Rate and Relative Growth Rate were studied. Results indicated that total phytomass, LAI, CGR, RGR were higher in sole cowpea than under trees. Whereas, greengram and black gram recorded higher total phytomass, CGR and RGR under trees at low density. The three crops suffered under trees at high density. The results clearly indicated that short duration legumes can be successfully intercropped with *F. albida* trees at low densities.

INTRODUCTION

Faidherbia albida (Syn. *Acacia albida*) is an important component of parkland agroforestry systems in sub-saharan and Eastern Africa (CTFT, 1989). It is reported as a multipurpose tree of great potential for India (Vandenbeldt, 1993). The positive effect of the species on crop growth (the *albida effect*) has been well documented and reported from natural stands on farmers fields (Poschen, 1986).

However, recently some on-station studies have shown decline in crop yields when associated with *F. albida*, especially under high tree densities (Jama and Getahun, 1991). The adverse effect was attributed to shading. It was concluded that understanding the mechanism regulating leaf fall/retention phenomena of *F. albida* is crucial. Some authors attributed *albida effect* partly due to the pre-existing islands of soil fertility rather than solely due to the fertility built up by the tree (Geiger et.al., 1992). Competition factors frequently negate beneficial fertility effects (Sanchez, 1995).

Studies at Hyderabad (ICRISAT-ICAR, 1993) indicated that, in semi-arid tropics (SAT), though the tree becomes leafless with the onset of monsoon (June month), yet the leaves reappear fast and by the end of July tree canopy is fully developed. This severely restricts light to associated sorghum crop thus reducing its yield (by 48%). It was then suggested that perhaps short duration crops could do well in association with *F. albida* in SAT region. The objective of this study was to examine the growth and yield of short duration pulses when grown under *F. albida* at high and low densities.

MATERIALS AND METHODS

A field experiment was initiated in 1995 at Hayathnagar research farm of Central Research Institute for Dryland Agriculture, Hyderabad (17° 27' N latitude and 78° 28' E longitude, altitude 515.6 m above MSL), on *F. albida* trees planted in June 1984 at a spacing of 4 □ 4 m. In 1989 in half of the area alternate trees and rows were cleared to get a spacing of 8 □ 8 m as well. The experiment was laid in strip plot design, with 3 horizontal factor treatments (*F. albida* at 4 □ 4 m, *F. albida* 8 □ 8 m, and no trees) and three vertical factor treatments of arable crops viz., cowpea (C-152), black gram (T-9) and green gram (ML-267). The experiment was replicated 3 times. The crops were spaced at 40 □ 10 cm, the net plot size was 64 m^2. A basal dose of 20 kg N and 40 kg P_2O_5 per hectare was given. The study was carried out in two seasons, viz., 1995 and 1996. The soil type of the experimental site was red loamy sand representing alfisol and related soil order. Average annual rainfall is 730 mm.

The data was recorded for crop yield in both control and with trees. In each plot five plants were selected at random at vegetative, flowering, pod formation and harvesting stages. Plant samples were dried at 70°C and then weighed. Mean CGR, RGR and LAI were calculated (Radford, 1967).

RESULTS AND DISCUSSION

Total Dry Weight

Most accurate and precise method of expressing growth is by the phytomass production. There was a reduction in total dry matter in all the three crops with high tree density. However, increase in phytomass of green gram and black gram was observed when grown with trees @ 8 □ 8 m spacing. However, in cowpea, sole crop recorded higher phytomass. This low phytomass in cowpea under the tree is mainly due to reduction of vegetative and reproductive growth which may be due to low light intensity. Similar observations were made by Neves *et. al.* (1982).

Growth Analysis

The photosynthetic efficiency of a crop depends not only on number of leaves, but also on leaf area per unit ground area. Hence, on optimum leaf area index (LAI) is required for higher yields. Higher crop LAI was observed under trees @ 8 □ 8 m spacing upto vegetative stage, later the LAI decreased with increase in shading. Similar results were observed in cowpea by Tarila et.al. (1977). Contrary to this in green gram and black gram higher LAI was observed @ 8□8 m tree spacing in initial stages, but at later stages the LAI increased with increase in shading. Such synergistic effect of shading on LAI was reported by Schoch and Santos (1974) in *Vigna sinesis*. The increase in LAI under shade at later stages of crop growth may be due to delay in the senescence of lower plant leaves.

The total dry weight (TDW) per unit leaf area (LA), an indicator of the photosynthetic efficiency, increased with the age of the crop upto harvest. Sole cowpea recorded maximum TDW per LA compared to the crop grown with close and wider spacing of trees. On the other hand, green gram and black gram grown under 8 □ 8 m spacing recorded higher TDW per LA than sole crop and the crop under 4 m □ 4 m spacing. This may be due to higher total dry matter production.

Crop growth rate (CGR) increased gradually upto flowering and decreased thereafter. Cowpea registered higher CGR in sole cropping compared to the crop grown under trees. This low CGR may be due to reduced light intensity. Similar results were recorded by Hiremath (1983). Whereas, in green gram and black gram the CGR was maximum at 8□8 m spacing compared to 4□4 m spacing and sole cropping.

Relative growth rates (RGR) of all the three crops increased upto 60-75 days and declined thereafter. In green gram higher RGR was recorded in 4 □ 4 m spacing upto 60-75 days. But at later stages intercropping of black gram and green gram at 8 □ 8 m spacing recorded maximum RGR, whereas sole cowpea recorded highest RGR.

Seed Yields

The seed yields of green gram and black gram improved at low densities of *F. albida* whereas, such yield improvement was not noticed under higher densities (Table 1). The improvement of seed yields under 8 □ 8 m may be due to improvement in soil nutrient status. The lower yields of black gram and green gram under 4 □ 4 m spacing, and cowpea in both the densities of trees might be due to reduction in light interception which in turn affected the growth of the pulse crops, viz. CGR, TDW, LAI under the trees.

TABLE 1: *Effect of F. albida on yield of intercrops*

Treatments	Crop yields (kg/ ha)		Haulm yields (kg/ ha)	
	1995	1996	1995	1996
625 Trees				
Greengram	*517*	*469*	*3060*	*3533*
Cowpea	*25*	*69*	*3157*	*3540*
Blackgram	*325*	*357*	*1951*	*2339*
156 Trees				
Greengram	*822*	*763*	*3543*	*4387*
Cowpea	*869*	*713*	*3835*	*4856*
Blackgram	*1146*	*1083*	*2341*	*3490*
Sole crop				
Greengram	*731*	*635*	*2708*	*3438*
Cowpea	*999*	*836*	*3872*	*4930*
Blackgram	*901*	*871*	*1661*	

The present study suggests that cowpea is more sensitive to light compared to green gram and black gram. Hence green gram and black gram can be successfully intercropped under *F. albida* in marginal lands of Telangana region of AP for higher productivity.

REFERENCES

1. CTFT. 1988. *Faidherbia albidai (Del.)* A. Chev. (Synonym: *Acacia albida* Del.) Monographie. Centre Technique Forestier tropical, Noget-sur. Marne, France.
2. Geiger S. C.; Vandenbeldt R. J. and Manu, A. 1992. In: Vandenbeldt R. J. ed., *Faidherbia albida* in the West African Semi-arid Tropics, ICRISAT, India: 121-124.
3. Hiremth, M. S. 1983. Thesis submitted to UAS, Bangalore.
4. ICRISAT-ICAR. 1993. In: ICRISAT-ICAR, proceedings of CRIDA-ICRISAT (RMP) Collaborative Research Projects planning meeting held at ICRISAT centre on 26/04/1993, ICRISAT, India: 10-13.
5. Jama B. and Getahun, A. 1991. Agroforestry Systems, 14: 193-205.
6. Karim, A. B.; Savil, P. S. and Rhodes, E. R. 1991. Agroforestry Systems, 16: 203-211.
7. Neves M. C. P.; Summerfield R. J. and Minchin F. R. 1982 Tropical Agric. 59: 248-253.
8. Poschen, P. 1986. Agroforestry Systems, 4: 129-143.
9. Radford P. J. 1967. Crop Sci. 7: 171.
10. Schoch P. G. and Santos C. L. 1974 Turrrialba, 24: 84-89.
11. Tarila A. G. J.; Ormod D. P. and Adedipe, N. O. 1977. Ann. Bot. 41: 75-83.
12. Vandenbeldt R. J. 1993 Range Mgmt. and Agroforestry. 14: 159-169.

Management of Fruit Trees Suitable for Agroforestry System in Arid Region

R. N. PRASAD AND B. B. VASHISHTHA
Central Arid Zone Research Institute, Jodhpur (Rajasthan)

ABSTRACT

Due to peculiar climatic conditions of Indian arid zone and continuous increase in biotic pressure the process of desertification gets accelerated. Frequent tillage operations associated with arable crops further aggravate the problems due to soil erosion. Failure of rainfall at prime time leads to the total crop failure, resulting in total loss of yield. Thus, the monoculture of arable crops is neither environment friendly nor economically viable for fragile desert ecosystem. Therefore, horticultural crops mainly fruit trees such as jujube (*Ziziphus mauritiana),* pomegranate (*Punica granatum*), aonla (*Emblica officinalis*), bael (*Aegle marmelos*), gonda (*Cordia myxa*) etc. are the alternate crops for such ecosystem to stablize the crop productivity and improve the economy. Further, the combination of fruit crops with silviculture, pasture grasses and with arable crops can form an ideal farming system for overall development of the arid region.

Proper management like choice of fruit trees, planting season, spacing, irrigation, manuring, plant protection, etc. and techniques such as water harvesting techniques and shelter belt plantation should be incorporated for initial establishment of the plants. Thus, farming system with fruit crops would maximise and stabilise the productivity under arid conditions.

INTRODUCTION

The resource poor farmer behind the plough is the most neglected person in the rural societies of India. Simple and high input package designed for monocrop systems do not fit well either with complexity and diversity of resource poor farming systems or with their poor access to agricultural services and risk prone environment. In the arid region of Rajasthan, the estimated forage availability (dry matter) from range land, top feeds, crop residues, etc. is about 15.41 million tonnes against the total dry matter requirement of 17.46 million tonnes for the existing livestock population, showing a shortfall of 2.05 million tonnes (Sharma, 1994). Similarly, the consumption of fruits in the daily diet of rural mass is almost negligible as against the recommended requirement of 60 gm fruit per day (Chadha, 1993). Cultivation of fruit crops has remained in its infancy in the arid and semi-arid regions of the country, in spite of its great potential in these areas. Drought hardy fruit crops can survive and provide income to the farmers even under severe drought.

Necessity of fruit production assumes greater importance because of low calorie and protein diet consumed by the desert dwellers, which results into malnutrition amongst the poor masses. To manage the arid lands in judicious way, combination of forestry with horticulture can be an ideal solution for fulfilling the need of desert dwellers in terms of balanced diet and fuel wood, timber and other forest produce.

SELECTION OF SPECIES

In an integrated farming system approach, selection of fruit plants, trees/bush plants/ grass-legume is to be done keeping in mind the geo-physical characters of the soil, climatic conditions and last but not the least, the farmer's choice. The fruit crops suitable for different locations have been listed in Table 1.

TABLE 1: *Fruit crops for different agro-climatic conditions of arid region*

< 500 mm rainfall	500-750 mm rainfall	Partially irrigated
Lasoda or Gonda (Cordia myxa)	*Aonla (Emblica officinalis)*	*Sour lime (Citrus aurantifolia)*
Ker (Capparis decidua)	*Bael (Aegle marmelos)*	*Pomegranate (Punica granatum)*

< 500 mm rainfall	500-750 mm rainfall	Partially irrigated
Pilu (Salvadora oleiodes)	*Wood apple (Feronia Limonia)*	*Guava (Psidium guajava)*
Ber (Zizyphus mauritiana)	*Phalsa (Grevia subinaequalis)*	
Karonda (Carissa carandas)	*Mulberry (Morus alba)*	
Custard apple (Annona squamosa)	*Tamarind (Tamarindus indica)*	
Drumstick (Moringa olefera)	*Khirni (Manikara hexandra)*	

PROPAGATION AND PLANTING

The usual methods of propagation in nursery and their planting in the field will not be suitable under agroforestry system. The taproot of such plants is often lost as a result of repeated transplanting, a common practice with the nursery man. Plants of dryland conditions must have a root architecture with a strong tendency to penetrate deep in the soil. Therefore, softwood grafting on in-situ raised seedlings in guava, aonla, etc. is advocated. *In situ* budding in ber should be done to avoid mortality. However, rooted cuttings to pomegranate, gonad, mulberry, phallus, etc. should be planted during the rainy season to get better establishment. The planting should be done in such a way that runoff water is available to the roots for establishment and for better fruit production.

MANAGEMENT OF FRUIT CROPS

Water remains the foremost constraint in dryland horticulture, therefore, its conservation and judicious use is an important aspect. This can be achieved by adopting water harvesting, use of mulches and sub-soil barriers.

PRUNING AND TRAINING

Only one branch should be allowed from a node and these should be balanced around the trunk. Shoots emerging below the bud union must always be removed.

Pruning is seldom done in tamarind, aonla, lime lemon, etc. except to regenerate old, less production trees. Ber trees are pruned every year in the second week of May to induce new shoots on which the flowering and fruiting take place. In case where the animals like sheep and goat are allowed for grazing in the pasture, the budding should be done at the height of 1 metre and the plants should be trained so that the animals do not harm the fruit trees. In case of jujube or ber, the leaves can be fed to the animals after the harvest of fruits is over. The plants of pomegranate should be trained on single stem and branching should be allowed at one metre height. Phalsa should be heavily pruned in the month of February. Mulbery is also pruned at this time for inducement of new bearing shoots.

NUTRIENT MANAGEMENT

The fruit plants should be provided required nutrition. The requirement of N, P and K will depend upon the species and age of the plant. The doses of N, P and K are constant after 6 year of age of the plants. The doses of nitrogen should invariably be given in 3 splits and that of phosphorus and potash should be given once a year. Table 2 provides the general recommendation of nutritional requirements of various fruit crops.

TABLE 2: *Nutrition for fruit crops (More than six year of age)*

Fruit Crops	FYM (kg)	N (g)	P_2O_5 (g)	K_2O (g)
Citrus fruits	75	500	200	250
Ber	50	500	200	150
Pomegranate	50	500	200	150
Aonla	75	500	280	225
Sapota	75	500	240	300
Annona	20	450	150	250
Bael	40	400	250	250
Guava	75	500	250	300

MANAGEMENT SYSTEMS

Agri-horti Systems

In this system short duration kharif crops can be taken as inter crop between the rows of tree plants. It is always better to select a legume crop such as moong, moth, guar, etc. which, in addition to production of grains, will also fix nitrogen in the soil. One more point to be kept in view while selecting the crop is that the height of the inter crop should not be more than tree plants of young age, otherwise the growth of fruit trees will be suppressed. The intercrops can safely be taken till the plants come into bearing after which it will largely depend upon the competition between fruit trees and crop for moisture and nutrition.

Horti-pasture System

This system is especially beneficial in areas where animal husbandry is the main occupation of the region as in western Rajasthan. The work conducted at CAZRI, revealed that the fruit trees should be established ahead of planting the grasses. When planted simultaneously, heavy mortality was observed even in *Z. nummularia*, which is otherwise a hardy species.

The grass (*Cenchrus ciliaris*) competed for the moisture and did not allow *Z. nummularia* to establish (Sharma and Vashishtha, 1985). The fruit trees in this system should be so trained that the goats do not cause harm. In case of ber, the budding should be done at a height of one metre and subsequently pruning should be done in such a way that the plant grows upright. By planting grass and ber, the farmers can get fodder (grass+ber leaves), fruit and fuel (pruned wood).

PLANT PROTECTION

Fruit plants are attacked by various pests and diseases. In addition to the polyphagus insects pests and disease, there are some specific pests and disease of fruit crops, which need managment. Ber is attacked by fruit fly (*Carpomyia vasuviana*) which can be controlled by two sprays of any systemic insecticide (*metasystox, rogor, dimecron, etc.*) 0.03% active ingredient. The first spray should be done when the fruits are in 'pea stage', followed by second spray two weeks later. It is also attacked by powdery mildew (*Oidium* sp.) which can be controlled by spray of 0.1% Karathan. Black or brown spots on pomegranate fruits are due to anthracnose disease which can be controlled by the repeated sprays of copper fungicide (0.21%) viz. Fyloton, blue copper, blitox etc. Lime fruits are spoiled by citrus canker, caused by *Xanthomonas citri* bacteria. This disease spreads more during rainy season and can be checked by repeated spray of *streptocyclin* (50-100 ppm depending upon the intensity of disease).

In aonla, rust (*Ravenelia emblicai*) can cause serious losses. It can be controlled by the sprays of 0.1% Bayleton or 0.2 Dithane M 45.

Fortunately no disease or insect pests cause serious damage in custard apple, tamarind, phalsa, jamun, karonda, etc. There are many diseases and insect/pests of guava, citrus fruits which should be controlled timely.

REFERENCES

1. Chadha, K. L. 1993. Golden Jubilee Symposium Souvenir: 1-4.
2. Sharma, S. K. and Vashishtha, B. B. 1985. Annals of Arid Zone 24 (4): 308-9.
3. Sharma, S. K. 1994. Training course on horticultural crops for Arid and Semi-arid areas (Jan. 24-31), CAZRI, Jodhpur.

Early Growth Pattern of Some MPTs in Association with Pasture Species

M. M. ROY, B. K. CHOUBEY AND M. P. RAI
Indian Grassland and Fodder Research Institute, Jhansi (UP)

ABSTRACT

Studies on survival and growth pattern of three MPTS in association with different pasture combinations was carried out during 1994 - 1996 at Indian Grassland and Fodder Research Institute, Jhansi under rainfed conditions on red and mixed sandy loam soils. Highest average survival was recorded in *Dalbergia sissoo* (89 %) followed by *Leucaena leucocephala* (86.9 %) and *Acacia nilotica* (82.5 %). However, highest average growth was attained by *L. leucocephala* followed by *D. sissoo* and *A. nilotica.* In general, higher survival and early growth rates were achieved by all three MPTS without pasture. Survival and early growth rates were significantly less in the pasture combination involving *Dichanthium annulatum* + *Cenchrus ciliaris* + *Stylosanthes hamata* in all the three MPTS.

INTRODUCTION

The Indian National Forest Policy 1988 aims at achieving a minimum one third of the total geographical area of the country under forest or tree cover with the basic objective of improving the ecological balance of the country. Presently, forests occupy hardly 12 per cent of the geographical area of the country. The average growing stock of wood (65 m^3/ha) is below the world standard (110 m^3/ha). Also, the average productivity of forests in India is just 0.5 m^3/ha against the world average of 2.1 m^3/ha (FSI, 1989). The grazing pressure is heavy

- about 99 million ACUs graze in forests which has a carrying capacity of only 39 million ACU (Lal, 1990).

The greening of degraded forests following silvopastoral approach developed at Indian Grassland and Fodder Research Institute (IGFRI), Jhansi has great potential in conserving the soil, water, plant genetic resources and the environment which are the prime need of the day (Pathak and Roy, 1994). Growing of multipurpose trees (MPTS) with pasture species on degraded lands, offer many advantages at virtually no expense. However, there may be competition between tree and pasture species for resources like moisture, light and nutrients, especially on degraded lands in arid/semi-arid regions. The intensity of competition is usually more in establishment/early growth phase of the tree (Harsh *et. al.*, 1993; Pathak and Roy, 1994).

Early growth pattern of trees is an important attribute of silvopastoral systems. It is an indication of the role of trees in producing desirable biomass besides determining compatibility of trees with understorey species. In this study survival and early growth pattern of three important MPTS was assessed in association with various pasture combinations at Jhansi.

MATERIALS AND METHODS

The study was conducted at the Central Research Farm of Indian Grassland and Fodder Research Institute at Jhansi (20□27'N latitude and 78°35'E longitude and about 275 m above mean sea level) during 1993 -1996. The soil of the experimental site was a red and mixed (red + black) sandy loam, neutral in reaction. The soil was low in nitrogen and phosphorus content and medium to rich in potassium content.

In July 1993 three promising MPTS *viz., L. leucocephala* var. K-8, *D. sissoo* and *A. nilotica* were planted in 4 □ 4 m spacing in a randomised block design with three replications. The plot size was 20 □ 16 m. The space available between two trees in a line was utilized by *L. leucocephala* var. Cunningham at 1 m distance, maintained at a height of 60 cm. The interspace between two tree-lines was utilized by different pasture combinations (sown in lines 50 cm apart) involving *Cenchrus ciliaris* (CC), *Dichanthium annulatum* (DA) and *Stylosanthes hamata* (SH). (T1 = CC; T2 = DA; T3 = SH; T4 = CC + SH; T5 = DA + SH; T6 = DA + CC; T7 = DA + CC + SH; T8 = Only Tree).

The monthly meteorological data for the study period were collected from the Agro-meterological observatory at IGFRI, Jhansi. The study period received an average total rainfall of 778 mm, with 84 per cent rain during July - October. May was the hottest month with mean maximum temperature (42.8□ C) followed by June (40.2□ C). December was the coldest month with mean minimum temperature (5.7□ C) followed by January (6.5□ C).

The survival of trees was estimated by actually counting the living trees/plot. The tree growth parameters viz., height (H), canopy spread (CS), collar diameter (CD), diameter at breast height (DBH) were measured by following standard procedures outlined by Mac Dicken *et. al.* (1991). The data were analysed following standard statistical procedures (Panse and Sukhatme, 1978).

RESULTS AND DISCUSSION

Survival

After three years, higher survival was found in *D. sissoo* (84.0 - 94.5 %). It was closely followed by *L. leucocephala* (83.1 - 95.3 %). The survival of *A. nilotica* (74.8 - 86.2 %) was less when compared to the other two species (Table 1).

Among the three MPTS, the survival of *L. leucocephala* was most affected with pasture species. The survival was significantly higher in control plots (without pasture species) than in plots having pasture combinations *viz.,* DA + CC + SH, DA + CC, DA + SH, CC + SH and DA. The survival of *D. sissoo* was comparatively less affected with pasture species. The survival was significantly higher in control plots than in plots having pasture combinations viz., DA + C, DA + SH, DA + SH and CC + SH. The survival of *A. nilotica* was least affected with the pasture species. The survival was significantly higher in control plots than in plots having only one pasture combination DA + CC + SH (Table 1).

Rai *et. al.* (1995) studied survival of several MPTS with range grasses on gravelly soil at Jhansi. They found higher survival of *A. nilotica* (93.7 %) followed by *L. leucocephala* and *D. sissoo* (both 87.5 %). The survival of *A. nilotica* was least affected with pasture species in this study also. However, overall survival of this species was lower. This may be attributed to different soil conditions and pasture species in this study.

TABLE 1: *Survival (%) of some MPTS in association with different pasture combinations*

Treatment	Tree Species		
	A. nilotica	*D. sissoo*	*L. leucocephala*
T1	82.5	93.0	90.8
T2	82.9	92.6	87.3*
T3	86.2	94.5	91.4
T4	83.6	86.1*	85.4*
T5	84.9	84.2*	85.2*
T6	80.5	84.0*	83.1*
T7	74.8*	85.4*	77.2*
T8	84.6	92.4	95.3
CD (<0.05)	7.3	4.6	5.3

Growth in Height and Canopy Spread

After three years, higher growth in height (7.14 - 8.43 m) and canopy spread (3.18 - 4.06 m) was attained by *L. leucocephala*. It was followed by *D. sissoo* (4.69 - 5.82 m and 1.41 - 2.19 m) and *A. nilotica* (3.12 - 3.93 m and 0.59 - 1.18 m) (Table 2).

TABLE 2: *Growth in height and canopy spread (m) of some MPTS in association with different pasture combinations*

Treatment	Tree Species					
	A. nilotica		*D. sissoo*		*L. leucocephala*	
	H	CS	H	CS	H	CS
T1	3.40*	1.12	5.27	2.14	8.12	3.62
T2	3.53	1.08	5.10	2.10	7.90	3.65
T3	3.67	1.18	5.82	2.06	8.09	3.77
T4	3.38*	1.18	5.32	2.04	7.83	3.47
T5	3.36*	1.03	5.05*	1.94	7.87	3.56
T6	3.18*	0.87*	5.14	2.04	7.86	3.52
T7	3.12*	0.59*	4.69*	1.41*	7.14*	3.18*
T8	3.93	1.15	5.69	2.19	8.43	4.06
CD (<0.05)	0.47	0.24	0.64	0.58	1.12	0.63

Unlike survival, in *L. leucocephala* the growth in height and canopy spread was least affected with pasture combinations. The growth was significantly higher in control plots than in plots having only one pasture combination DA + CC +SH. In *D. sissoo*, the growth in height was significantly affected in DA + CC + SH and DA + SH pasture combinations. However, the growth in canopy spread was affected only in the former combination. In *A. nilotica*, the growth in height was significantly affected in several pasture combinations viz. DA + CC + SH, DA + CC, DA + SH, CC + SH and DA + CC. The growth in canopy spread was, however, affected only in two pasture combinations Da + cc + SH and DA + CC (Table 2).

The height growth attained by trees in silvopastoral/ agrisilvicultural systems has been studied by several workers. Deb Roy *et. al.* (1982), reported no marked difference in height growth of *Acacia tortilis* in association with *Cenchrus ciliaris* or *Cenchrus setigerus* at Jhansi. Rai *et. al.* (1995) reported peak height growth in *L. leucocephala*, followed by *D. sissoo* and *A. nilotica*, with range grasses on gravelly soils at Jhansi. Similar trend of growth has been found in this study.

Growth in Diameter

After three years, higher growth in CD (12.1 - 14.4 cm) and DBH (8.1 - 9.6 cm) was attained by *L. leucocephala*. It was followed by *D. sissoo* (6.8 - 9.4 cm and 4.7 - 6.2 cm) and *A. nilotica* (4.6 - 5.5 cm and 2.6 - 3.1 cm) (Table 3).

The growth in CD was not significantly affected in *A. nilotica* and *L. leucocephala* with different pasture combinations. In *D. sissoo,* however, the CD growth was significantly less in three pasture combinations DA + CC + SH, DA + CC and DA. The growth in dbh was not significantly affected in *A. nilotica* with different pasture combinations. The DBH growth in *L. leucocephala* was significantly low only in one pasture combination DA + CC + SH. In *D. sissoo*, DBH growth was significantly low in several pasture combinations namely, DA + CC + SH, DA, CC + SH and DA + SH (Table 3).

TABLE 3: *Growth in collar diameter and diameter at breast height (cm) of some MPTS in association with different pasture combinations*

Treatment	Tree Species					
	A. nilotica		*D. sissoo*		*L. leucocephala*	
	CD	DBH	CD	DBH	CD	DBH
T1	5.1	2.9	8.6	5.7	13.5	9.5
T2	5.1	2.7	7.8*	5.0	13.4	9.3
T3	5.2	3.0	8.7	6.0	13.7	9.3
T4	5.0	2.7	8.1	5.3	12.3	8.9
T5	4.8	2.6	8.0	5.3	12.1	8.7
T6	4.9	2.7	7.6*	5.4	12.4	8.9
T7	4.6	2.6	6.8	4.7	12.1	8.1*
T8	5.5	3.1	9.4	6.2	14.4	9.6
CD (<0.05)	NS	NS	1.6	0.9	NS	1.3

Comparative MPTS Growth

After three years, highest average survival was found in *D. sissoo* (89 %) followed by *L. leucocephala* (86.9 %) and *A. nilotica* (82.5 %). The differences were not statistically significant. However, there was a marked difference in average growth rates of these species at this site. Highest average growth, in all the parameters for this study, was attained by *L. leucocephala* followed by *D. sissoo* and *A. nilotica*. The differences in all the parameters were statistically significant both at 5 and 1 per cent levels of probability (Table 4).

TABLE 4: *Comparative growth characteristics of some MPTS in association with pasture species at Jhansi*

Species	Survival (%)	Growth Characteristics			
		Height (m)	Canopyspread (m)	CD (cm)	DBH (cm)
A. nilotica	82.5	7.91	3.61	13.00	9.04
D. sissoo	89.0	5.30	1.99	8.13	5.45
L. leucocephala	86.9	3.45	1.03	5.03	2.79
CD (0.05)	NS	0.33	0.11	0.88	0.53

Similar trend of tree growth has been reported by Rai *et. al.*, (1995), in association with range grasses on gravelly soils at Jhansi. However, unlike in this study they reported higher survival of *A. nilotica* as compared to *D. sissoo* and *L. leucocephala.*

ACKNOWLEDGEMENTS

The authors are grateful to Director, IGFRI, Jhansi for encouragement and providing facilities for this work.

REFERENCES

1. Deb Roy, R.; Patil, B. D.; Pathak, P. S. and Gupta, S. K. 1982. Indian J Range Mgmt., 3: 45-51.
2. FSI 1989. State of Forest Report. Forest Survey of India, Dehra Dun.
3. Harsh L. N.; Tewari, J. C. and Puri, S. 1993. In: Agroforestry in South Asia (Eds.: W. R. Bentley, P. K. Khosla and K. Seckler). Oxford & IBH Publishing Co., New Delhi,: 315-328.
4. Lal, J. B. 1990. Indian Forester, 116: 431-441.
5. Mac Dicken K. G.; Wolf, G. V. and Briscoe, C. V. 1991. Standard Research Methods for Multipurpose Trees and Shrubs. WIIAD, Arlington (USA),: 91.
6. Panse, V. G. and Sukhatme P. V. 1978. Stastical Methods for Agricultural Workers. ICAR, New Delhi,: 347.
7. Pathak P. S. and Roy M. M. 1994. Silvipastoral System of Production. IGFRI, Jhansi, pp. 55.
8. Rai, P.; Deb Roy R. and Rao, G. R. 1995. Range Mgmt. and Agrof., 16: 103-113.

Effect of Chirpine on above Ground Biomass of Herbage Layer and Carrying Capacity of the Natural Silvipastoral System in North-West Himalaya

G. R. RAO, V. K. MISHRA*, B. GUPTA*, K. R. SOLANKI AND ANIL KUMAR
National Reserach Centre for Agroforestry, Jhansi (UP)
** Dr. Y S Parmar University, Solan (HP)*

ABSTRACT

The study explored the effect of widely spaced naturally grown trees of chirpine on aboveground biomass productivity and carrying capacity of herbage layer in North-West Himalaya. It included three crown classes, four directions and three distances as treatments. Sampling of herbage was carried out at monthly intervals starting from July 15th to October 15th. The size of the quadret was 30 □ 30 cm and number of quadrats per replicate were twelve under tree crown and four (one each/direction) outside the tree, acting as control. The aboveground biomass of herbage increased from July to September, in both environments and declined thereafter in October. In both environments total aboveground biomass was higher in open grasslands as compared to that under tree canopies. The variation in aboveground biomass of herbage with respect to the crown class and direction effect was not perceptible, however with increase in distance from tree base it enhanced markedly. Statistical analysis of the aboveground biomass in dominant species evinced similar kind of response as observed for total aboveground biomass of the herbage i.e. it attained significantly higher values in open grassland and a substantial enhancement with increase in distance from tree base. With regards to grazing capacity and carrying capacity, they fallowed similar trend as was observed in case of aboveground biomass.

INTRODUCTION

During the evolutionary history of man since its beginning in the pleistocene period, forests and grasslands have been important. There is a long history of development of grasslands and coexistence of grazing animals through their evolution. Silvipastoral systems involving a large number of trub (trees + shrubs) species and various management intensities, ranging from extensive nomadic silvipastoralism to very high intensity cut-and-carry fodder systems, have been practiced in various countries.

In India grazing in forests has risen from 35 million ACU (Adult cattle Units) in 1958 to 60 million ACU in 1970, which further increased to 90 million ACU in 1990. Production of dry grass in such grasslands ranges from 0.5 to 6.0 tones/ha/annum depending on conditions of grasslands (Dabadghao and Shankarnarayan, 1973). Improvement of these grasslands spread under tree canopy calls for an immediate attention. However, before launching any comprehensive programme of improvement, information on their present yield level appears an essential pre-requisite for implementing a concerted management of grasslands.

So far, very little work is done on these aspects, in India or abroad. General convention is that grass yield decreases under tree (Rajvanshi *et. al.*, 1987) and so far little but inconclusive information is available in this regard. Looking at the paucity of data base on current productive potentials of natural silvipastoral systems, the present study was carried out with following objectives:

1. Estimate herbage productivity as a function of crown spread and direction from the tree trunk.
2. Determine the grazing capacity and carrying capacity of the system.

MATERIALS AND METHODS

The present investigation was carried out in natural grasslands located in the Dr Y S Parmar University of Horticulture and Forestry, Solan, Himachal Pradesh, during two consecutive years i.e. 1996 and 1997. The study site was situated about 1300 m above m.s.l. The study area represents a transition between sub-tropical to sub-temperate climate. The area receives an average annual rainfall of 1100-1350 mm most of it concentrating during monsoon (mid June to mid September). The average annual temperature ranges from 3°C to 32° C with mean annual temperature (MAT) around 18°C The soil is loamy in texture derived from inferakarol series rocks mainly of Shimla group comprised by carbonaceous shales, calcareous shales, dolomitic limestone within bands of intermittent shales.

The site was protected from grazing and harvesting of grasses was practiced over it. To explore the effect of individual trees on herbage dynamics, selected sample trees at each site were divided arbitrarily into three crown classes on the basis of crown width. Each crown class comprised of four trees with individual tree acting as a replicate. Under each tree four directions, viz., North, East, South and West were marked with the help of prismatic compass and ranging rods. In every direction, three distances from the trunk of each sample tree viz. upto 33%, 33 to 66% and 66 to 100% of portion of crown spread (crown radius) were selected. Four controls (open area), one each per direction, were selected away from periphery of the tree canopy, where there was no effect of the tree shade. The design fallowed was RBD. The observations were recorded for herbaceous vegetation and its above ground biomass production. Carrying capacity and grazing capacity of two herbage production systems under study were expressed in Adult Cattle Units (ACU) and were worked out on the basis of body weight of adult cattle unit (320 kg av. body weight) using the formula given by Suresh and Vinod (1987). The data generated from the present investigation was statistically analysed by Randomized Block Design in accordance with procedure outlined by Gomez and Gomez (1984).

RESULTS AND DISCUSSION

Aboveground Biomass

The aboveground biomass increased from July to September and declined thereafter in both the environments. It exceeded by 73.1, 83.2, 85.2 and 46.8 per cent in open environment as compared to that under chirpine, in the months of July, August, September and October, respectively. Similar pattern of increase in biomass yield from July to September and decline thereafter in October was also demonstrated by the data for dominant species as well as the total aboveground biomass of all the species for the year 1996 and 1997. However, less dominant species and other grasses did not evince any particular pattern in biomass yield with regard to change in months.

The increase in aboveground biomass production may be related to spurt of growth with advent of monsoon in June as rainfall, relative humidity and soil moisture increased from June onwards and attained maximum in August-September. Significant role played by climate on growth and development of vegetation has been advocated by Singh and Yadava (1974) and Trivedi (1994), for monsoonal grasslands of India. A marked increase in aboveground biomass can also be related to increase in phytosociological characters of grasses like density and basal area during this period.

The appreciable decline of aboveground biomass under tree canopies, in comparison to their respective open grasslands is the manifestation of reduction in rate of transpiration, leaf temperature and stomatal conductance of grasses under canopy caused by lower relative illumination, which ultimately led to low biomass production (Bhatt *et. al.*, 1994). Allelopathy and interference of needle biomass to grass species under chirpine have also been advocated by Dass (1995).

The aboveground biomass in relation to crown class was recorded to be maximum in open grassland (27.50 q ha^{-1}) and reduced by 38.7, 41.8 and 40.6 per cent at C_1, C_2 and C_3 compared to C_0. The trend was almost identical in case of dominant species as well as total aboveground biomass during the year 1996 and 1997 (Table 1). The direction effect did not exercise a discernible variation in aboveground biomass, nonetheless, the yield in north was, in general, relatively lower as compared to other directions.

TABLE 1: *Effect of crown class on aboveground biomass (q/ha) of grasses under chirpine and in open grassland*

Species	Year	Crown Class			
		C0	C1	C2	C3
C. montanus	1996	20.06	12.38	10.49	8.19
	1997	8.27	6.40	7.10	6.11
P. maximum	1996	13.16	6.08	5.40	3.68
	1997	4.05	1.69	1.86	3.14
H. contortus	1996	0.19	2.00	0.66	0.59
	1997	1.54	0.53	0.37	0.03
A. mutica	1996	0.25	0.99	0.26	1.16
	1997	2.26	0.71	0.12	1.01
E. binata	1996	0.25	0.00	1.61	3.37
	1997	2.46	0.23	1.38	2.39

Species	Year	Crown Class			
		C0	C1	C2	C3
Other grasses	*1996*	*0.19*	*1.80*	*2.50*	*1.90*
	1997	*2.41*	*0.62*	*0.33*	*1.09*
Total	*1996*	*34.10*	*23.50*	*20.92*	*18.89*
	1997	*20.98*	*10.18*	*11.06*	*13.77*
Pooled		*27.50*	*16.94*	*15.99*	*16.33*

The aboveground biomass responded positively to increase in distance from tree base (Table 2). It showed a hike of 27.8 per cent from S_1 to S_2 and 20.2 per cent from S_2 to S_3 when data werw polled. Similar trend of increase in biomass yield as distance increased from tree trunk was also visible, for the individual grass species as well as total aboveground biomass of all the grasses for the year 1996 and 1997.

TABLE 2: *Effect of distance on aboveground biomass (q/ha) of grasses under chirpine.*

Species	Year	Distance		
		S1	S2	S3
C. montanus	*1996*	*8.19*	*10.20*	*12.02*
	1997	*5.46*	*6.29*	*7.79*
P. maximum	*1996*	*3.12*	*4.92*	*5.29*
	1997	*1.12*	*1.81*	*2.45*
H. contortus	*1996*	*0.46*	*1.04*	*1.75*
	1997	*0.42*	*0.41*	*0.10*
A. mutica	*1996*	*0.88*	*0.95*	*0.57*
	1997	*0.56*	*0.54*	*0.73*
E. binata	*1996*	*1.40*	*1.61*	*1.98*
	1997	*0.85*	*1.58*	*1.58*
Other grasses	*1996*	*1.58*	*1.80*	*2.83*
	1997	*0.70*	*0.53*	*0.80*
Total	*1996*	*15.63*	*20.52*	*24.44*
	1997	*9.07*	*11.03*	*13.47*
Pooled		*12.35*	*15.78*	*18.96*

This can be ascribed to the fact that as one moves away from tree trunk the solar influx increased. These findings in the present study are in consonance with observations of Hazra and Patil (1986). The reduction in root competition with increasing distance may have also contributed for the same.

Grazing Capacity and Carrying Capacity of the System

The data on grazing capacity and carrying capacity revealed that the pooled values of above attributes increased from July to September and declined thereafter. Similar trend was also recorded for individual years. Open environment registered higher values of these two attributes as compared to those under chirpine in all the months during both the years.

The probable reason may be referred to increase in dry matter production in these months in the system. Further, the grazing capacity as well as carrying capacity were recorded to be remarkably higher in open grassland compared to that under tree cover in the system which is again a function of difference in biomass production in two environments. The values for both these characters are well in agreement with those reported by Dass (1995) in different grassland of Himachal Pradesh.

CONCLUSIONS

The following conclusions were emerged out of the present study:

1. Open grasslands exhibited superiority over that under trees with respect to aboveground biomass of herbage layer.
2. Effect of crown class and direction did not evince any discernible response as regards to aboveground biomass.
3. As distance from tree increased, in general, biomass demonstrated appreciable enhancement.
4. The study reveals that productivity of grassland is low i.e. around 2 to 4 t/ha/year, imparting that there is a need to introduce germplasm of high yielding improved grasses in them. *Chrysopogon montanus* and *Panicum maximum* can be given due priority owing to their good performance.

REFERENCES

1. Bhatt, R. K.; Misra, L. P. and Pathak, P. S. 1994. Range Mgt. Agrof. 15 (1): 87-93.
2. Dabadghao, P. M. and Shankarnarayan, A. K. 1973. The grass cover of India, ICAR, New Delhi, 713.
3. Dass, B. 1995. M.Sc. Thesis, Dr Y S Parmar University, Nauni, Solan.

4. Gomez, K. A. and Gomez, A. A. 1984. Statistical procedures for Agricultural Research. John Wiley and Sons, New York, Toronto and Singapore: 99-107.
5. Hazra, C. R. and Patil, B. D. 1986. Indian J. Rang. Mgmt. 7 (1): 33-36.
6. Rajvanshi, R.; Kumar, V. and Singh, R. 1987. Ind. For. 113 (9): 599-608.
7. Singh, J. S. and Yadava, P. S. 1974. Ecological Monograph 44 (3): 360-375.
8. Trivedi, B. K. 1994. Range Mngt. and Agroforestry 15 (2): 123-129.

Micro-Climate Variation in *Dalbergia sissoo* based Agroforestry System

ARUN K. SHANKER, RAM NEWAJ AND A. K. HANDA
National Research Centre for Agroforestry, Jhansi (UP)

ABSTRACT

Leaf Physiological parameters viz., Diffusive resistance, Transpiration rate, Leaf temperature, Canopy temperature constituting the micro-climate of *Dalbergia sissoo* based Agroforestry system was studied in the farm of National Research Centre for Agroforestry, Jhansi. Root barrier, Deep basin, Deep ploughing, Pruning and Ga-100 ppm spray treatments were imposed on four year old *Dalbergia sissoo* trees. Under storey crop Mustard was sown in Rabi season of 97-98. The above mentioned parameters were recorded at three different stages viz., Aug. 97 (without crop), Nov. 97 (with crop at vegetative stage) and Jan.98 (with crop at podfilling stage). Transpiration rate and diffusive resistance were inversely related. Root barrier and pruning treatments showed significant difference in all the parameters in comparison with control. General trend of decrease in canopy temperature in all treatments was seen in Aug.97 and Nov.97, but leaf temperature reduced in Jan.98. There was a slight increase in canopy temperature in winter. Root barrier and pruning treatment were best as far as water conservation was concerned. Deep basin treatment showed high water loss in terms of transpiration, which was reflected in growth. GA-100 ppm treatment and control were at par exhibiting no significant difference in all leaf physiological parameters.

INTRODUCTION

Dalbergia sissoo is a medium sized deciduous tree, 10-30 m tall, with trunk 2-4 m in girth at base. Leaves are imparipinnate and alternate facilitating light transmission though canopy (Tiwari 1994). These characteristics can be effectively tapped in an agroforestry situation with various agricultural crops as Under storey vegetation. Much work has been done in the direction of evaluating the potential of *D. sissoo* as useful agroforestry component (Sheik *et. al.*, 1985; Mishra and Prasad, 1980). Limited literature is available on microclimate interaction in an Agroforestry system comprising of *D. sissoo*. Hence the present study was undertaken with the objective of elucidating the micro-climate variation in *D. sissoo* based Agroforestry system.

MATERIALS AND METHODS

The field chosen for the micro-climate observations was planted with *D. sissoo* in which six treatments were imposed viz., Root barrier, Deep basin, Deep ploughing, Pruning, GA-3 100 ppm spray and Control. Mustard was the Under storey crop sown in Rabi season of 97-98. Micro-climate readings (Diffusive resistance, Transpiration rate, Lead temperature, Canopy temperature) were taken at three different stages viz., Aug.97 (without crop), Nov.97 (with crop at vegetative stage) and Jan.98 (with crop at pod filling stage). The above parameters were measured using steady state parameter model LI-1600 fitted with narrow-leaf aperture cap (1-2 cm □ 0.2-0.6 cm). The data was analyzed statistically in RBD design in SYSTAT for Windows software.

RESULTS AND DISCUSSION

Leaf physiological parameters viz., Diffusive resistance, Transpiration rate, Lead temperature, Canopy temperature showed distinct trends in relation to treatment and season. Diffusive resistance decreased through Aug.97-Jan.98. Root barrier, Pruning and GA-3 100 ppm treatments differed significantly from control (Table 1) in all stages of observation. A similar trend was observed in transpiration rate. Diffusive resistance and transpiration rate showed inverse relationship. Leaf temperature was maximum in Aug.97 and reduced gradually till Jan.98 by 3.34°C ±0.67°C. The treatments did not exhibit significant variations in leaf temperature, whereas in the case of canopy temperatures, as they proceeded towards maturity, the gradual decrease was less compared to leaf temperature.

TABLE 1: *Leaf physiological parameters as affected by different management practices in Dalbergia sissoo with mustard as intercrop*

Treatment	Diffusive Resistance ($s.m^{-2}$)			Transpiration Rate ($ugm.\ cm^{-2}.sec.^{-1}$)		
	Aug.97	Nov.97	Jan.98	Aug.97	Nov.97	Jan.98
Control	*0.873*	*0.884*	*0.851*	*15.81*	*15.87*	*14.11*
Deep Ploughing	*0.812*	*0.823*	*0.871*	*16.18*	*16.03*	*15.62*
Root Barrier	*0.918*	*0.931*	*0.963*	*14.23*	*14.16*	*14.08*
Deep Basin	*0.861*	*0.873*	*0.912*	*15.32*	*15.64*	*15.13*
Pruning	*0.913*	*0.942*	*0.921*	*14.68*	*14.51*	*14.22*
GA-100 ppm	*0.901*	*0.911*	*0.892*	*14.59*	*14.31*	*14.18*
CD At p=0.05	*0.048*	*0.041*	*0.052*	*0.81*	*1.21*	*1.08*

Treatment	Leaf Temperature (°C)			Canopy Temperature (°C)		
	Aug.97	Nov.97	Jan.98	Aug.97	Nov.97	Jan.98
Control	*30.26*	*27.18*	*24.23*	*28.17*	*27.08*	*25.31*
Deep Ploughing	*30.78*	*26.93*	*25.14*	*28.87*	*26.54*	*25.01*
Root Barrier	*31.24*	*28.13*	*25.21*	*28.91*	*27.63*	*24.94*
Deep Basin	*31.01*	*28.31*	*25.76*	*28.90*	*27.51*	*25.01*
Pruning	*30.52*	*29.31*	*25.66*	*29.32*	*28.41*	*25.00*
GA-100 ppm	*31.21*	*30.01*	*26.42*	*29.64*	*28.51*	*25.38*
CD At p=0.05	*1.03*	*1018*	*1.02*	*1.41*	*1.37*	*1.26*

The decrease in leaf and canopy temperature in successive observations was mainly due the reduction in ambient air temperature as the season progressed from Monsoon to Winter. The reduction in canopy temperature was less compared to the corresponding reduction in leaf temperature and this can be explained in terms of adaptive mechanism of the tree to changes in air temperature. Transpiration rate and diffusive resistance were inversely related. This can be explained by the fact that due to increases in resistance to water vapor movement through stomata there is a consequent reduction in transpiration rate. The study showed that optimum temperatures was provided for understorey crop growth by *Dalbergia sissoo* trees and Root barrier and pruning treatment were best as far as water conservation was concerned. Deep basin treatment showed high water loss in terms of transpiration which was reflected in growth.

REFERENCES

1. Mishra, J. and Prasad. V. N. 1980. Indian Forester 106 (10): 675-695.
2. Sheikh, M. I.; Hussain, R. W. and Khan, M. 1985. Pakistan J. Forestry 35 (1): 13-14.
3. Tiwari, D. N.1994. A monograph on *Dalbergia sissoo* Roxb.-International Book Distributers, Dehradun, India.

Ber in Agroforestry System

B. B. VASHISHTHA AND VISHAL NATH
National Research Centre for Arid Horticulture, Bikaner (Rajasthan)

ABSTRACT

Arid region is characterised by extremes of temperature, low and erratic rainfall, high vapour pressure deficit which ultimately creates the conditions of drought quite often. Under such circumstances monoculture of arable crops is highly risk prone and therefore, planting of perennial woody plants along with annual crop is now advocated for sustained development of the region. Owing a drought hardy, multipurpose species with quick economic returns, ber, very much fits in agro-forestry system of the arid region. It is compatible to various plant species including legumes, oil seeds, grasses and some woody plants.

INTRODUCTION

The land and water are two basic resources of the earth and their complementary interaction raises the biotic systems that sustain all forms of life. Increase in production can either be achieved by increased productivity of individual crop on available cultivated land or by exploiting untapped areas being left out in form of arid, semi arid and hilly regions. Moreover, an integrated approach consisting of agriculture, horticulture, agrostology and forestry would ensure multiple production pattern and can maximise the production of food, fodder and fuel. These systems aim at synthesis of processes reducing the negative interactions and adding of positive useful ones leading to high system efficiency with conserved land resources (Pathak, 1989).

Ber or Jujube (*Zizyphus mauritiana* Lamk.) a drought hardy, xerophyte produces nutritious fruits for human, fodder for animals and fuel for domestic use. An estimate made at Jodhpur on hectare basis, shows that ber produces 25000 kg fruits containing 37.5 kg vitamin C and 4500 kg sugars with normal cultivation practice of arid region (Vashishtha and Prasad, 1997). Ber plant produces the nutritive leaf fodder (fresh and dry) for camels and small ruminants in desertic region. Leaves of bordi (*Z. mauritiana* var. *rotundifolia)* and jharber *(Z. nummularia)*, a closely related species of ber, found extensively in desert and river basins (Vishal Nath *et. al.,* 1996) are used as goat and sheep feed after lopping the branches and even cutting them from ground level. These plants have very good sprouting capacity and produce dense foliage every year. The cut-wood of these plants are used for fencing of fields in rural areas. Ber bears on current season's growth. To obtain new growth, pruning is a common practice during summer (Pareek and Vishal Nath, 1996).

The concept of ber based farming system depends both on biological and socio-economic considerations. The biological premises include all its advantages of fruit as well as natural forest plant on the soil and environment, such as maintenance of organic matter, prevention of run-off and soil erosion, regulation of micro-climate, and above all adaptability of plants to soils that are incapable of sustaining annual agricultural crops as such. The socio-economic factors that substantiate the potential value of ber based agroforestry system are the mounting population pressure and lack of resources due to which the poor farmers are forced to utilize unproductive areas for food production. Thus, the ber based agroforestry system has great potential to provide a sustainable land use system which would maintain an acceptable level of production of fruits, vegetable, fuelwood, fodder, etc. and at the same time, conserve the basic resources in the arid and semi arid areas (mainly soil) on which production depends.

BER UNDER VARIOUS AGROFORESTRY SYSTEMS

Apart from various uses and importance, ber has been found to have an efficient and positive interaction with several crop and plant species like cereals, pulses, oil seeds, vegetables, grasses, forest trees and fruit plants in a composite system under arid and semi arid ecosystem of the country. Followings are the different ber based agroforestry approaches being adopted in the region.

Agri-horti System

In rainfed areas of arid and semi arid region, kharif and rabi crops have been grown successfully with ber plantations. Ber is planted at 6 □ 6 m distance and inter space between the two rows is being utilized to raise the crops. Legumes and non legumes both were tried but the former has an edge over the later in terms of biological nitrogen fixation by its root. Work conducted at Central Arid Zone Research Institute (CAZRI) Jodhpur has revealed that a farmer can get an additional production of 7.82 q/ha clusterbean (*Cymopsis tetragonaloba*) when grown with jujube plantation in kharif (Vashishtha, 1997). Singh and Vishal Nath (1997) reported that under semi arid conditions of Pali (Rajasthan) 4.36 q/ha mung bean *(Vigna radiata)*, 10.23 q/ha clusterbean and 1.88 q/ha sesame *(Sesamum indicum)* can be produced under ber (cv. Seb) based farming system during kharif. I n aridisols at National Research Centre for Arid Horticulture (NRCAH) Bikaner nearly 45 per cent area between the ber rows have been utilized for raising the crops under rainfed conditions during both the seasons. During kharif leguminous crops for various purpose (Table 1) were raised while during rabi only mustard *(Brassica campestris)* was taken. Table clearly indicates that the agri-horti system has a tremendous potential in arid parts of thc country which is almost rainfed.

TABLE 1: *Production of crops under ber based agri-horti system*

Name of Crops	Cultivar	Final Produce	Total yield (q/ha)*	Cropping Season
Clusterbean	*Durga Bahar*	*Green pods*	67	*Kharif*
Clusterbean	*Durga Bahar*	*Seed*	2.25	*Kharif*
Clusterbean	*RGC-936*	*Grain*	6.3	*Kharif*
Moth bean	*RMO-40*	*Grain*	5.4	*Kharif*
Mustard	*Pisa Bold*	*Grain*	9.14	*Rabi*

* *Yield from only 4500 m^2 area available between ber rows. Annual rainfall 350-450 mm.*

Horti-pastoral System

The economy of the people in arid region is animal husbandry based. Ber based horti-pastoral system in the region provides fodder for the animals and fruits for additional income. Jujube, in combination with grasses, can form an ideal combination in animal husbandry based arid ecosystem. The system provides, fodder from grasses and leaves of ber (after pruning), fruits and fuel (after pruning) from ber plants. At CAZRI, Sharma and Vashishtha (1985) concluded that combination of jujube and buffel grass *(Cenchrus ciliaris)* under rainfed condition in Barmer district of western Rajasthan has provided 8.4 q/ha grass fodder, in addition to fruit yield. They also noticed that after 5 years of establishment of jujube, growth of plant was more than double while production of dry leaves and pruned wood was more than six times under horti-pastoral system as compared to jujube grown as monocrop. They suggested that to avoid damage of plants by animals, budding in jujube should be done at 1 m height.

Horti-silvi System

In this system, forest species are planted in association with fruit species. Under ber based horti-silvi system in arid regions selection of suitable tree species should be given major thrust. Now a days, a new concept of horti-silvi system is coming into effect where both, fruit species as well as forest species are planted with the aim to act as complimentary to each other. In both the cases, drought hardy species are selected for extreme arid conditions.

Ber based horti-silvi system has been found more beneficial at the places facing the problem of very high wind velocity in arid region. Plantation of tree species viz. *Acacia tortilis, Prosopis juliflora, Cassia siamea, Albizzia lebbeck, Z. nummularia* etc. as wind break all around the orchard (Vashishtha, 1995) and provision of woody tree species preferably leguminous trees *(Prosopis cineraria)* inside the system in a pre defined layout can give better out put from the system in adverse soil and climatic conditions. The tree species planted around the system provide fuel or some marginal timber wood, at the same time protect the orchard from hazards of cold and hot winds, as well as generate more income through fruit production while the production of fruits and timber can be achieved from the main plantation. Selection of the tree species can be made as per local demands and its suitability in the region. *Rohira (Tecomela undulata)* has been found well in arid region and therefore, it can be a component in ber based horti-silvi system. The fodder trees *Leucaena leucocephala, Ailanthus excelsa* etc. should be selected in predominantly animal husbandry areas.

Horti-silvi-pastoral System

This is an ideal system of farming under arid and semi arid agroclimatic zone. All the three components can be successfully utilized to generate more output for the local habitants under adverse soil and climatic conditions of the region. Besides, fruit production the system checks the degradation of land by high wind velocity and rains on one hand and helps in stabilising the already degraded lands on the other hand. The selection of forest trees and pasture grasses for ber based horti-silvi-pastoral system entirely depends upon soil and climatic conditions of the particular place. At Central Horticultural Experiment Station Godhara (Gujarat) having about 700 mm annual rainfall, heavily eroded wasteland having a network of two deep and three medium gullies were brought under agroforestry system and planted with various fruit trees (ber, mango, guava), grasses (*C. ciliaris* and *S. hamata)* and forest trees (*Leucaena, Eucalyptus*). The area between the gullies was planted with ber, mango and guava using staggered contour trenches to harvest rain water and to check surface runoff. The highly degraded land within a distance of 6 m from the gully rims was planted with *L. leucocephala* and *Eucalyptus* with a population density of 3000 trees/ha on either side. *C. ciliaris* was planted along the gully rims and on the bunds of contour trenches. Besides, fruit trees were intercropped with vegetable like cow pea (*V. unguiculata*), okra (*Abelmoschus esculentus*), clusterbean, brinjal (*Solanum melongena*) and a tropical pasture legume *S. hamata*. The collected surface run-off was used for life saving irrigation at critical crop growth stages. *In situ* grafting and budding respectively in mango and ber gave good establishment even during drought years (Raturi and Chadha, 1993).

The system started giving returns from the very first year with a production of 4 q/ha from ber which increased upto 16 in second year and upto 49 q/ha in the third year with a net return of Rs. 1705/ha in the second year and Rs. 11599/ha in the third year. Besides, the ber trees also provided fuelwood of about 2.2 and 7.5 q/ha in the second and third year, respectively from the annual pruning. The other fruit trees started giving returns from the third year. The production of pasture grasses from the system is given in Table 2.

TABLE 2: *Yield of fodder in q/ha (fresh weight basis)*

Crop species	First year	Second year	Third year
C. ciliaris	60	75.2	107.84
S. hamata	31.91	38.88	92.05
L. leucocephala	45.75	56.25	80.5

Source: Raturi and Chadha, 1993

After 5 years, the value of timber from *Eucalyptus* was nearly Rs. 90,000/ha. A comparison of the different intercropping system with the local practice of growing maize *(Zea mays)* plus pigeon pea *(Cajanus cajan)* by the tribal farmers revealed that ber and clusterbean combination gave the highest net return of Rs. 14,630/ha (Table 3) with the lowest return of Rs. 616/ha from the maize and pigeon pea combination (Raturi and Hiwale, 1990).

TABLE 3: *Comparative performance of different cropping systems*

Intercropping system (ber planted in 1984)	Yield q/ha (1986)		Net returns (Rs./ha)	Yield q/ha (1987)		Net returns (Rs./ha)
	Ber	Intercrop		Ber	Intercrop	
Ber	16.1	-	1705	49	-	11599
Ber + Cowpea	16.24	66.12	5685.4	52.16	31.98	13808.2
Ber + Okra	17.08	69.97	5497.4	47.28	33.91	12581.2
Ber + Clusterbean	16.2	58.77	4502	51.22	42.95	14630.6
Ber + Brinjal	17.1	82.9	2121.2	50.08	-	11899
Ber +Stylosanthes	15.97	31.91	2091.2	47.65	38.88	11943.68
Maize+ Pigeonpea	8.00*	3	1126	5.00*	2.5	616

**Yield of maize. (Source: Raturi and Hiwale, 1990)*

Thus, it is possible to optimise integrated use of land, water and vegetation not only to obtain sustained productivity from the arid and semi arid wastelands but also to manage the situation of drought and soil erosion by adopting the ber based agroforestry systems.

TIPS FOR BER PRODUCTION IN AGROFORESTRY SYSTEM

Selection of Cultivars

The choice of cultivars in ber under different agroforestry systems vary with existing agro-climatic conditions. In extreme arid conditions (rainfall <350 mm) early maturing cultivars (Gola, Seb, Mundia) are most suited to the system while in areas having little higher rainfall (350-500 mm) cultivars like Katha, Umran, Kaithali, Maharwali can be taken (Vashishtha, 1991).

Orchard Establishment

1. **Shelter belt plantation:** Orchard establishment is a great problem in arid and semi arid areas. To ensure higher establishment success of ber under agroforestry system in these regions, 2-3 row plantation with fast growing tree species in staggered fashion protect the plants from hot and cold winds. Besides having the shelter belts all around the periphery of the farm, the individual block of the system can have a mini shelter belt of tall type castor *(Ricinis communis)*. Sowing of 2-3 rows of castor staggered manner at 1-2 feet will not only help in developing microclimate but also provides additional income to the grower.

2. ***In situ* plantation:** Owing to higher evapotranspiration from newly planted ber sapling (nursery raised budded plants) the survival under field is very less. Cut and coil roots of plant in nursery aggravate the problems upto greater extent. Therefore, under arid and semi arid zone *in situ* plantation of *boradi* seedling or sowing of boradi seeds and budding them there after one year (Pareek, 1978) have been recommended. Method provides deeper and extensive root system to the plant on one hand and eliminates the age gap in case the failure of budding success in first year on the other hand.

Water Conservation

1. **Treatment of pit:** The problem of deep percolation losses of water is acute in light soils (15-20 cm/hr). It causes problem at the time of initial establishment of plants when water requirement at the upper 30 cm layer is maximum. To conserve the water from deep percolation and improve the establishment of plants, the pits should be treated with materials which would easily allow the penetration of roots but at the same time minimises the percolation/seepage losses of water. Clay can be used in the bottom of pits for this purpose but availability of materials in the regions of sandy soils, is very difficult. Bentonite, a natural mineral available in abundance in Thar desert of western

Rajasthan has been found very effective to minimise the downward movement of water. To make it much effective treatment of pit should be done in such a manner so that it not only reduces the percolation of water but also minimises the seepage from the sides of pit. For this purpose, bentonite layer of 2-3 cm thickness is laid at the bottom of the pit while sides can be filled with bentonite/soil mixture in the space provided by putting a hollow iron cylindrical frame having 2-3 cm less diameter than that of pit. The pit should be irrigated after filling the sides and then the cylindrical frame should be taken out. Such a pit forms a pot like structure.

2. **Use of double walled pot *(Jaltripti)*:** To reduce the water loss and assure the plant establishment in desertic soil, a double walled clay pot in which inner part is hollow and outer part (Gupta *et al*, 1989) can be is sealed at the bottom with the inner part, is used. The plant can be planted in the inner part and whole pot is kept in the pit. About 2-3 litres of water is filled in the space between inner and outer part and through seepage the water reaches the root zone of plant. Thus, the water requirement for the establishment of plant is drastically reduced. This pot has been named as '*Jaltripti*'.

3. ***In situ* water harvesting:** A vast area of arid and semi arid region does not have irrigation facility and therefore the production system only based on natural precipitation which ranges from 100-159 mm in Jaisalmer (Rajasthan) to 700-800 mm in Aruppukottai (Tamil Nadu). Average annual rainfall in the *Thar desert* of western Rajasthan is however, very low and makes fruit growing very difficult. Owing to light soil and high solar radiation, quick loss of rain water takes place. Nevertheless, there is about 25% rainfall to runoff ratio even in light soils with high infiltration rate under 250-300 mm rainfall zone (Ramakrishna and Singh, 1974) which indicates the prospects of *in situ* water harvesting in fruit tree species. The runoff can, therefore be used for growing fruit trees in such a way that each tree has its own micro-catchment area. The layout of micro catchment in relation to the location of tree would depend upon the topography of area. Studies carried out at CAZRI, Jodhpur on the effect of runoff concentration on production ber have established that the runoff yield is a composite function of rainfall, slope, catchment/planted area ratio, contributing area and capacity of soil to store moisture.

More than a decade studies on ber have revealed that 54 m^2 catchment area (in addition to canopy area) with 5 per cent slope has been found to be ideal for optimum fruit production (Vashishtha, 1988).

4. **Mulching:** It is an age old practice used by the farmers to reduce the evapotranspirational loss of water from soil surface. The mulching material can be either from organic sources (grass, straw or any locally available biomass) which improves the soil after decomposition or from inorganic (polythene) material. Use of black polythene (100 gauge) has been found to be very much effective in ber (Pareek and Vishal Nath, 1996). Besides, mulches also check the emergences weeds near the plant.

5. **Use of antitranspirants:** Owing to intense radiation and high wind velocity in arid and semi arid region, considerable amount of water is lost from the leaf surface. Spray of radiation reflectants (Kaoline, PMA, Power Oil, Waxol, etc.) have been found effective in reducing the transpiration losses of water by lowering the leaf temperature. (Vashishtha, 1988). Scarcity of water, particularly during fruiting period in ber reduces the yield drastically. Spray of 4-5% Kaoline and 10% PMA have been found effective (Pareek and Vishal Nath, 1996) to conserve plant water for a longer time which helps in development and maturity of the fruits.

Agrotechniques

1. **Training and pruning:** For proper utilization of natural resources and optimisation of tree canopy under agroforestry system, it is best to train ber plants to be low except if a horti-pastoral management system is envisaged. This reduces the cost of orchard management. Only one branch should be allowed from a node and these should be balanced around the trunk. Shoots emerging below the bud union must always be removed.

   Since ber, bears on current season growth therefore, the trees should be pruned every year in the second week of May to induce new shoots on which the flowering and fruiting take place during monsoon. In arid areas, light pruning upto 6th to 8th secondary (20-25 buds) have been found effective while in semi arid regions having little moderate climate, pruning after 4th secondary (at 15th bud) has been recommended. Pre pruning spray of thiourea (3.0%) have also found effective in ber to reduce the large number of unfruitful branches on base (Pareek and Vishal Nath, 1996)

2. **Plant protection:** The plant protection measures for ber vary in different regions depending upon the extent of aridity, atmospheric humidity and temperature. Nonetheless, there are some polyphagous insect pests which need to be controlled. Termite causes more damage in rainfed areas particularly with sandy soils. Pre-treatment of pits before planting should be done by adding 100 g endosulfan dust. However, after planting only water soluble insecticide (chlorophyriphos) can be used so as to penetrate over the root zone. Painting of ber trunk upto 1 feet height with the paste prepared by mixing slacked lime +chloropyriphos+water+mancozeb (4:1:2:1) reduces the attack of termite and any other fungal diseases related to collar portion of the plant. The paste can also be applied on the cut ends after pruning of plants. Fruitfly (*Carpomyia vesuviana)* is a serious pest of ber in North India which can be effectively controlled with the spray of systemic insecticide (0.03%) when fruits are in 'pea' stage. Spray should be repeated after three weeks after the first spray. In South India, fruit borer (*Maridarchis scyrodes*) attacks the ber fruits which should be controlled before the damage has been done. To control bark eating caterpillar, insert cotton swab, absorbed with fumigant insecticide, in the hole and plug it with clay. Mites, bark eating caterpillars, leaf chaffers and white grub beetle also attack ber plants which need to be timely controlled. Powdery mildew (*Oidium* sp) and black leaf spot (*Isariospsis* sp.) are the major diseases of ber in comparatively humid areas which can be controlled by three sprays of 0.1% karathane and carbendazim, respectively at 10 day interval starting from initiation of symptoms.

REFERENCES

1. Anonymous. 1997. Annual Report 1996-97. AICRP on Arid Zone Fruits-NRC for Arid Horticulture, Bikaner:1-56.
2. Gupta, I. C.; Singh, P. M.; Yadav, N. D and Sharma, B. D. 1989. Agriculture International. 42 (5):176-78.
3. Pareek, O. P. 1978. Indian Horticulture 23:5-6.
4. Pareek, O. P. and Vishal Nath 1996. Coordinated Fruit Research in Indian Arid Zone. NRC for Arid Horticulture, Bikaner: 1-104.

5. Pathak, P. S. 1989. In: Perspective in Ecology (J. S. Singh and Brij Gopal eds.). Jagminder Book Agency, New Delhi, India: 169-179.
6. Ramakrishna, Y. S. and Singh, R. P. 1974. Crop production strategy in the arid zone in relation to rainfall distribution models. Seminar on Desert Technology, CAZRI, Jodhpur.
7. Raturi, G. B. and Chadha, K. L. 1993. Horti-Silvi Pastoral System. Advances in Horticulture Vol.2 - Fruit Crops (K. L. Chadha and O. P. Pareek eds.). Malhotra Publishing House, New Delhi: 805-811.
8. Raturi, G. B. and Hiwale, S. S. 1990. Ber (*Z. mauritiana* Lamk.) based cropping system for marginal lands under rainfed conditions. Seminar on Rainfed Cultivation Technology, WALMI, Bhopal, 14 January, 1990: 1-7.
9. Singh R. S. and Vishal Nath 1997. Intensive Agriculture 34 (11 & 12): 14-17.
10. Sharma, S. K. and Vashishtha, B. B. 1985. Annals of Arid Zone 24 (4):303-309.
11. Vashishtha, B. B. 1988. In: Desert Ecology, (Ishwar Prakash ed.). Scientific Publisher, Jodhpur:135-148.
12. Vashishtha, B. B. 1995. In: Land Resources and their Management for Sustainability in Arid Regions (A. S. Kolarkar, D. C. Joshi and A. Kar eds.). Scientific Publishers, Jodhpur: 197-203.
13. Vashishtha, B. B. 1991. In: Summer Institute on Fruit Production and utilization in wasteland. June 24-July 14, 1991. N. D. University of Agriculture and Technology, Kumarganj, Faizabad.
14. Vashishtha B. B. and Prasad, R. N. 1997. Jujube or Ber: A multipurpose plant for desert. Souvenir on Symposium-R&D priorities for Arid Zone, CAZRI, Jodhpur:59-61.
15. Vishal Nath; Singh R. S. and Pareek, O. P. 1996. Current Agriculture 20:83-86.

Performance of MPTs and their Effect on Forage Production under Rangelands

R. C. SINGH AND SUDHIR KUMAR
Indian Grassland and Fodder Research Institute, Jhansi (UP)

ABSTRACT

A field experiment on collection and evaluation of trees/shrubs spp. was initiated at C.R. Farm of IGFRI, Jhansi in July 1987. After 7-1/2 years trees were felled and it was revealed that *Acacia tortilis* recorded the maximum survival (93.7%) while the minimum survival (5.1%) could be recorded in case of *Albizia lebbeck*. Growth in terms of total height and bole height was maximum (9.90 m and 4.25 m respectively) in case of *Eucalyptus* while *A. nilotica* showed the maximum diameter at collar and breast height (8.92 cm). *A. amara* produced the maximum number of branches (7.12). *Dichrostachys cineria* being shrub showed comparatively poor growth among all. As regard the production of natural pasture, it was noticed that the higher rate of tree survival reduced the pasture yield. However, among all trees/shrubs *A. nilotica* itself performed well but showed more suppressive effect on pasture production and only 1.3 q/ha dry forage could be harvested. Contrary to this *A. tortilis* seems to be the best because of its higher survival (93.7%) and forage yield of range lands (5.2 q/ha) under semi-arid conditions.

INTRODUCTION

In semi-arid regions silvipasture system is gaining importance rather than crop production because of many constraints. The selection of suitable tree/shrub is of more significance for higher yield/unit time/unit of land. Hence, 11 species of trees/shrub were planted in

Sehima-Dichanthium dominated grass cover of Bundelkhand area in July 1987.

MATERIALS AND METHODS

According to productive life cycle and stature four to six month old seedlings of 11 species were planted at 3 different spacing. *L. leucocephala* (k-8, and Silva-4) and *E. camaldulensis* at 2 □ 2 m; *A. tortilis, A. nilotica, Prosopis juliflora, D. cineria* at 3 □ 3 m and *Dalbergia sissoo, Holoptelia integrifolia, A. lebbeck and A. amara* at 4 □ 4 m spacing in compact family block design and repeated three times. After 7 ½ years, data for 25 trees of each species were recorded for their survival and growth in terms of height, thickness, canopy diameter etc and felled. The production of under storey grass cover was also recorded and weighed simultaneously to see the effect of tree canopy on pasture yield in association.

RESULTS AND DISCUSSION

Survival Percentage

Trees planted at 3 □ 3 m spacing showed their superiority for survival over trees planted at 2 □ 2 m and 4 □ 4 m spacings. *A. tortilis* (3 □ 3 m) gave the maximum survival (93.7%) and was closely followed by *P. juliflora* and *A. amara*. The minimum survival (5.1%) was recorded in *A. lebbeck*. The higher survival of *A. tortilis* may be attributed to its suitability for this region under rainfed conditions.

Height

Among 3 groups trees planted at 2□2 m spacing i.e. *L. leucocephala* spp. and *E. camaldulensis* attained maximum average height of 8.86 m followed by trees with 3 □ 3 m spacing. *E. camaldulensis* attained the maximum height (9.90 m) at 7 ½ years age among all the tree spp. while *D. cineria* attained the minimum height of 3.34 m.

Diameter

A. nilotica attained the maximum diameter (12.68 cm) at collar region among all species followed by *A. lebbeck* and *D. sissoo*. Again *D. cineria* remained the lowest in this respect which could attain only 6.10 cm dia. at collar region. The same trend was recorded for diameter growth at breast height also. This may be attributed due to suitability of *Acacia spp*. for this region.

Number of Branches

Among all tree species *A. amara* produced the maximum number of branches/plant (7.12) where as the minimum number (1.00) was produced by *E. camaldulensis*. This may be by virtue of nature of producing more branches at lower part and upper part of stems of *A. amara* and *E. camaldulensis* respectively.

Clear Bole Height

It was seen that larger clear bole of 4.25 m was produced by *E. camaldulensis* and the smallest clear bole (1.09) was observed in case of *D. cineria*.

Canopy Spread

It was observed that *D. cineria* being shrub maintained the smallest canopy spread of 2.98 m/plant in its group as well as over all. The largest canopy cover (6.04 m) could be recorded in *A. lebbeck*. This may be because of poor tree population/ha and less compact canopy of *A. lebbeck*.

Pasture Production

It is revealed that trees exhibiting higher survival (Av. 86.6%) like *Acacias, P. juliflora* and *D. cineria* showed more suppressive effect on pasture grasses (Av. 2.9 q/ha) as compared to those with low survival percent in general and *A. nilotica* in particular planted at 3 □ 3 m spacing (1.3 q/ha). The findings support the earlier findings of Singh and Pathak (1990; 1993), Lowell (1973) and Donald (1967).

Contrary to this *A. tortilis* planted at the same spacing as of *A. nilotica* and survived better than *A. nilotica* even than produced 5.2 q/ha dry fodder. The highest forage production (11.0 q/ha) was recorded under *L. leucocephala* (K8) closely followed by *A. lebbeck* (9.2 q/ha). This may be attributed to the less number of trees/ha (51-27.1) which could not cast sufficient shade on pasture grasses.

ACKNOWLEDGEMENT

The authors are thankful to the Director, IGFRI, Jhansi for providing facilities.

REFERENCES

1. Donald, A. J. 1967. Range Mgmt. and Agroforestry. 20 (4): 247-249.
2. Lowell, K. H. 1973. J. Forestry. 72 (12): 752-757.
3. Singh, R. C. and Pathak, P. S. 1990. Rang. Mgmt. and Agroforestry. 11 (2): 189-198.
4. Singh, R. C. and Pathak, P. S. 1993. Range Mgmt. and Agroforestry. 14 (2): 171-178.

Ber Based Agroforestry System in Marginal Lands under Rainfed Conditions

RAM NEWAJ, K. R. SOLANKI AND S. K. SHUKLA
National Research Centre for Agroforestry, Jhansi (UP)

ABSTRACT

A field experiment was conducted during 1989-90 with four varieties of ber (Seo, Gola, Banarasi karaka and Deshi) two spacings (12 □ 6 m and 6 □ 6 m) and five crop sequences [fallow, sesame (*Sesamum indicum* L.), pigeonpea (*Cajanus canjan* Milsp), blackgram (*Vigna mungo*) and greengram (*Vigna radiata*)]. The experiment was laid out in split plot design with three replications. Ber based agroforestry system in marginal lands under rainfed conditions has potential to produce on an average 8.94 to 10.27 q/ha fruit, 0.68 to 3.74 q/ha grain from different crops 3.85 to 4.75 q/ha fuel wood and 0.63 to 0.72 q/ha fodder (leaves) every year. The blackgram was found to be the most compatible intercrop among all the crop sequences under marginal lands with ber. This crop produced on an average 3.74 q/ha grain yield every year.

INTRODUCTION

The land available with farmers is not suitable to produce sufficient food grain to meet out their increasing demands. Besides that, due to lack of irrigation facilities and low rainfall, farmers are not able to meet their multiple needs. Under such conditions, perennial drought hardy crop like ber (*Zizyphus mauritiana*) can be grown with agricultural crops and grasses to meet out the increasing demand of human and livestock population.

Ber is compatible to various kharif crops (legumes and oil seeds) and intercropping with ber also reduces the risk of crop failure under adverse weather conditions. In rainfed ecosystem growing of kharif crops with ber having an opportunity to get employment income during whole of the year because in ber, fruit set begins in September and completes by the end of October, by this time intercrops are ready to harvest and after that, fruits are ready to harvest in month of Feb to March. Agrihorticultural system with ber is very relevant, viable and sustainable for semi-arid environments (Singh, 1995). Ber fruit contains 14-16% sugar and 160 mg per 100 g of vitamin C, as well as other minerals and vitamins. After the fruit harvesting, leaves can be fed to sheep or goat. For fruit production, *Zizyphus* requires annual pruning and pruned wood is a good source of fencing material. A well grown plant of ber produces 40-50 kg pruned wood every year depending upon intensity of pruning (Anon., 1997)

MATERIALS AND METHODS

The study was undertaken during 1989-90 with four varieties of ber (Seo, Gola, Banarasi Karaka and Deshi), two spacings (12 □ 6 m and 6 □ 6 m) and five crop sequences [fallow, sesame (*Sesamum indicum* L.), pigeonpea (*Cajanus cajan* Milsp.), blackgram (*Vigna mungo*) and greengram (*Vigna radiata*)]. The soil of experimental field was red gravelly having pH 7.7, organic carbon 0.15 %, EC 0.17 m mhos/cm, ESP 0.62 me/ 100 g, available N, P, K (110.0, 12.2 and 120.5 kg/ha respectively).

RESULTS AND DISCUSSION

Fruit bearing in ber had started after four years (1992-93) of planting. The performance of individual variety with different crop combinations and spacings indicated that Banarasi karaka produced on an average 10.93 q/ha fruit every year followed by Seo 10.57 q/ha. The performance of individual variety of ber varied during different years which indicated that in a particular year due to seasons effect the yield of a variety has increased or reduced (Table 1). For example the yield of Banarsi karaka was 17.80 q/ha during 1993 but it was reduced during 1994 to 1996 and further the yield was increased during 1997. The lowest fruit yield was achieved from Deshi (on an average 4.73 q/ha).

TABLE 1: *Fruit yield of ber (q/ha) in marginal lands under rainfed conditions*

Varieties	Years						
	1992	1993	1994	1995	1996	1997	Average
Seo	6.65	11.4	5.85	7.74	18	13.8	10.57
Gola	8.05	13.1	5.45	6.56	10.6	16	9.95
Banarasi Karaka	9.17	17.8	4.16	6.52	8.85	19.1	10.93
Deshi	4.25	7.42	4.16	3.91	4.54	4.11	4.73

The production potential of agrihorticulture system with ber under rainfed conditions on marginal lands indicated that the system provided 8.94 to 10.27 q/ha fruit from ber, 0.68 to 3.74 q/ha grain from different crops, 3.85 to 4.75 q/ha wood and 0.63 to 0.72 q/ha fodder (leaves) every year. Among different crop sequences, sesame could not give the satisfactory level of grain production because of crops is severely affected by the climatic conditions (some time heavy rainfall just after sowing and some time continuous rainfall). The black gram was found the most compatible intercrop under marginal lands with ber. This crop produced on an average 3.74 q/ha grain yield every year (Table 2). The production potential of greengram was next to the blackgram Singh and Vishal Nath (1997) also reported that under semi arid conditions of Pali (Rajasthan) 4.36 q/ha mung bean (*Vigna radiata*), 10.23 q/ha cluster bean and 1.88 q/ha sesame can be produced under ber (cv. Seb) based farming system during *Kharif*.

TABLE 2: *Production potential of ber + intercrops in marginal land under rainfed conditions*

Ber + intercrop	Yield (q/ha)							
		Years						
	Item	1992	1993	1994	1995	1996	1997	Average
Ber + fallow	Grain	-	-	-	-	-	-	-
	Fruit	7. 38	11.1	4.76	5.86	11.2	13.8	9.01
	Fuel wood	0.71	3.4	3.5	10.6	7.59	3.17	4.25
	Fodder	0.27	0.33	0.62	1.24	1.1	0.34	0.63
Ber + sesame	Grain	1.6	1.72	0.81	0	0	0	0.68
	Fruit	7.79	12.6	3.06	4.91	9.83	13.1	10.27
	Fuel wood	1.02	3.96	5.05	10.5	5.25	3.18	4.75
	Fodder	0.27	0.41	0.64	1.04	1.93	0.44	0.78
Ber + pigeonpea	Grain	2.6	2.84	0	0.8	1.59	1.67	1.58
	Fruit	6.12	10.8	3.18	5.31	10.9	11.6	9.79

Ber + intercrop	**Yield (q/ha)**							
	Years							
	Item	**1992**	**1993**	**1994**	**1995**	**1996**	**1997**	**Average**
	Fuel wood	*0.88*	*2.84*	*5.93*	*9.2*	*6.7*	*2.98*	*4.75*
	Fodder	*0.24*	*0.37*	*0.52*	*1.17*	*0.96*	*0.38*	*0.6*
Ber + blackgram	*Grain*	*9.37*	*1.21*	*0.85*	*2.92*	*5.52*	*2.6*	*3.74*
	Fruit	*6.91*	*13*	*4.01*	*5.92*	*11*	*12.8*	*8.94*
	Fuel wood	*0.71*	*3*	*2.53*	*8*	*7*	*3.2*	*4.07*
	Fodder	*0.23*	*0.4*	*0.53*	*0.96*	*1.1*	*0.35*	*0.59*
Ber + greengram	*Grain*	*8.11*	*0.38*	*0.65*	*1.92*	*4.22*	*0.76*	*2.67*
	Fruit	*6.96*	*14.6*	*4.58*	*4.58*	*9.63*	*13.4*	*9.58*
	Fuel wood	*0.81*	*3.25*	*6.9*	*9*	*7.5*	*2.82*	*3.85*
	Fodder	*0.19*	*0.35*	*0.69*	*1.11*	*1.24*	*0.37*	*0.65*

CONCLUSIONS

Ber + blackgram intercropping (agrihorticultural system) system was found very relevant, viable and sustainable for rainfed areas of marginal lands and it also reduces the risk of crop failure under adverse weather conditions.

REFERENCES

1. Anon. 1997. Annual Report 1996-97. AICRP on Arid Zone Fruit- NRC for Arid Horticulture, Bikaner: 1-56.
2. Singh, R. P. 1995. In: Regional training-cum-workshop on silvi-pastoral systems in arid and semi-arid ecosystems at CAZRI, Jodhpur: 311-332.
3. Singh R. S. and Vishal Nath 1997. Intensive Agriculture 34 (11 & 12): 14-17.

Agri-Horticultural System: A Sustainable Way of Meeting Diverse Needs of Farming Community

S. K. SHUKLA, A. K. BISARIA, A. S. GILL, R. P. DWIVEDI AND C. K. BAJPAI

National Research Centre for Agroforestry, Jhansi (UP)

ABSTRACT

The study was conducted at National Research Centre for Agroforestry, Jhansi to assess the performance of agri-horticultural system under irrigated condition, with the treatments comprising three tree spacings i.e. 10□10 m, 10□5 m and 5□5 m, in the main plot and three fruit tree species, namely guava (*Psidium guajava*), ber (*Zizyphus mauritiana*) and mandarin (*Citrus reticulata*) in the subplots. A common crop rotation of Urd-Wheat was followed in the interspaces of fruit trees. Data recorded on crop yield revealed maximum urd (5.6 q/ha) and wheat (36.4 q/ha) yield in control (sole crop). There was drastic reduction in crop yields with the introduction of tree component. Data with respect to growth of fruit trees exhibited maximum tree height in ber (6.27 m) followed by mandarin (4.11 m) and guava (3.94 m). Ber trees in control (sole tree) attained maximum canopy size (7.14 m) which decreased with the increase in the density of ber trees with crop in the interspaces. Increase in tree spacing reduced the foliage production from ber markedly. Maximum fruit yield of mandarin (18.45 tonnes/ha) was recorded in 5 □ 5 m spacing in association with crop followed by sole tree, 10 □ 5 m and 10 □ 10 m. Guava gave maximum fruit production in sole tree with similar trend of reduced yield with the reduced tree density.

INTRODUCTION

Agroforestry, an age old practice in our country, is a way of life. Agri-horticultural system of agroforestry refers to the growing of fruit trees with understorey agricultural crops for concurrent production of fruits and food. In fact, tree growing and agriculture have always been conceived as diagonally opposite propositions but dwindling forest resource coupled with increasing food, fuel and fodder demands of growing population has necessitated bringing these two identities nearer. Agroforestry promises to help farming community in increasing productivity, profitability and sustainability of production on their land. In spite of all these benefits, farmers are often reluctant to introduce trees with their agricultural crops because of possible reduction in crop yields. Farmers when given choice for selection of agroforestry tree species prefer fruit trees to other MPTS owing to the early returns from fruit trees in comparison to other MPTS. A careful selection of fruit tree species and tree density are two major criteria which should be taken into account in an agri-horticultural system of agroforestry. In the present study, attempt has been made to find out suitable fruit tree species and optimum number of trees per unit area to be planted in agroforestry system.

MATERIALS AND METHODS

The experiment was laid out in Split Plot Design with two replicates. Treatments comprised of three fruit tree spacings i.e. 10×10 m, 10×5 m and 5×5 m, in the main plot and three fruit tree species, namely guava (*Psidium guajava*), ber (*Zizyphus mauritiana*) and mandarin (*Citrus reticulata*) in the subplots. Sole tree (spacing 5×5 m) and sole crop were maintained as controls. A common crop rotation of Urd (*Vigna mungo*)-Wheat (*Triticum aestivum*) was followed in the interspaces of fruit trees. Data were recorded on crop yield, growth parameters of fruit trees, fruit yield, foliage and fuel wood yield of ber.

RESULTS AND DISCUSSION

Crop Yield

Maximum urd yield was recorded in control i.e. sole crop (5.6 q/ha). Comparison of different fruit tree spacings showed maximum urd yield in 10×5 m spacing (3.65 q/ha). Among fruit trees, urd gave maximum yield in association with ber (2.49 q/ha) followed by mandarin and guava (Table 1). The interaction between tree spacing and fruit species revealed highest yield of 4.2 q/ha in 10□5 m spacing in association with mandarin. Among different tree densities, relative urd yield was maximum in 10□5 m spacing (65.2 % of control).

TABLE 1: *Effect of different fruit tree species and spacing on intercrop yield*

Name of fruit tree	Fruit tree spacing			Mean
	10□10 m	10□5 m	5□5 m	
Urd yield (q/ha) and Relative urd yield (in % in parentheses)				
Guava	*0.80 (14.3)*	*3.65 (65.2)*	*2.00 (35.7)*	*2.15 (38.4)*
Ber	*1.92 (34.3)*	*3.10 (55.3)*	*2.45 (43.7)*	*2.49 (44.4)*
Mandarin	*1.6 0 (28.6)*	*4.20 (75.0)*	*1.45 (25.9)*	*2.42 (43.2)*
Mean	*1.44 (25.7)*	*3.65 (65.2)*	*1.96 (35.1)*	
Wheat yield (q/ha) and relative wheat yield (in % in parentheses)				
Guava	*11.92 (32.7)*	*12.25 (33.6)*	*6.50 (17.8)*	*10.23 (28.1)*
Ber	*13.77 (37.8)*	*13.70 (37.6)*	*7.32 (20.1)*	*11.59 (31.8)*
Mandarin	*14.07 (38.6)*	*12.97 (35.6)*	*8.15 (22.4)*	*11.73 (32.2)*
Mean	*13.25 (36.4)*	*12.97 (35.6)*	*7.32 (20.1)*	

Similarly, maximum wheat yield was recorded in control (36.4 q/ha) which was reduced drastically in association with fruit trees, i.e. 13.25 q/ha in 10□10 m spacing, 12.97 q/ha in 10 □ 5 m and 7.32 q/ha in 5□5 m spacing (Table 1). As regards the effect of respective fruit trees, the yield reduction was more pronounced in association with guava (10.23 q/ha) followed by ber (11.59 q/ha) and mandarin (11.73 q/ha).

Growth of Fruit Trees

Data recorded on growth parameters exhibited maximum tree height in ber (6.27 m) followed by mandarin (4.11 m) and guava (3.94 m) among different fruit trees. Fruit trees raised in different spacings revealed maximum tree height in ber (6.52 m) in control (sole tree) followed by ber planted at 10□5 m (6.44 m), 10□10 m (6.33 m) and 5□5 m (6.04 m) with crop in the interspaces (Table 2). Data recorded on tree canopy revealed maximum canopy in ber (6.64 m) followed by guava (4.94 m) and mandarin (4.15 m). Ber trees in control (sole tree) attained maximum canopy size (7.14 m) which decreased with the increase in the density of ber trees in association with crop.

TABLE 2: *Effect of spacing on height and canopy of fruit trees*

Name of fruit tree	Fruit tree spacing				Mean
	Control (5 □ 5 m)	10 □ 10 m	10 □ 5 m	5 □ 5 m	
Tree height (m)					
Guava	*3.82*	*4.28*	*3.81*	*3.73*	*3.94*
Ber	*6.52*	*6.33*	*6.44*	*6.04*	*6.27*
Mandarin	*4.11*	*4.06*	*4.22*	*4.07*	*4.11*
Mean	*4.81*	*4.92*	*4.82*	*4.61*	
Tree canopy (m)					
Guava	*4.78*	*5.20*	*5.15*	*4.47*	*4.94*
Ber	*7.14*	*6.98*	*6.48*	*6.47*	*6.64*
Mandarin	*4.11*	*3.95*	*4.34*	*4.16*	*4.15*
Mean	*5.34*	*5.37*	*5.32*	*5.03*	

Foliage and Fuel Wood Production

As a result of pruning in ber during summer, yield of ber leaves was recorded. Maximum production of ber leaves (both fresh and dry) was recorded in 5 □ 5 m spacing and with crop in the interspaces. Increase in tree spacing reduced the foliage production from ber markedly. A production of 105.0 q/ha of fresh wood and 50.85 q/ha of dry wood was obtained from ber in control (sole tree). Slight increase in the production of fresh wood of ber (106.7 q/ha) was observed at 5□5 m spacing and with crop in the interspaces. The reduction in plant density resulted in decrease in the production of both fresh and dry wood of ber (Table 3).

TABLE 3: *Effect of spacing on foliage and fuel wood production from ber (q/ha)*

Wood and Leaf	Fruit tree spacing								Mean	
	Control (5□5 m)		10□10 m		10□5 m		5□5 m			
	Fresh	Dry	Fresh	Dry	Fresh	Dry	Fresh	Dry	Fresh	Dry
Wood	*105.0*	*50.85*	*30.45*	*14.77*	*57.77*	*28.57*	*106.7*	*50.45*	*64.97*	*31.26*
Leaf	*17.1*	*5.39*	*5.02*	*1.51*	*9.41*	*2.93*	*20.49*	*6.32*	*11.64*	*3.58*

Fruit Yield

In case of guava, maximum fruit yield of 143.5 q/ha was obtained from tree control followed by 5□5 m, 10□5 m and 10□10 m spacing in association with crop. Maximum yield of mandarin (184.5 q/ha) was obtained in case of 5□5 m spacing with the crop in the interspace followed by tree control (148.8 q/ha), 10□5 m and 10□10 m spacing respectively. No fruit yield was recorded in ber during the year.

TABLE 4: *Effect of spacing on fruit yield in different tree species*

Fruit tree species	Control (5□5 m)	Fruit tree spacings (m)			Mean
		10□10 m	10□5 m	5□5 m	
Guava	143.5	40.6	68.0	78.1	62.2
Ber	0.0	0.0	0.0	0.0	0.0
Mandarin	148.8	55.5	82.8	184.5	107.6

A cursory glance over the data recorded in tenth year of the study reveals that there has been drastic reduction in the yields of understorey crops with the introduction of fruit trees. Although the yield loss was compensated by the gain from the fruit trees in terms of fruit yield and other benefits in case of mandarin and guava. The gains from ber trees were not proportional to the loss caused by it. Increase in tree density reduced the crop yields, although on the other hand it had increased the fruit yields principally because of accommodation of more number of trees/unit area. Guava interferes more with crop yields probably because of its spreading canopy. Too much of vegetative growth in ber resulted in increased fodder and fuel wood production but poor flowering. Poor flowering coupled with severe occurrence of powdery mildew converted an on year into an off year for ber from the viewpoint of fruit production. Excessive vegetative growth in ber may be because of application of fertilizers and irrigation to the under storey crop. In higher tree densities such as 5 □ 5 m and 10 □ 5 m, intermingling of tree canopies was noticed which is not a healthy sign for growth and fruiting of trees. Increased foliage and fuel wood production with reduced tree spacing was because of more number of trees planted per unit area. Guava+ginger gave the highest net returns and without any adverse effect of intercrops on the growth of guava (Anonymous, 1988). Agri-horticultural system increased the economic returns considerably as compared to the monoculture of either fruit or crop (Chandra *et. al.*, 1981; Raturi and Hiwale, 1988; Reddy and Sudha, 1988).

CONCLUSIONS

On the basis of the study it can be stated that the performance of mandarin and guava was better than that of ber in agri-horticultural system under irrigated condition with the tree spacings viz 5 □ 5 m and !0 □ 5 m taking into account the crop and fruit yields. Ber was not found a successful crop for this system.

REFERENCES

1. Anonymous.1988. Annual Progress Report, ICAR Research Complex for NEH region, Shillong.
2. Chandra, S.; Sharma, A. K. and Shukla, D. 1981. Annual Report, CS&WCR&TI, Dehradun.
3. Raturi, G. B. and Hiwale, S. S. 1988. Annual Report, I.I.H.R., Banglore.
4. Reddy, Y. V. R. and Sudha, M. 1988. Annual Report, CRIDA, Hyderabad: 62-63.

Allelopathic Effect of *Leucaena leucocephala* on Germinability and Seedling Growth of Wheat

NEELAM KHARE, A. K. BISARIA AND K. R. SOLANKI
National Research Centre for Agroforestry, Jhansi (UP)

ABSTRACT

This paper addresses the allelopathic influence of *Leucaena leucocephala* Lam de vit. (*subabul*) on seed germination and seedling growth of wheat. The root length increased at 10-30 % concentration and reduced at higher concentrations. Similar trend was exhibited by shoot length and seedling length. The per cent germination was accelerated at 20, 40, and 60% concentration of decomposed leaf extract and reduced at 80 and 100%. It appears from the present observation, that the decomposed leaves of *Leucaena leucocephala* contain some active compound which promotes the seed germination and seedling growth of wheat. The negative influence on seed germination is perhaps due to presence of mimosine a non-protein amino acid. The active substance which promotes seed germination remains to be isolated and identified.

INTRODUCTION

Subabul (Leucaena leucocephala (Lam) de vit) belonging to family leguminosae and sub family mimosaceae is known as the "*Miracle tree*" due to its paramount economic importance. It is a fast growing thornless evergreen NFT. It is a plant of tropical and subtropical region, growing up to an elevation of 500 m. The best growth of *L.*

leaucaena has been observed in area with 600-1000 mm annual rainfall, though it can grow in area with 250 mm rainfall (NAS, 1977). Low temperature limits its growth, since it is a thermo sensitive tree. It performs well on deep fertile soils (Gray, 1968). It is extensively cultivated in Uttar Pradesh, Maharashtra, West Bengal, Andhra Pradesh, Orrisa, Karnataka, Punjab and Himachal Pradesh.

Among legumes, *L. leucocephala* is the most popular MPTS in the country, providing a wide assortment of cultivars for various uses. It can grow on marginally productive soil and wastelands. It has some important economic uses such as in pulp and paper industry, timber and lumber wood, reforestation, ornamental, fuelwood, in therapeutic use, soil improvement and green manure.

MATERIALS AND METHODS

The fresh and decomposed leaves were collected in the first week of August, 1997 and flowers, pods were collected in the second week of March, 1998 from the experimental farm of National Research Centre For Agroforestry, Jhansi. The location has an elevation of 300 m above the sea level and situated between 24□11' 26°27' N latitude and 78°17'-81°34'E longitude.

Five hundred gram fresh leaves and 100 grams of flowers and pods were kept in 1000 ml and 100 ml distilled water, respectively (for 24 hours), then blended for two minutes for obtaining a paste. The paste was diluted with equal amount of glass distilled water, filtered through Whatman No. 1 filter paper and centrifuged at 10,000 rpm for 5 minutes. The supernatant liquid was filtered through Whatman No. 1 filter paper and further diluted with distilled water to obtain 10, 20, 30, 40 and 50% concentration. In case of decomposed leaves 500 g sample was collected beneath the old plantation of *L. leucocephala* and same procedure was repeated to obtain 20, 40, 60, 80 and 100% concentrations. Distilled water was considered as control. One hundred seed of wheat (*Triticum aestivum*) were soaked in each concentration of each extract for twenty four hours. Twenty seeds of wheat were placed in petriplates (15 cm diameter) lined with two layer of filter paper moistened with glass distilled water. Three petriplates of twenty seeds for each treatment were taken into account. The experimental set was repeated thrice.

Regular observations were recorded on seed germination from the second day onwards. The root length, shoot length, and fresh weight were measured on fifth day. Dry matter in seedling was determined by keeping them in hot air oven at 80°C for 48 hours. The data was analysed statistically as per design, analysis of variance and correlation among environmental factor and growth parameters were worked out using SPSS.

RESULTS AND DISCUSSION

Pre-treatment of wheat seeds with *L. leucocephala* fresh leaf, flower and pod extract at different concentrations viz., control 10, 20, 30, 40 and 50% influenced the seed germination. The germination percent was 81, 83, 82, 58, 52, and 35% respectively on second day. Third day the germination percentage was 96.2, 96.6, 96.6, 96, 92 and 80% respectively. These values remained constant up to fifth day. The seed germination was promoted at 10 and 20% of concentration and decrease at higher concentration in progressive manner.

Shoot length of wheat increased significantly at 10 and 20% while it was reduced at subsequent concentration. This reduction was directly proportional to the concentration of extract. Similarly, root length also exhibited the same trend. Therefore, aqueous extract significantly increased the seedling length at 10 and 20%, reduced significantly at 40-50% and remained unchanged at 30% concentration, when compared with control.

The fresh weight and dry matter production of seedling showed the trend like seedling length. The fresh weight increased at lower concentrations over control and was maximum at 20% whereas, it was significantly reduced at 40 and 50% concentration over control.

These observations are in accordance with those reported for *Lectuca sativa, Oryza sativa, Raphanus sativa* and *Daucus carota* (Tawata and Hongo, 1987). In addition to the aforesaid findings inhibitory influence on seed germination has been reported in *T. aestivum, Zea mays, Pisum sativum* and *Brassica compestris* (Joshi and Prakash, 1992). Earlier, Suresh and Rai (1987), have reported inhibitory effect of leaf extract of leaves of *L. leucocephala* in *Sorghum bicolor, Helianthus annus* and *Vigina ungiculata*. There are contradictory reports pertaining to the influence of aqueous extract of leaves of *L. leucocephala*. According to Chou (1983), aqueous leaf extract stimulates seed germination and growth of seedling in *O. sativa*, contrarary to these observations Kaul (1991), has observed the inhibitory effect on germination in *O. sativa*.

CONCLUSIONS

The present experimental significance suggests that leaf extract inhibit germination at higher concentration and stimulate it at lower concentration. It may be due to the fact that leaves of *L. leucocephala* contain mimosine, which is a non-protein amino acid. It has been observed to inhibit seed germination in Letuce, rice, kidney bean, Turnip, Reddish, Carrot (Tawata and Hongo, 1987). It appears that the effect of aqueous extract of *L. leucocephala* on *T. aestivum* is mediated through mimosine present in the extract of *L. leucocephala* leaves, flowers and pods.

ACKNOWLEDGEMENT

The authors are grateful to Director, National Research Centre for Agroforestry, Jhansi for laboratory facility. Thanks are also due to Dr. Rajeev Tiwari for his assistance in these studies.

REFERENCES

1. Chou, C. H. 1983. In: Allelochemicals and pheromones. (ed. Chou, C. H. and Waller, G. R.) Taipai, Taiwan, Institute of Botany, Academia Sinica. 27-64.
2. Joshi, P. C. and Prakash, O. 1992. In: Proc. First National symposium Allelopathy in agroecosysytems. (ed. Tauro, P. and Narwal, S. S.) 127-128.
3. Koul, V. K.; Raina, A.; Khanna, Y. P.; Tickoo, M. L. and Singh, H. 1991. Indian J. of Forestry.14: 54-57.
4. Suresh, K. K. and Rai, R. S. V. 1987. International Tree Crops J. 4: 109-115.
5. Tawata, S. and Hongo, F. 1987. Leucaena Research Report. 8: 40-41.
6. Casal, J. F.; Reigosa, M. J. and Carbelleira, A. 1985. Ecol. Biol. Sol. 22: 1-12.
7. Gray, S. G. 1968. Trop. Grasslands. 2 (1): 19-30.
8. NAS, 1997. Leucaena: Promising Forage and Tree Crop for the Tropics. Washington, D.C. 115.
9. Jadhav, B. B. and Gaynar, D. G. 1992. Indian J. Plant Physiology. 35: 86-89.
10. Sundramoorthy, S.; Kalra, N. and Sen D. N. 1992. In: Proc. First National symposium Allelopathy in agroecosysytems. (Eds.: Tauro, P. and S. S. Narwal): 123-124.

Performance of Black Gram under Neem Based Agroforestry System in Semiarid Region

A. K. PANDEY, K. R. SOLANKI AND V. K. GUPTA
National Research Centre for Agroforestry, Jhansi (UP)

ABSTRACT

Studies on *Azadirachta indica* A. Juss (neem) based agroforestry system was conducted at National Research Centre for Agroforestry in randomised block design. In general, depressing effect of neem on crop yield and yield components in black gram was observed. Depressing effect was more near to trees compared to away from the trees. Maximum negative interaction (53-64% reduction) was observed for grain yield, while neem had positive effect on test weight.

INTRODUCTION

Indian soils, in general, are facing severe problem of degradation owing to unscientific management and intensive land use practices. This poses a great threat to sustainable agriculture and requires alternative management with low input technology to meet the increasing food grain demand of the human population. It is well documented that presence of trees in cultivated lands enhance soil fertility through nitrogen fixation, greater accumulation of organic matters in the form of litter fall, nutrient recycling, provide fuelwood, fodder, fibre, timber etc. for village communities. In spite of several advantages to plant trees, farmers have relatively less liking owing to depressing effect of trees on crop yields and enhance bird damage to the under storey crops. In this context, study was initiated to examine the effects of neem trees on grain yield and yield components in black gram.

MATERIALS AND METHODS

The experiment was conducted at research farm of National Research Centre for Agroforestry, Jhansi during *Kharif* in rainfed conditions. The soil of the experimental site was alfisol with poor water holding capacity. Sowing of black gram was done in Kharif 1997 in the alleys of 8 year old neem plantation in randomised block design having seven replications at a spacing of 50□15 cm. Neem stand was present at 5□5 m spacing having mean plant height of 6.3 m, DBH 10.7 cm, and canopy diameter 4 m. Crop was fertilized at the @ 20 kg N and 40 kg P_2O_5 per hectare. Plot size was 15 □ 15 m having 9 neem trees and 10 rows of annual crop. Observations were recorded on stem girth, plant height, number of leaves per plant, number of primary branches per plant, leaf area, 100-seed weight, number of pods per plant, number of grains per pod, grain yield per plant in 5 randomly selected plants at a distance of 50, 100, 150 and 200 cm from trees. Grain yield per running meter row, photosynthetic active radiation and leaf temperature were also taken at respective distances. In trees, height, DBH, collar diameter and canopy diameter were recorded.

RESULTS AND DISCUSSION

Vegetative Characters

Studies on neem trees (8 years old) and black gram crop interaction indicated that trees significantly affected the vegetative growth of under storey crop (Table 1).

Depressing effect on all the characters under study was more pronounce to the plants near to the tree base (50 cm). As the distance from the trees increased, the depressing effect decreased. Reduction in the magnitude for all the characters ranged from 29.5% for plant height to 38.5% for leaf area in the plants which were 50 cm away from the trees. In contrary to this, reduction percentage for all the vegetative characters of the plants which were 200 cm away from the trees ranged from 7.7 for number of leaves per plant to 19.9 for number of primary branches per plant. Intermediate distances had intermediate depressing effects.

TABLE 1: *Effect of neem trees on grain yield and yield components of black gram at various distances under agroforestry system*

Distancefrom trees	50	100	150	200	Control (No tree)	SEm+	CD (5%)
			Vegetative characters				
Girth (cm)	0.29 (65.4)	0.34 (77.4)	0.35 (79.5)	0.36 (81.9)	0.44 (100)	0.02	0.07
Plant ht. (cm)	27.4 (70.5)	30.1 (77.6)	32.7 (84.0)	32.5 (83.7)	38.9 (100)	1.13	3.30
No. of prim. Bran./pl.	5.8 (64.6)	6.4 (71.5)	7.1 (79.7)	7.1 (80.1)	8.9 (100)	0.25	0.74
No. of leaves	9.5 (69.0)	11.4 (83.0)	14.0 (101.5)	12.7 (92.3)	13.8 (100)	0.77	2.26
Leaf area (cm^2)	85.9 (61.5)	111.6 (79.9)	108.9 (77.9)	124.4 (89.0)	139.8 (100)	6.38	18.82
Leaf temp. (°C)	32.7 (94.2)	32.6 (93.9)	32.2 (92.8)	32.1 (92.5)	34.7 (100)	NS	
PAR (µ mole $S^{-1}m^2$)	1031 (73.6)	922 (65.9)	677 (48.4)	656 (46.9)	1400 (100)		
			Yield components				
No. of pods/ plant	13.9 (72.9)	13.7 (72.2)	15.6 (82.3)	16.7 (87.9)	19.0 (100)	1.23	3.60
No. of grains/ pod	6.3 (96.9)	6.3 (98.3)	6.5 (100)	6.2 (95.4)	6.5 (100)	NS	
100-seed wt. (g)	3.8 (105.0)	3.9 (108.0)	3.9 (108.0)	3.9 (107.2)	3.6 (100)	NS	
Grain yield/ plant (g)	4.5 (37.0)	4.1 (34.0)	4.6 (37.4)	5.9 (48.4)	12.2 (100)	0.61	1.78
Grain yield/ m line (g)	24.5 (39.8)	22.4 (36.3)	23.1 (37.4)	28.8 (46.7)	61.7 (100)	2.96	8.65

Yield Components

Effect of tree distance on yield components of under storey crop was also negative (suppressing) as in the case of vegetative characters except test weight which had positive trend. Degree of reduction for all the characters except test weight was more to the plants near to trees as compared to away from the trees. Maximum reduction

(43.3 - 61.3%) were observed for grain yield per running meter row followed by number of pods per plant. Distance from trees had no significant effect on number of grains per pod. This suggested that production in grain yield was due to reduction in number of pods per plant among reproductive characters. Low yield under agroforestry systems and near to trees were also reported by several workers in various tree species and crop combinations (Ong *et. al.*, 1991, Khybri *et. al.*, 1992, Newaj *et. al.*, 1996, Sreemannarayan *et. al.* 1996). However, further study is needed to identify the real causes of grain yield reduction in black gram under neem based agroforestry system so that suppressing effect can be reduced to a great extent.

REFERENCES

1. Khybri, M. L.; Gupta, R. K.; Ram, S. and Tomar, H. P. S. 1992. Agrofor. Systems. 17: 193-204
2. Newaj, R.; Roy, Deb R.; Bisaria, A. K. and Singh, B.1996. Range Mgmt and Agrofor. 17:69-74
3. Ong, C. K.; Corlett, J. E.; Singh, R. P. and Black, C. R. 1991. For. Ecol. Manage. 45: 45-57.
4. Sreemannarayana, B.; Joseph, S. and Rao, G. L. G. 1996. Range Mgmt. Agrofor. 17: 179-186

Evaluation of *Anogeissus pendula* Genotypes under Agrisilvicultural System in Rainfed Conditions

A. K. RAI, K. R. SOLANKI AND V. K. GUPTA
National Research Centre for Agroforestry, Jhansi (UP)

ABSTRACT

Studies conducted on evaluation of five genotypes (AP-12, AP-28, AP-35, AP-52 and AP-S2) with two pruning managements (pruning up to 25% and 50% height from ground level) in degraded land revealed that growth w.r.t. height, CD, DBH and canopy diameter was maximum with AP-28 followed by AP -35. The minimum growth was recorded with AP -52. Pruning up to 25% and 50% height did not show significant differences in plant height and collor diameter. However, differences were significant in case of DBH and canopy diameter. The dry leaf fodder production (446.67 kg/ha) recorded with AP-S2 was significantly higher than other genotypes except AP-28. Similarly, maximum fuel wood production (1606.93 kg/ha) was obtained with AP-S2 followed by AP -35. Pruning up to 50% height gave significantly higher leaf fodder and fuel wood production than pruning up to 25% height.

INTRODUCTION

Anogeissus pendula Edgew, locally known as kardhai is grown in dry hot regions of the country. It is naturally distributed between Aravally hills in Rajasthan and Bundelkhand regions and down to Narbada river and Panchmahal in deccan plateau (Gupta, 1967). Thus it grows mainly in Rajasthan, southern Uttar Pradesh and Madhya Pradesh. It grows in mixture with *Acacia catechu, Zizyphus zyloprus, Butea monosperma etc.* (Singh, 1982). The maximum temperature varies from 46 to 49°C and minimum from about-1 to 3°C and annual rainfall from 425 to 875 mm (Troup, 1921) in its natural zone.

Anogeissus pendula is a slow growing tree with a short crooked bole. It attains a maximum height of 12 m and girth of 1.5 m. It has been identified as excellent firewood and it sold on higher rates as compared to other fuel wood tree species. The leaves of the tree are palatable and contain crude protein 13.16 per cent (Rai *et. al.*, 1995).

Looking to the wide range of adaptability and excellent fuel wood value of this species, studies were conducted to evaluate the different genotypes of *A. pendula* to find out a fast growing genotype for higher above ground biomass production.

MATERIALS AND METHODS

Four genotypes were collected from Haryana and Rajasthan (AP-28, AP-35, AP-52, AP-S2) and tissue culture raised seedlings of these genotypes were obtained from Tata Energy Research Institute, New Delhi. One genotype raised through seeds was received from Jhansi. Thus five genotypes were planted in degraded land at the spacing of 3 □ 2 m during September 1994 at National Research Centre for Agroforestry, Jhansi, in randomized block design with three replications. The soil of the site was neutral in reaction, low in nitrogen and phosphorus and medium to high in potassium. During 1997 two pruning (25% and 50% from ground level) treatments were imposed. Thus from 1997 a total of eleven treatments (Five genotypes, two pruning management and control i.e. crop without trees) were evaluated in agrisilvicultural systems under rainfed conditions. During kharif season the blackgram was sown in the interspaces of genotypes of as well as in control. The crop was provided 20 kg N+ 40 kg P_2O_5 /ha at the time of sowing. The data on growth parameter were recorded during June and December every year. The pruning was done during December, 1997.

RESULTS AND DISCUSSION

Growth Attributes

Effects of genotypes - Data recorded on growth attributes of A. pendula genotypes at 46 months of age (June 98) revealed that these genotypes significantly differed for their plant height, collor diameter and diameter at breast height. However, the difference were non significant in case of canopy diameter (Table 1). The plant height of AP-28 (2.47 m) and AP-35 (2.37 m) was significantly higher as compared to other three genotypes. The growth in collar diameter of three genotypes such as AP-28, AP-35 and AP-S2, did not differ significantly. The growth in DBH of AP-28 and AP-35 was significantly higher than AP-12, AP-52, and AP-S2.

Effect of pruning- There was no significant differences in plant height and CD due to different pruning intensities. However, pruning up to 50% height from the ground level exhibited slightly higher plant height and CD. As regard to DBH pruning up to 50% height gave significantly higher DBH than pruning up to 25%. In case of canopy diameter pruning up to 25% height showed significantly higher (1.80 m.) than pruning up to 50% height.

TABLE 1: *Growth attributes of Anogeissus pendula as influenced by genotypes and pruning at 46 months of age.*

Treatments	Growth attributes			
	Plant height (m)	Collor diameter (cm)	DBH (cm)	canopy diameter (m)
		Effect of genotypes		
AP-35	2.37	3.53	1.01	1.71
AP-S2	2.07	3.62	.71	1.55
AP-52	1.97	2.96	.61	1.53
AP-28	2.47	3.69	1.02	1.90
AP-12	2.09	3.12	.75	1.56
SEm±	0.08	0.11	0.06	0.15
CD at 5%	0.24	0.33	0.18	NS
		Effect of pruning intensities (%)		
25	2.14	3.33	.76	1.80
50	2.25	3.45	.87	1.48
SEm±	0.05	0.07	0.04	0.09
CD at 5%	NS	NS	0.12	0.28

Biomass Production

Green and dry leaf fodder production: Data in Table 2 revealed that AP-S2 gave significantly higher green (690.67 kg/ha) and dry (446.67 kg/ha) leaf fodder yield as compared to all the other genotypes except AP-28 for dry leaf fodder. The minimum green and dry leaf fodder was recorded with AP-12 (local Jhansi).

Pruning up to 50% height from the ground level gave significantly higher green (434.67 kg/ha) and dry (272.91 kg/ha) fodder yield as compared to pruning up to 25% height (Table 2).

TABLE 2: *Green and dry leaf fodder and fuel wood yields (kg/ha) of different genotype of Anogeissus pendula at different pruning intensities at 40 months age*

Treatments	Green leaf fodder	Dry leaf fodder	Green fuel wood	Dry fuel wood
		Effects of genotypes		
AP-35	*264.83*	*165.22*	*2371.17*	*1557.63*
AP-S2	*690.67*	*446.67*	*2144.67*	*1606.93*
AP-52	*103.83*	*62.69*	*1283.83*	*829.23*
AP-28	*317.83*	*196.02*	*2461.17*	*1413.51*
AP-12	*84.50*	*52.58*	*1414.67*	*856.97*
Sem ±	*70.00*	*90.00*	*60.00*	*170.00*
CD at 5%	*210.00*	*280.00*	*180.00*	*510.00*
		Effect of Pruning (%)		
25	*150.10*	*96.20*	*1358.67*	*841.89*
50	*434.67*	*272.91*	*2511.33*	*1663.81*
SEm ±	*40.00*	*60.00*	*20.00*	*110.00*
CD at 5%	*130.00*	*180.00*	*60.00*	*320.00*

Green and dry fuel wood production: Perusal of data in Table 2 showed that maximum green fuel wood production of 2461.17 kg/ha was recorded with AP-28 which is significantly higher than all the other genotypes except AP-35. The minimum green fuel wood production was recorded with AP-52. As regards to dry fuel wood the maximum production was obtained with AP-S2 which is significantly higher, than AP-52 and AP-12. The minimum dry fuel wood was noted with AP-52. Pruning up to 50% height from the ground level produced significantly as compared to pruning up to 25% height (Table 2).

REFERENCES

1. Gupta, A. C. 1967. Proc. II Silvi Conf., 5 vol II, 160-164. FRI, Deharadun.
2. Rai, P.; Roy, Deb R. and Rao, G. R. 1995. Range Mgmt. Agroforestry 16 (2): 103-113.
3. Singh, R. V. 1982. Fodder trees of India. Oxford and IBH Publishing Co. Pvt. Ltd., New Delhi.
4. Troup, R. S. 1921. The Siliviculture of Indian tress Vol. I. II, III, Clarendon Press, Oxford.

Ecological Studies on Multipurpose Woody Species of *Iseilema* Grasslands

B. K. TRIVEDI, J. N. GUPTA AND D. K. PARASHAR
Indian Grassland and fodder Research Institute, Jhansi (UP)

ABSTRACT

Herbaceous species contribute the major portion of forage to our large livestock population while woody species which include trees and shrubs also play an important role in meeting forage and fuelwood demand to a greater extent. For getting the sustained production from shrub infested grasslands an appropriate management technique needs to be evolved based on the regeneration potential of trees and shrubs. The performance of woody species under two management practices viz. reserved and felled was studied with regard to plant density, growth attributes and biomass production of these woody species.

INTRODUCTION

The vegetation of grasslands/rangelands is comprised of herbaceous and woody species. Herbaceous species contribute the major proportion of forage to our large livestock population while woody species which include trees and shrubs also play an important role in meeting forage and fuelwood demand to a greater extent. Trees and shrubs have been providing valuable forage to herbivores probably since the time to their domestication (Robinson, 1985). Apart from this, importance of woody vegetation increases with the onset of drought season particularly in draught years. For getting the sustained production from shrub infested grasslands an appropriate management technique needs to be evolved based on the regeneration potential of trees and shrubs. With this view the ecological studies on multipurpose woody species of natural grassland were carried out under two systems of management.

MATERIALS AND METHODS

The experiment was carried out in a natural grassland dominated by *Iseilema laxum* from 1986-91 at Central Research Farm of Indian Grassland and Fodder Research Institute, Jhansi (78° 37' E long. and 25° 27' N latitude and 275 m MSL). The climate is tropical monsoonic with three distinct seasons in a year viz. summer (March to June), rainy (July-October) and winter (November to February). Long term meteorological records revealed a mean total annual precipitation of 876.1 mm with mean minimum and maximum temperature of 6.8 and 42.6 °C, respectively. For the study period the total precipitation varied from 700.5 mm in 1989 to 1092.9 mm in 1990. Soil of experimental site was black in colour and sandy clay to sandy clay loam in texture with 8.4 pH.

The performance of woody species under two management practices viz. reserved and felled was studied on plant density, growth attributes (number of branches and diameter at basal height) and biomass production (fuelwood and top-feed). These observations were recorded in alternate years from 1987 onwards. Biomass data of felled treatment were recorded by actual felling of trees as per method outlined by Newbould (1967). In reserve treatment sampling from three average plants of each species was done.

The herbaceous vegetation of the site comprised of 39 species which included 6 perennial, 10 annual grasses, 10 legumes and 13 forbs. Among these *Iseilema laxum, Dichanthium annulatum and Heteropogon contortus* were the prominent perennial grasses and *Atylosia scarabaeoides* and *Alysicarpus longifolius* legumes and *Borreria sttrictta* and *Evolvulus alsinoides* as forbs.

RESULTS AND DISCUSSION

As a result of this study 21 woody species were found on the site and out of these, there were 7 trees and 14 shrubs. The dominant tree species were *Butea monsperma* and *Phoenix silvestris* which *Acacia catechu, A. leucophloea, Carissa spinarum, Flacourtia indica,*

Securinegga virosa, Zizyphus mauritiana, Z. oenoplia and *Z. xylopyrus* were the prominent shrubs.

Plant Density

During study period *B. monsperma* showed a regular decrease in its density in both the treatments while *P. silvistris* exhibited about 90 percent increase in reserved and 27 per cent decrease in felled treatment. The density of four shrubs viz. *A. catechu, A. leucophloea, Z. oenoplia* and *Z. xylopyrus* decreased and contrary to this the density of *S. virosa and Z. mauritiana* increased regularly in both treatments. The effect of these management practices was very conspicuous on the density/population of two shrubs namely *C. spinarum* and *F. indica*, where both the shrubs increased considerably in reserved treatment while decreased in felled treatment.

Felling treatment reduced the total tree density from 306 to 181 plants/ha over a period of 5 years and same was true for most of the shrubs. Yearwise trend in average total density of woody vegetation was almost same in both the treatments.

Growth Attributes

Average number of branches of *A. leucophloea, C. spinarum, F. indica and Z. xylopyrus* increased in both management systems but however, this increase was higher under felling system in *A. leucophloea* and *F. indica*. The effect of these treatments is clear on the DBH of trees. In case of *B. monsperma* DBH remained unchanged under reserved treatment and increased from 3. 6 to 4. 7 cm under felling treatment while that of *P. silvistris* increased under reserve treatment and decreased considerably under felled treatment. Among the shrubs DBH of *A. leucophloea* increased in both the treatment. Reverse to this DBH of *C. spinarum* and *S. virosa* decreased in these treatments. Apart from this *A. catechu, Z. mauritiana* and *Z. oenoplia* exhibited decreased trend in their DBH under felling treatment while *Z. xylopyrus* showed considerably decrease in its DBH under reserved treatment.

Fuelwood Production

Among tree component only two species viz. *B. monosperma* and *P. silvistris* produced the fuelwood in all the years. But it is interesting to note that the response or trend of these species in wood production was reverse in these management systems. *P. silvistris* showed an increasing trend in fuelwood production in reserved treatment. Fuelwood production of *B. monosperma* got increased regularly under felling system whereas that of *P. silvistris* decreased from 12. 1 to 4. 6 q/ha at the end of experiment. Although, total fuelwood production of tree component did not reveals any trend in any of the treatment but about 47. 1 per cent increase in total fuelwood production was recorded in these treatments over a period of 5 years.

As regards the shrub component the fuelwood production of *A. catechu, A. leucophloea* and *Z. xylopyrus* declined regularly in both the treatments while contrary to the fuelwood of *F. indica, S. virosa, Z. mauritiana* and *Z. oenoplia* increased in both system and this increase was maximum in *S. virosa*. The total wood production of shrubs exhibited a linear increase (76. 5%) in reserved treatment while that of felled treatment showed about 4. 9% decrease against the initial wood production (32. 3 q/ha).

Top Feed Production

The shrub component contributed about 95 per cent of total top feed and the top feed of shrub component showed a regular increase in reserved treatment over the years. In most of the shrub species the trend of top-feed production was similar to fuelwood production in both the treatment. The maximum average feed was produced by *C. spinarum* followed by *Z. oenoplia* and *Z. xylopyrus*.

Different species have differential response to the types of management. Many trees and shrubs have the capacity to regenerate new shoots from stumps, roots or branches after being cut. This is the most desired trait in a fodder tree. This capacity enables us to obtain fodder and other products repeatedly without destroying the trees and shrubs. An increasing trend in the density, fuelwood and top feed production of *S. virosa* and *Z. mauritiana* over the years in both the treatment shows their good regeneration potential. *S. virosa, H. isora, Ehretia aspera, Flacourtia indica, G. spionsa* and different species of *Zizyphus* are also identified as promising species of grassland by Gupta and Singh (1995). Decreasing trend in the density of *P. silvistris* and *B. monosperma* in felled treatment also play an important role in controlling these undesired plants in the grassland.

REFERENCES

1. Gupta, J. N. and Singh, J. P. 1995. Annual Report, IGFRI, Jhansi.
2. Newbould, P. J. 1967. Methods of Estimating the Primary Production of Forests. IBP Hand Book, Blackwell Scientific Publication, Oxford, UK.
3. Robinson, P. J. 1985. In: (Cannel M. G. R. and J. E. Jackson (Eds.) Attributes of Trees as Crop plants. Institute of Terrestrial Ecology, Huntington, UK: 281-300.

Intercropping with MPTs based Industrial Plantation and its Beneficial Effects

M. S. MALIK AND C. SURENDRAN
SKUA&T, Regional Research Station, R. S. Pura, Jammu (J&K)

ABSTRACT

The practices and beneficial effects of intercropping potato, beans and maize in multipurpose tree species *Eucalyptus globulus* based industrial plantation in the hilly zone of Tamil Nadu are summarized in the paper. The *E. globulus* plantation was raised for production of pulp wood. The results of the field experiment revealed that the growth traits of tree viz. tree height, diameter, and volume increased considerably with intercrops. Among the three intercrops studied, potato was most compatible with *E. globulus* and triggered fast growth of the tree.

INTRODUCTION

The intercropping in multipurpose tree species based industrial agroforestry plantations is becoming popular because of increased demand for wood based raw materials. The Forest Dept. can supply only a fraction of the expanded requirement of the industries. Hence, local industries are encouraging the farmers to take up multipurpose tree species as a monocrop and integrating food crops as intercrops as a means of food supply and enhancing their living conditions. Hofstad (1978) held that the increased revenue from a food crop in the first one or two years of plantation made intercropping an attractive proposition. In Pulney hills of Tamil Nadu, wood based industries have large scale plantation especially of *E. globulus* for rayon pulp.

MATERIALS AND METHODS

Pulney hills of Tamil Nadu of South India receive a mean annual precipitation of 1455 mm with maximum temperature of 19 to 28°C and a minimum temperature of 8°C and annual average relative humidity of 62% in the forenoon and 87% in the afternoon. *Eucalyptus globulus* a multipurpose tree species plantations were raised for production of pulpwood. Espacement of 3 □ 3 m was adopted for the tree. Split plot design was adopted for the field experiments with three replications. Tree assigned to the main plot and three crops, weeded and unweeded experiment were carried out under sub-plots. The sub-plots size was 12 □ 12 m. The following five treatments were given.

- *E. globulus* clean weeded and intercropped seasonally with potato. Potato was planted on the hill on the ridges at a spacing of 60 □ 15 cm.
- *E. globulus* clean weeded and intercropped seasonally with beans. Beans were planted at a spacing of 60 □ 40 cm.
- *E. globulus* clean weeded and intercropped seasonally with maize. Maize was planted at a spacing of 20 cm between plants in the rows which were 60 cm apart.
- *E. globulus* monoculture maintained free of weeds (clean weeded).
- *E. globulus* monoculture given no weeding after planting (unweeded).

Unweeded Eucalyptus and monocropping of agricultural crops served as control. Mean of each of the seasonal crop grown twice a year were taken for study. The crops were sown adopting the standard agronomic practised.

RESULTS AND DISCUSSION

Growth of Eucalyptus

The increment in height of one year old *E. globulus* with intercropping of crops was 10.3 times higher than control (Table 1). Between the three crops, culturing with potato resulted in highest increment for height and diameter growth. Intercropping in two year old plantation resulted in better height and diameter growth as influenced by the crops and was superior to weeded and unweeded eucalyptus (Samraj *et. al.*, 1982; Ghose *et. al.*, 1989; Redhead, 1993). The impact of intercropping on tree growth at three and four year old plantation. was not as pronounced as with the younger plantation, showing that the beneficial effect of intercropping was diminishing in older plantation. Thus all intercropped treatments presented higher eucalyptus yield than the eucalyptus monoculture treatments (Couta and Gomes, 1995).

TABLE 1: *Data regarding average height and DBH of E. Globulus*

Treatment	Years after planting							
	1		2		3		4	
	Height (m)	DBH (cm)	Height (m)	DBH (cm)	Height (m)	DBH (cm)	Height (m)	DBH (cm)
Intercropped with potato	6.52	3.58	3.59	5.97	9.29	10.74	11.66	12.15
Intercropped with beans	2.54	1.95	2.81	4.48	8.85	10.45	11.39	11.52
Intercropped with maize	3.1	2.01	2.79	2.87	8.37	9.75	11.34	10.65
Monoculture weeded	1.08	1.94	2.34	2.77	7.82	8.59	11.14	10.74
Monoculture unweeded	0.63	1.16	1.73	2.03	7.61	8.15	11.01	9.8

Potato, although being solanaceous crop, exerted greater influence in accelerating initial growth as compared to the leguminous beans in the first year, followed by the maize crop. Potato required better site preparation through deep soil working, formation of ridges and furrows to conserve moisture on one hand and this facilitated better tuber growth and also put up a faster growth of eucalyptus, since the species is reported to be an efficient user of water (Chaturvedi, 1984). Higher fertilizer application in potato also accelerated the tree growth.

In general, net primary production of roots was substantially more than the standing biomass of trees, owing to the turn over of finer roots of trees. Roots are good store house of nutrients and by such turn over of roots, these nutrients are returned to the soil, which in turn stimulates finer root growth thus resulting in nutrient recycling. The absolute biomass of roots in agroforestry was more than twice that of all agricultural system reported (Ewel et. al, 1982).

The ridges formed during potato intercropping also resulted in shallow placement of roots in potato compared to the roots in eucalyptus in deeper substratum of soil; thereby lessening mutual competition, between the tree and then intercrop, which becomes advantageous both to the intercrop and the deep rooting trees (Buck, 1986). Further,

E. globulus has most of the lateral roots arising from the upper portion of the tap root (Mathur, *et. al.*, 1996), an advantage for nutrient mobilization.

Crop Performance

There was some increase in yield of potato in first year of intercropping compared to control, while it was at par with control in second year of intercropping. During the third year of intercropping there was reduction in yield of potato to the extent of 25.1 and 72.38 percent, respectively compared to the control. Hence, even with lower yield of potato in the third year of plantation, it has been found advantageous to grow potato during the first three year of plantation (Samraj *et. al.*, 1982).

The decline in yield of beans at all age group of trees was obvious. Control plots yielded 358 kg per ha of seed while the most promising intercropped plot at first year could only record 291.34 kg per ha of yield. The plants in the oldest plantation three and four years had a significant yield and reduction of 59.39 and 91.68 per cent respectively compared to control. Similar observation was reported by Redhead; 1993 in Tanzania.

The yield of maize in first year was at par with control but there was reduction in yield with increase in age of plantation. From the third year onwards, trees cost much shade for the maize to flower properly (Redhead, 1993).

Economic Evaluation

The cost of intercropping in a plantation of *E. globulus* with potato, beans and maize was Rs. 20,231; Rs. 11,399; and Rs. 7912, respectively. The revenue from intercrop was calculated on the prevailing market price for potato at Rs. 5, beans at Rs. 80 and maize at Rs. 5 per kg. The profit from intercrop and pure crop (control) are mentioned in Table 2. When compared with control first year intercrop of potato with one year old eucalyptus plantation have higher yield and consequently higher profit. But profit from beans and maize as compared to control was slightly better. In two year old plantation, the trend of profit over control was practically of same order.

TABLE 2: *Profit/loss analysis of intercropping (in rupees)*

	Years after planting (Intercropping)				Control
	1	2	3	4	
Potato	*64603*	*62076*	*41853*	*2658*	*62653*
Beans	*11907*	*4747*	*254*	*- 9013*	*17293*
Maize	*8691*	*6538*	*977*	*- 4699*	*9206*

In third year, beans and maize showed steep decline in yield and consequently profit, while in fourth year plantation intercrop of beans and maize showed a loss.

CONCLUSIONS

Tree cropping under this practice has conferred a wide variety of direct and indirect benefits to the farmers and society. It, ensures seasonal income through intercropping, repeated crop harvest minimises cost of weeding, there is also reduction in rotation period of trees and it also minimise soil erosion and employment generation (Malik, 1994).

REFERENCES

1. Buck, M. G. 1986. Agroforestry Systems 4: 191-203.
2. Chaturvedi, A. N. 1984. Paper for National Seminar on application of Science and Technology for Afforestation, ACT Jaipur, Rajasthan, India.
3. Couto, L and Gomes, J. M. 1995. Int. Tree Crops Journal 8:83-93.
4. Ewel, J.; Benadict, F.; Barish, C.; Brown, B.; Gliessmen, S.; Amador, M.; Bermudez, R., Martinez, A.; Miranda, R. and Price, N. 1982. Agroecosystems 7: 305-26.
5. Ghose, S. P.; Kumar, B. M,; Kabeerathumma, S. and Nair, G. M., 1989. Agroforestry Systems; 8:67-82.
6. Hofstad, O. 1978. Preliminary evaluation of the taungya system for combined wood and food production in north-eastern Tanzania. Record No.2, Division of Forestry, University of Dares Salaam, Morogoro, Tanzania.
7. Malik, M. S. 1994. FAO- APAN Report No. 15, Indonesia.
8. Mathur, H. N.; Rajau, K. and Rajagopal, K. 1986. KFRI, Peechi, Kerala. 225-228.
9. Redhead, J. F. 1993. Growth and water use of forest plantation. Eds. Calder, Hall and Aollard. John Wiley & Sons, U. K. 128-137.
10. Samraj, P.; Haldorai, B. and Henry, C. 1982. Annual Report, CSWCR&TI, Dehradun- 49.

Income Generation from Tree and Crop Based Systems: A Farmer's Experience

P. S. TOMER
Indian Grassland and Fodder Research Institute, Jhansi (UP)

ABSTRACT

Out of the four systems tried, growing of Eucalyptus at a spacing of 3.0 □ 1.5 m with intercropping of seasonal crops during first three years (intercropping on rental basis) come out to be thc most remunerative system under irrigated areas of Bulandshehar district of Uttar Pradesh.

INTRODUCTION

In present days the labour is a very scarce and costly input. In the absence of labour, intensive cropping is getting a setback. As a result of this, the latest thinking developed in the minds of land owners is to adopt the system which is least labour consuming and more income generating. Keeping this consideration in view, different systems of income generation were tested at farmers's field in village Garhia Sultanpur of Bulandshehar district of Uttar Pradesh.

MATERIALS AND METHODS

A field trial consisting four systems i.e. (I) silviculture having sole crop of Eucalyptus at a spacing of 1.5□1.0 m, (ii) agroforestry having Eucalyptus at a spacing of 3.0 □ 1.5 m and intercropped with seasonal crops during first three years, (iii) seasonal crops i.e. Maize-Wheat and

(iv) rental basis, was conducted in village Garhia Sultanpur of Bulandshehar district, UP during the period 1982-83 to 1993-94.

In silviculture, standing Eucalyptus tree were sold during 1992-93. Since it was harvested one year earlier than completion year (1993-94), one year interest on the amount was worked out and added. In agroforestry, standing Eucalyptus trees were sold during 1993-94 and additional income received on account of giving the land on rent for first three years alongwith interest was added. The third system in which maize-wheat was grown consistently upto 12 years gave income in every season. Therefore interest worked out on the seasonal income was also added in total income of the system. In fourth system the land was given on rent upto 12 years. The income was received every year, therefore interest was worked out and added in the income of this system.

RESULTS AND DISCUSSION

The data presented in Table 1 revealed that highest net income was generated from the system in which Eucalyptus was grown at a spacing of 3.0 □ 1.5 m and intercropping was done during first three years. It gave net income of Rs. 18,450/ha/yr. The second in merit was maize-wheat sequence which gave net income of Rs. 8,180/ha/yr. The third in merit was silviculture yielding a net income of Rs. 6,989/ha/yr, while the lowest income of Rs. 5,648/ha/yr was obtained when land was given on rent. It is worthwhile to point out that expenditure on inputs was more with the system of raising the seasonal crops as compared to giving the land on rent, growing trees alone or in combination with crops. However, gross income obtained from growing of seasonal crops alone or in combination with trees was much higher. Since the input expenditure on tree-crop combination was low and gross income was high, the net income obtained with this system was highest.

TABLE 1: *Net income from different systems*

Systems	Cumulative net income/ha (Rs.)	Net income/ ha/yr (Rs.)
Silviculture	*83,875*	*6989*
Agroforestry	*221410*	*18450*
Maize-Wheat	*98170*	*8180*
Rent Basis	*67785*	*5648*

Farmers' Preference for MPTs in Agroforestry

R. P. DWIVEDI, S. K. SHUKLA, P. RAI AND C. K. BAJPAI
National Research Centre for Agroforestry, Jhansi (UP)

ABSTRACT

A study was conducted at farmer's field at village Karari in Jhansi district of Uttar Pradesh. Plantation was done in 1990 to establish agri-horticultural system in the field and trees of *Leucaena leococephala* (subabul) on the field boundaries. The present study was conducted to know the farmers' preference for MPTS in agroforestry. The data were collected by personal interviews with 100 farmers at village Karari with the help of an interview schedule which contained mostly open-ended questions. Data were also recorded on yield of fuel and fodder obtained from the *L. leucocephala* on farmer's field. Looking to the economic gains from *L. leucocephala*, majority of farmers were tempted to have the same on their farmlands. The sequential preference of trees indicated *L. leucocephala* as most preferred MPTS. Other popular MPTS included *Azadirachta indica* (Neem), *Zizyphus mauritiana* (Deshi Ber), *Acacia nilotica* (Babul), *Anogeissus pendula* (Kardhai), *A. latifolia* (Dhau), *Ficus religiosa* (Peepal), *Madhuca latifolia* (Mahua), *Butea monosperma* (Dhak), *Inga dulcis* (Jangal jalebi), *Syzigium cuminii* (Jamun) and *Cordia myxa* (Lasoda) in a descending sequence. The leaves of these trees are used for feeding animals like cattle, buffalo, goat and sheep. The other uses of these MPTS are as firewood, timber, agricultural implements and fruits. The farmers' perception indicated that the potential of cash income formed a major criterion in selection of MPTS.

INTRODUCTION

Agroforestry encompasses land use systems where woody perennials are intentionally grown on the same piece of land with agricultural crops and/or animals either in form of spatial arrangement or temporal sequence. A good deal of work has been done on the subject during last one decade. Any innovation starts generally from the research done by the scientists and whether or not the innovation would be useful for the clients or users (farmers) would depend upon how the farmers' problems or needs considered by scientists before developing technology and how it is perceived by the farmers. Therefore, the success of agroforestry technologies largely depends upon how it is accepted by farmers. Despite its usefulness farmers are reluctant to introduce woody perennials in any form in their cultivated fields. This is the reason why our agroforestry technologies are not being adopted by the farming community. Taking the fact into account a study was initiated in 1990 at village Karari district Jhansi at farmer's field to show the impact of agroforestry system amongst the farmers. In this study *L. leucocephala* (Subabul) was planted around the cultivated field on the boundary in order to get production of fodder and fuelwood. During the period, farmers of the village used to visit the demonstration site. The present study was conducted with the objective to identify MPTS of farmers' choice as a component of agroforestry.

In our country, ruminant species, especially buffaloes, cattle, goat and sheep contribute significantly to the productivity, stability and sustainability constitute important component of farming system. Inadequate animal feed is widely regarded as the major constraint to animal production in Bundelkhand region. The research efforts in animal feed are commonly focused on fodder crops or pasture grasses and legumes primarily because these constitute the main component of diet in traditional feeding system for ruminants. However, in Bundelkhand region tree fodder is widely available and used by farmers in addition to grasses and legumes. The potential contribution of such shrubs is well recognised. Their role is considered especially important in our country where small land holdings and large ruminant population result in a severe problem of feed availability from more conventional sources. In semi-arid and arid region browse shrubs and tree fodder are becoming increasingly important as feeds because of harsh environment and the limited fluctuating feed supplies. Also in such environments, there are large areas of marginal lands that can support good pasture growth consequently the browse and leaves of shrubs and trees are often the main source of feeds. Under such circumstance, value of shrubs and tree fodder is considerable in meeting nutrient needs and in maintaining the productivity of ruminant population. A great deal of work has been done with regard to identification and management of tree fodders and for various agro-ecological regions of the country (Patil and Pathak, 1977; Singh, 1982; Pathak, 1993).

MATERIALS AND METHODS

Sample

The sample for the present study consisted of 100 farmers of village Karari of Jhansi district (U.P.). The reason of selection of respondents from village Karari was that in this village there are demonstration trials at three farmers' fields and the villagers have been in touch with the agroforestry activities of the centre such as Kisan Diwas and Kisan Gosthis, Aonla Diwas, Kisan Mela, etc. Most of the farmers got themselves acquainted with MPTS and agroforestry.

Data Collection

Since the present study is of an exploratory nature, the questions pertaining to the problem were mostly of an open ended nature avoiding pre-categorization. This was done to allow for maximum freedom of responses and to include all shades of opinion. Data were collected by individual personal interviews with the farmers by using interview schedule to know the preference of the farmers for the Multipurpose Tree Species (MPTS) in agroforestry. The data were analyzed on a nominal level, using numbers and percentages to express the responses.

RESULTS AND DISCUSSION

Performance of *L. leucocephala* at Demonstration Site

Initially 200 subabul plants were planted on the boundaries. Trees were irrigated up to third year. They were lopped for fodder purpose from third year onwards. In 1996, first selective felling of 178 trees was done which resulted in the total income of Rs 5340/- @ Rs 30/-

per tree. In spite of the felling, about 415 plants are growing naturally on the boundary as a result of seed fallen from the trees in the vicinity. In addition, trees gave the fodder yield of 21 kg per day for 3 months from December to February every year. Feeding of milch animals with 35 per cent of Leucaena leaves of their total requirements per day increased the milk yield by 20 per cent.

FARMERS' PREFERENCE FOR MPTS

From farmers' responses to the question that "*if you have a choice, which MPTS would you prefer to grow*?" *L. leucocephala* (used for fodder and fuelwood) was found to be most preferred species (73 %). Sixteen per cent of the respondents expressed their response in negative i.e. they would not plant this tree in any form, rest 11 per cent were indecisive about this species. Similarly, *A. indica* was preferred by 69 per cent farmers, 29 per cent expressed no-choice while two per cent did not give any response. It was found that 67 per cent respondents would accept *Z. mauritiana* as fodder species grown for its fruits, fuelwood and small timber. Farmers giving negative response and undecided response for *Z. mauritiana* were 16 and 17 per cent respectively. *A. nilotica* was preferred as MPTS by 63 per cent of the respondents, while 28 per cent farmers did not give any response, only nine per cent gave no-choice (Table 1).

It was revealed by 59 per cent respondents that they prefer *A. pendula* as MPTS for the purpose of nutritious fodder and good quality firewood. while 13 per cent farmers expressed no- choice and the rest 28 per cent were indecisive. About *A. latifolia*, 55 per cent respondents preferred this as MPTS, six per cent no-choice and remaining 39 per cent were in undecided category. *F. religiosa* is another species used for fodder purpose by 51 per cent farmers, 34 per cent respondents replied that they would not plant this species as MPTS, while 15 per cent respondents were in undecided category *M. latifolia* is also a common MPTS liked by 48 per cent of the respondents, it was not liked by 27 per cent respondents, while 25 per cent farmers were indecisive (Table 1).

The tree species *B. monosperma*, *I. dulcis*, *S. cuminii* and *C. myxa* were least preferred MPTS among the respondents (29%, 23%, 21% and 17%). Most of the respondents would not prefer these tree species to grow as MPTS (51%, 14%, 67% and 72%). On an average 27 per cent respondents were indecisive regarding these four tree species (Table 1).

TABLE 1: *Farmers' choice for MPTS*

S. No	Name of MPTS	Response of farmers regarding acceptability of MPTS			Uses
		Yes	No	Undecided	
1	*L. leucocephala*	73	16	11	*fodder, fuelwood*
2	*A. indica*	69	29	2	*timber, fodder, medicine*
3	*Z. mauritiana*	67	16	17	*fruits, fuelwood, small timber, agricultural tools*
4	*A. nilotica*	63	9	28	*fuelwood, fodder, small timber, agril implements*
5	*A. pendula*	59	13	28	*fodder, firewood*
6	*A. latifolia*	55	6	39	*fodder, firewood*
7	*F. religiosa*	51	34	15	*fodder*
8	*M. latifolia*	48	27	25	*fodder, timber, fuelwood*
9	*B. monosperma*	29	51	22	*leaves for plate making, fodder, fuel*
10	*I. dulcis*	23	14	63	*live fence, fruits*
11	*S. cuminii*	21	67	12	*fruits, fodder*
12	*C. myxa*	17	72	11	*fruits, fodder, fuelwood*

A large majority of the farmers (51 to 73 per cent) are quite interested in MPTS but social desirability might also have coloured their responses. It is also clear that physical facilities and manpower for agroforestry need strengthening; this is also supported by the study of Samanta (1985). It is noteworthy that about three-fourth of the farmers have perceived the potential of cash income as a major criterion in selection of MPTS.

Farmers have perceived the MPTS as an important source of fuel, fodder, small timber and fruits for creating job opportunities for the rural masses. There is a need to find out suitable MPTS based agroforestry system in order to prove that the economic returns can be enhanced considerably from specific tree-crop combinations. Moreover, intensive extension efforts of government, state departments and non-government organizations will go a long way in popularizing and exploiting the tree in suitable agro-ecological zones for the upliftment of rural poor.

REFERENCES

1. Pathak, P. S. 1993. In: Agroforestry - Indian Perspective (Eds.: L. K. Jha, P. K. Sen Sarma) Ashish Publishing House, New Delhi.
2. Patil, B. D. and Pathak, P. S.1997. Invention Intelligence 12:79-87.
3. Samanta, R. K.1985. Indian J. Extension Edu. 21 (1 and 2): 53-61.
4. Singh, R. V. 1982. Fodder trees of India. 663. Oxford & IBH, New Delhi.

Recommendations of the Symposium

The major recommendations of the symposium are as follows:

I POLICY

1. Patenting rights of agroforestry products and technologies need to be strengthened and popularise among scientists and other professionals.
2. Multipurpose trees species (MPTS) present in the farm lands should be free from the application of forest laws. Farmers should be allowed to take their own decisions on choice of species for planting and felling according to their needs.
3. Transparency in the implementation of legislative measures for production and marketing of agroforestry products specially tree and shrub based group.
4. Human resource development in agroforestry should be strengthened by establishing close linkages of SAUs with the Ministry of Environment and Forest, ICRAF and other International Organisations.
5. Participation of women in plantation, care and production of MPTS under agroforestry systems should be encouraged.
6. There should be close linkage of producers with tree based industries particularly paper and pulp industries to ensure the marketing and realise the benefits.
7. Activities related to plantation of MPTS in degraded lands, range lands, problematic soils etc. should be strengthened for reclamation of these lands as well as for fulfilling the fuel wood need of the villagers.

8. Close linkages between research and development institutions, farmers and tree based industries should be strengthened for efficient transfer of technology and quick marketing.
9. Establishment and strengthening of infrastructure for mass multiplication of MPTS through NSC and State Seed Corporations.
10. Mechanism for releasing varieties in MPTS should be developed at Government level.
11. Teaching on tree farming principles into environmental conservation should begin at primary school level and continue up to the college level to popularise tree based farming systems.
12. Views and suggestions of experienced farmers in the field of agroforestry should be invited before planning agroforestry programme for adaptation.

II SCIENTIFIC AND TECHNICAL

1. Inventories of MPTS should be made based on farming system preference, tree-crop association, farm holding size, suitability, edaphic preferences and socio-cultural and economic suitability.
2. The data and information generated on the work done on MPTS for agroforestry including the bench mark/diagnostic and design surveys conducted must be secured, compiled and updated on 15 years of work must be brought.
3. A separate inventory of unexploited indigenous fruit and medicinal trees should be prepared. Their domestication as future MPTS resource possibilities should be explored.
4. Plantation of neem trees in farm lands should be encouraged because it has good economic and environmental value and global level.
5. *Acacia nilotica* based agroforestry system is quite successful particularly in rice growing area. Extension activities to popularise this system should be strengthen.
6. Plantation of tree species namely *Dalbergia sissoo, Ficus religiosa, Ailanthus excelsa, Cassia fistula, Calotropis procera, Moringa oleifera, Ficus bengalensis, Azadirachta indica, Pithecolobium dulce, Mangifera indica, Tectona grandis* and *Ficus infectoria* in polluted area should be encouraged since these species are resistant to the

automobile exhaust pollutants, air pollutants and work as dust filtering agents.

7. Research is needed on optimum tree density including its geometry.
8. Silvicultural and management aspects (below ground and above ground) need to be worked out to reduce competitive effect between components of a tree crop system.
9. Intensified research should be initiated to reduce competition for nutrients and soil moisture and develop more knowledge on tree behaviour and functions as well as on shading effect on understorey crops.
10. There is urgent need to conserve bio-diversity to MPTS in situ and ex-situ.
11. Economic and socio-economic analysis of MPTS in agroforestry systems should be done on priority basis.
12. Collection and evaluation of promising MPTS should be done under various agroforestry systems rather than in isolation.
13. Analysis of social, anthropological and economic relations of MPTS with particular reference to women, children, tribals and landless farmers and small holders should be done.
14. National network should be established for exploration, collection, multiplication, distribution and conservation of germplasm of MPTS for agroforestry species.
15. In first, genetic improvement work should be started at national level in *Populous deltoids, Alnus nepalensis, Ailanthus excelsa, Azadirachta indica, Dalbergia sissoo, Prosopis cineraria, Tamarindus indica, Emblica officinalis, Anthocephlus cadamba, Tamarindus indica* and *Acacia nilotica.*
16. Mode of pollination of reproductive biology of important MPTS should be studied.
17. Simple and feasible techniques should be standardised for vegetative propagation.
18. Seed productive zone should be identified to supply quality seeds for immediate needs of farmers.

19. Clonal seed orchards of important MPTS should be established for agroforestry systems.

20. Seed standards should be prepared for MPTS.

21. Nursery techniques for MPTS should be developed to reduce the cost of seedlings.

22. Research on tree physiology including nutritional requirements, biochemistry of nitrogen fixation of trees, mycorrhizal association and prolonging seed viability should be investigated to increase production.

23. Research on nutrient cycling in relation to the rentention/uptake ratio, redistribution and nutrients use efficiency in respect of MPTS is required.

24. More emphasis should be given to silvipastoral system for livestock farming on degraded lands.